F L A G S O F T H E W O R L D

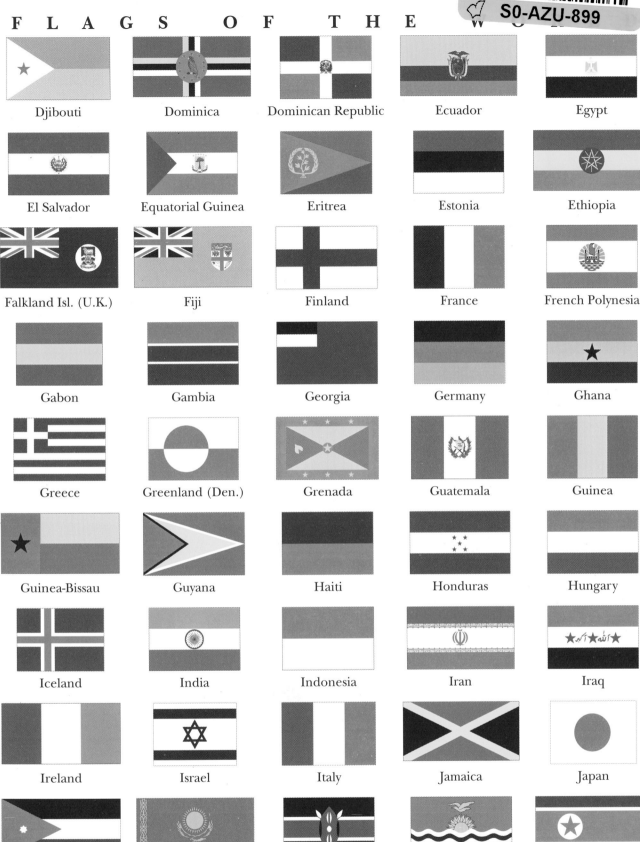

Djibouti	Dominica	Dominican Republic	Ecuador	Egypt
El Salvador	Equatorial Guinea	Eritrea	Estonia	Ethiopia
Falkland Isl. (U.K.)	Fiji	Finland	France	French Polynesia
Gabon	Gambia	Georgia	Germany	Ghana
Greece	Greenland (Den.)	Grenada	Guatemala	Guinea
Guinea-Bissau	Guyana	Haiti	Honduras	Hungary
Iceland	India	Indonesia	Iran	Iraq
Ireland	Israel	Italy	Jamaica	Japan
Jordan	Kazakhstan	Kenya	Kiribati	Korea, North
Korea, South	Kuwait	Kyrgyzstan	Laos	Latvia

WORLD GEOGRAPHY

WORLD GEOGRAPHY

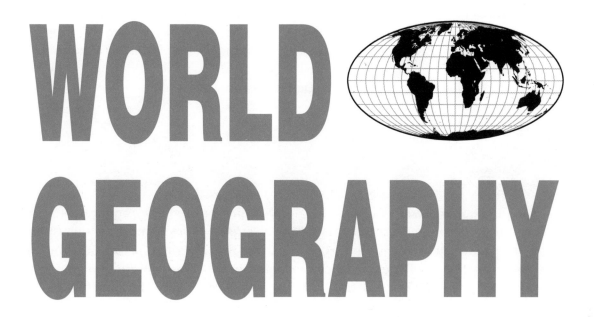

Volume 1

The World

Editor

Ray Sumner

Long Beach City College

Managing Editor

R. Kent Rasmussen

SALEM PRESS, INC.

Pasadena, California   Hackensack, New Jersey

Editor in Chief: Dawn P. Dawson

Managing Editor: R. Kent Rasmussen   *Research Supervisor:* Jeffry Jensen
Manuscript Editor: Irene Struthers Rush   *Acquisitions Editor:* Mark Rehn
Production Editor: Cynthia Beres   *Page Design and Layout:* James Hutson
Photograph Editor: Philip Bader   *Additional Layout:* William Zimmerman
Assistant Editors: Andrea Miller, Heather Stratton   *Graphics:* Electronic Illustrators Group
Cover Design: Moritz Design, Los Angeles, Calif.

Frontispiece: Earthrise seen from the Moon. *(PhotoDisc)*

∞ The paper used in these volumes conforms to the American National Standard for Permanence of Paper for Printed Library Materials, Z39.48-1992 (R1997).

Library of Congress Cataloging-in-Publication Data

World geography / editor, Ray Sumner ; managing editor, R. Kent Rasmussen.
    p. cm.
    Contents: v. 1. The World. — v. 2. North America and the Caribbean. — v. 3. Central and South America. — v. 4. Africa. — v. 5. Asia. — v. 6. Europe. — v. 7. Antarctica, Australia, and the Pacific. — v. 8. Glossary and Appendices.
    Includes bibliographical references (p.).
    ISBN 0-89356-024-3 (set : alk. paper) — ISBN 0-89356-276-9 (v. 1 : alk. paper) — ISBN 0-89356-277-7 (v. 2 : alk. paper) — ISBN 0-89356-335-8 (v. 3 : alk. paper) — ISBN 0-89356-336-6 (v. 4 : alk. paper) — ISBN 0-89356-399-4 (v. 5 : alk. paper) — ISBN 0-89356-650-0 (v. 6 : alk. paper) — ISBN 0-89356-699-3 (v. 7 : alk. paper) — ISBN 0-89356-723-X (v. 8 : alk. paper)
    1. Geography—Encyclopedias. I. Sumner, Ray.

G133.W88 2001
910′.3—dc21

2001020281

First Printing

PRINTED IN THE UNITED STATES OF AMERICA

CONTENTS

THE WORLD

THE NATURE OF GEOGRAPHY

PHYSICAL GEOGRAPHY

BIOGEOGRAPHY AND NATURAL RESOURCES

HUMAN GEOGRAPHY

ECONOMIC GEOGRAPHY

POLITICAL GEOGRAPHY

GAZETTEER OF OCEANS AND CONTINENTS 321

PUBLISHER'S NOTE

North Americans have long thought of the field of geography as little more than the study of the names and locations of places. This notion is not without a basis in fact: Through much of the twentieth century, geography courses forced students to memorize names of states, capitals, rivers, seas, mountains, and countries. Both students and educators eventually rebelled against that approach, geography courses gradually fell out of favor, and the future of geography as a discipline looked doubtful. Happily, however, the field underwent a remarkable transformation during the 1990's, as Dr. Ray Sumner explains in her introduction to this set, and geography now has a bright future at all levels of education.

While learning the locations of places remains an important part of geography studies, educators recognize that place-name recognition is merely the beginning of geographic understanding. Geography now places much greater emphasis on understanding the characteristics of, and interconnections among, places. Modern students address such questions as how the weather in Brazil can affect the price of coffee in the United States, why global warming threatens island nations, how preserving endangered animal species can conflict with the economic development of poor nations, and why other parts of the world can never be the same as North America.

World Geography addresses these and many other questions. Designed and written to meet the needs of high school students, while being accessible to both middle school and undergraduate college students, these eight volumes take an integrated approach to the study of geography, emphasizing interconnections of every kind. The set's 236 core essays examine aspects of physical, human, and economic geography and

range in length from three to ten pages each. All these essays are in the first seven volumes; the first volume covers geography as a field and examines the world as a whole. Each of the next six volumes focuses on a major world region. The eighth volume contains a detailed glossary of geography terms, appendices and a general index.

Volume 1 introduces the field of geography, examines basic concepts and issues, and explores the entire world, with special attention to the world's oceans. Volume 2 covers North America and the islands of the Caribbean Sea; volume 3 covers South and Central America; volume 4 covers Africa; volume 5 covers Europe; volume 6 covers Asia; and volume 7 covers Antarctica, Australia, and the islands of the Pacific. This regional division of the volumes makes the set easier to use for students studying specific countries or parts of the world, while not hindering those studying topics affecting even larger regions. As readers familiarize themselves with the contents of the volumes, they will appreciate the fact that each regional volume is organized on the same principles.

While each regional volume has its own index and can stand alone as a reference tool, it also fits into a larger scheme for the overall set. Information in the regional volumes is amplified and supplemented by the broad essays in volume 1 and the detailed glossary and appendices of volume 8. Moreover, the index to volume 8 combines all the individual volume indexes; it should be consulted by readers seeking information in addition to what they find in the regional volumes.

To make this set as easy as possible to use, its first seven volumes are similarly organized. The first volume opens with a section on the nature of geography. The next six volumes open with sections on regional geography. All seven volumes follow with sections on physical, human, and economic geography. The numbers of subregions covered in each volume vary. Volume 2, for example, opens with separate essays on Canada, the United States, Mexico, Greenland and the Arctic, and the Caribbean. By contrast, Volume 3 has overviews only on Central America and South America, and volume 4 has essays on the continent of Africa and Africa's offshore islands. Each overview essay surveys its region as a whole and explores connections among the physical and human dimensions of its geography.

Regional overview essays are followed by sections on Physical Geography, which include essays on physiography, hydrology, and climatology. Some volumes have separate essays on these topics for their subregions. For example, volume 2 covers North America and the Caribbean separately because they are physically very distinct regions. By contrast, volume 4 has single essays on the physiography, hydrology, and climatology of Africa.

The next section in each regional volume, Biogeography and Natural Resources, has essays on resources, flora, and fauna. This section is followed by Human Geography, with essays on the people, population distribution, culture regions, urbanization, and political geography of the region. Apart from volume 5, on Europe, all the regional volumes also have essays on exploration. The final essay section in each regional volume covers Economic Geography. These sections have essays on agriculture, industries, engineering projects, transportation, trade, and communications.

The last part of every regional volume is an extensive gazetteer, which contains descriptive entries on important places, especially those mentioned in the volume's essays. Volume 1 also has a gazetteer, with entries on the world's major land masses and inland waterways. The first seven volumes contain more than 2,200 gazetteer entries—an average or more than 300 entries per volume. A typical entry gives a place's name and location and tells what category of place it is, such as a mountain, river, city, or country. Entries

also provide statistics relevant to the categories of place—such as altitudes of mountains, lengths of rivers, areas of regions, and populations of cities and countries. Variant names and spellings are cross-referenced within each gazetteer, and all entries are indexed.

Volume 8 stands somewhat apart from the rest of the set. The bulk of its space is devoted to a glossary of basic geographical terminology that contains 1,600 entries, as well as cross-references. The volume also contains a comprehensive annotated bibliography and a selection of appendices summarizing global geographical data. These include rankings of the world's rivers, lakes, oceans, deserts, land masses, islands, and countries by size, and lists of the world's most populous cities and countries and most and least densely populated countries.

Volume 8's comprehensive index incorporates all the information in the individual volume indexes, which cover all the essays, sidebars, gazetteer entries, glossary entries, appendices, and illustrations. Thoroughly cross-referenced, the index contains detailed listings of places, issues, and terminology.

Both a physical and a social science, geography is unique among social sciences in the demands it makes for visual support. For this reason, *World Geography* contains more than 100 maps, more than 1,300 photographs, and scores of other graphical elements. In addition, essays are punctuated with more than 500 textual sidebars and tables, which amplify information in the essays and call attention to especially important or interesting points.

All maps and more than 800 photographs are printed in full color in four sixteen-page gallery sections in each volume. Like the black-and-white pictures in the text pages, the maps and pictures in the gallery pages illustrate the essays and gazetteer and glossary entries. Their arrangement parallels that of the text they illustrate. Green "camera" and "globe" icons in the outside margins of text pages direct readers to relevant photographs and maps in the galleries. Placed as close to the text to which they pertain as possible, the icons are labeled with the subjects they illustrate. Icons are linked only to gallery pages within their own volumes, but readers can find additional pictures of many subjects by following the entries in the comprehensive index in volume 8.

Both English and metric measures are used throughout this set. In most instances, English measures are given first, followed by their metric equivalents in parentheses. It should be noted that in cases of measures that are only estimates, such as the areas of deserts or average heights of mountain ranges, the metric figures are often rounded off to estimates that may not be exact equivalents of the English-measure estimates. In order to enhance clarity, units of measure are not abbreviated in the text, with these exceptions: "Kilometers" are rendered as "km." and "square kilometers" as "sq. km." This exception has been made because of the frequency with which these measures appear.

Reference works such as this would be impossible without the expertise of a large team of contributing scholars. This project is no exception. Salem Press would like to thank the more than 175 people who wrote the signed essays and contributed entries to the gazetteers. Their names and affiliations can be found on the contributor list that follows this note. We also wish to thank the many photographers who contributed pictures to this set, particularly Joan Clemons, Tom McKnight, and Clyde L. Rasmussen. Finally, we are especially pleased to express our thanks to Dr. Ray Sumner, of California's Long Beach City College, for the expertise and insights that she has brought to this project as Editor.

INTRODUCTION

When Henry Morton Stanley of the *New York Herald* shook the hand of David Livingstone on the shore of Central Africa's Lake Tanganyika in 1871, the moment represented the high point of geography to many people throughout the world. A Scottish missionary and explorer, Livingstone had been out of contact with the outside world for nearly two years, and European and American newspapers had buzzed with speculation about his disappearance. At that time, so little was known about the geography of the interior of Africa that Stanley's finding Livingstone was acclaimed as a brilliant triumph of exploration.

The field of geography in Stanley and Livingstone's time was—and to a large extent still is—synonymous with exploration. Stories of epic journeys, both historic and contemporary, continue to exert a powerful attraction on readers. Mountains, deserts, forests, caves, and glaciers still draw intrepid explorers, while even more armchair travelers are thrilled by accounts and pictures of these exploits and discoveries. We all love to travel—to the beach, into the mountains, to our great National parks, and to foreign countries. In the need and desire to explore our surroundings, we are all geographers.

Numerous geographical societies welcome both professional geographers and the general public into their membership, as they promote a greater knowledge and understanding of the earth. The National Geographic Society, founded in 1888 "for the increase and diffusion of geographical knowledge," has funded more than 6,500 field expeditions and now has seven million dues-paying members. Each year the society invests more than five million dollars in expeditions and scientific field research related to environmental concerns and global geographic issues. The findings are recorded in the pages of the familiar yellow-bordered *National Geographic* magazine, which circles the globe with ten million copies in fifteen different languages, bringing readers up-to-date scientific information and memorable images of both familiar and exotic people and places. The National Geographic International television network reaches out to more than eighty million subscribers in 111 countries, broadcasting in sixteen languages.

An even older geographical association is Great Britain's Royal Geographical Society, which grew out of the Geographical Society of London, founded in 1830 with the "sole object" of promoting "that most important and entertaining branch of knowledge—geography." Over the century that followed, the Royal Geographical Society focused on exploration of the continents of Africa and Antarctica. In the society's London headquarters adjacent to the Albert Hall, visitors can still view such historic artifacts as David Livingstone's cap and chair, as well as diaries, sketches, and maps covering the great period of the British Empire and beyond. Today the society assists more than five hundred field expeditions every year.

With the aid of satellites and remote-sensing instruments we can now obtain images and data from almost anywhere on Earth. However, remote and inaccessible places still invite the intrepid to visit and explore them in person. Although the outlines of the continents have now been completed, and their interiors filled in with details of mountains and rivers, cities and political boundaries, remote places still exert a fascination on modern urbanites.

The enchantment of tales about strange sights and courageous journeys has been with us since the ancient voyages of Homer's Ulysses, Marco Polo's travels to China, and the nautical expeditions of Christopher Columbus, Ferdinand Magellan, and James Cook. While those great travelers are from the remote past, the age of exploration is far from over—a fact repeatedly demonstrated by the modern Norwegian navigator Thor Heyerdahl. Moreover, new journeys of discovery are still taking place. In 1993, after dragging a sled wearily across the frigid wastes of Antarctica for more than three months, Sir Ranulph Twisleton-Wykeham-Fiennes announced that the age of exploration is not dead. Six years later, in 1999, the long-missing body of British mountain climber George Mallory was found on the slopes of Mount Everest, near whose top he had mysteriously vanished in 1924. That discovery sparked a new wave of admiration and respect for explorers of such courage and endurance.

How many people have been enthralled by the bravery of Antarctic explorer Robert Falcon Scott and the noble sacrifice his injured colleague Lawrence Oates made in 1912, when he gave up his life in order not to slow down the rest of the expedition? There can be no doubt that the thrills and the dangers of exploring find resonance among many modern readers.

The struggle to survive in environments hostile to human beings reminds us of the power of our planet Earth. Recent best-selling books on this theme have included Jon Krakauer's *Into Thin Air* (1998), an account of a disastrous expedition climbing Mount Everest, and Sebastian Junger's *The Perfect Storm* (1997), the story of the worst gale of the twentieth century and its effect on a fishing fleet off the East Coast of North America. *Endurance* (1998), the epic of Sir Ernest

Henry Morton Stanley (left) meets Dr. Livingstone in Central Africa. (H. M. Stanley, *How I Found Livingstone,* 1872)

Shackleton's survival and leadership for two years on the frozen Arctic, attracts the same people who avidly read *Undaunted Courage* (1996), the story of Meriwether Lewis and William Clark's epic exploration of the Louisiana Purchase territories in the early nineteenth century. In 1997 *Seven Years in Tibet* premiered, a popular film about the Austrian Heinrich Harrer, who lived in Tibet in the mid-twentieth century. The more urban people become, the greater their desire for adventurous, remote places, at least vicariously, to raise the human spirit.

There are, of course, also scientific achievements associated with modern exploration. In November, 1999, the elevation of Mount Everest, the world's tallest peak, was raised by 7 feet (2.1 meters) to a new height of 29,035 feet (8,850 meters) above sea level; the previously accepted height had been based on surveys made during the 1950's. This new value was the result of Global Positioning System (GPS) technology enabling a more accurate measurement than had been possible with land-based earthbound surveying equipment. A team of climbers supported by the National Geographic Society and the Boston Museum of Science, was equipped with GPS equipment which enabled a fifty-minute recording of data based on satellite signals. At the same time, the expedition was able to ascertain that Mount Everest is moving northeast, atop the Indo-Australian Plate, at a rate of approximately 2.4 inches (10 centimeters) per year.

In 2000, the International Hydrographic Organization named a "new" ocean, the Southern Ocean, which encompasses all the water surrounding Antarctica up to 60 degrees south latitude. With an area of approximately 7.8 million square miles (20.3 million square kilometers), the Southern Ocean is about twice the size of the entire United States and ranks as the world's fourth largest ocean, after the Pacific, Atlantic, and Indian Oceans, but just ahead of the Arctic Ocean.

Despite the humanistic and scientific advantages of geographic knowledge, to many people today geography is a subject where one merely memorizes long lists of facts dealing with "where" questions (Where is Andorra? Where is Prince Edward Island? Where is Kalamazoo?) or "what" questions (What is the highest mountain in South America? What is the capital of Costa Rica?) This approach to the study of geography has been perpetuated by the annual National Geographic Bee, conducted in the United States each year for students in grades four through eight. Participants in the competition display an astonishing recall of facts but do not have the opportunity of showing any real geographic thought. To a geographer, such factual knowledge is simply a foundation for investigating and explaining the much more important questions dealing with "why"—"Why is the Sahara a desert?"

Geographers aim to understand why environments and societies occur where and as they do, and how they change. Geography must be seen as an integrative science; the collection of factual data and evidence, as in exploration, is the empirical foundation for deductive reasoning. This leads to the creation of a range of geographical methods, models, theories, and analytical approaches that serve to unify a very broad area of knowledge—the interaction between natural and human environments. Although geography as an academic discipline became established in nineteenth century Germany, there have always been geographers, in the sense of people curious about their world. Humans have always wanted to know about day and night, the shape of the earth, the nature of climates, differences in plants and animals, as well as what lies beyond the horizon. Today, as we hear about and actually experience the sweeping effects of globalization, we need more than ever to develop our geographical skills. Not only are we connected by economic ties to the countries of the world, but we must also appreciate the consequences of North America's high standard of living.

Political boundaries are artificial human inventions, but the natural world is one biosphere. As concern over global warming escalates, national leaders meet to seek a solution to emission of greenhouse gases. Are we connected to our environment? At a time when the rate of species extinction is a hundred times above normal, and the human population is crowding in increasing numbers into huge urban centers, we have, nevertheless, taken time each year in April to celebrate Earth Day since 1970. We need now to realize that every day is Earth Day.

Geography languished in the United States in the 1960's, as social studies was taught with a history emphasis in schools. American students became alarmingly disadvantaged in geographic knowledge, compared with most other countries. Fortunately members of the profession acted to restore geography to the curriculum. In 1984 the National Geographic Society undertook the challenge of restoring geography in the United States. The society turned to two organizations active in geographic education: the Association of American Geographers, the professional geographers' group with more than 6,500 members, mostly in higher education in the United States; and the National Council for Geographic Education, with some one thousand members. The council administers the Geographic Alliances, found in every state of the United States, with a national membership of about 120,000 schoolteachers. Together they produced the "Guidelines in Geographic Education," which introduced the Five Themes of Geography, to enhance the teaching of geography in schools. Using the themes of Location, Place, Human/Environment Interaction, Movement and Regions, teachers were able to plan and conduct lessons in which students encountered interesting real-world examples of the relevance and importance of geography. Continued research into geographic education led to the inclusion of geography in 1990 as one of the core subjects of the National Education Goals, or "Goals 2000," along with English, mathematics, science, and history.

Another milestone was the publication in 1994 of "Geography for Life," the national Geography Standards. The earlier Five Themes are subsumed under the new Six Essential Elements: The World in Spatial Terms, Places and Regions, Physical Systems, Human Systems, Environment and Society, and The Uses of Geography. Eighteen Geography Standards are included, describing what a geographically informed person knows and understands. States, schools, and individual teachers have welcomed the new prominence of geography, and enthusiastically adopted new approaches to introduce the geography standards to new learners. The rapid spread of computer technology, especially in the field of Geographical Information Science, has also meant a new importance for spatial analysis, a traditional area of geographical expertise. No longer is geography seen as an outdated mass of useless or arcane facts; instead geography is now seen, again, to be an innovative and integrative science, which can in the twenty-first century contribute to solving complex problems associated with the reciprocal human-environmental relationship.

Geographers may no longer travel across uncharted realms, but there is still much we long to explore, to learn, and seek to understand, even if it is only as "armchair" geographers. This reference work, *World Geography*, will help carry readers on their own journeys of exploration.

Ray Sumner

CONTRIBUTORS

Emily Alward
Henderson, Nevada Public Library

Earl P. Andresen
University of Texas at Arlington

Debra D. Andrist
St. Thomas University

Charles F. Bahmueller
Center for Civic Education

Timothy J. Bailey
Pittsburg State University

David Barratt
Nottingham, England

Maryanne Barsotti
Warren, Michigan

Thomas F. Baucom
Jacksonville State University

Michelle Behr
Western New Mexico University

Alvin K. Benson
Brigham Young University

Cynthia Breslin Beres
Glendale, California

Nicholas Birns
New School University

Olwyn Mary Blouet
Virginia State University

Margaret F. Boorstein
*C.W. Post College of Long Island
University*

Fred Buchstein
John Carroll University

Joseph P. Byrne
Belmont University

Laura M. Calkins
Palm Beach Gardens, Florida

Gary A. Campbell
Michigan Technological University

Byron D. Cannon
University of Utah

Steven D. Carey
University of Mobile

Roger V. Carlson
Jet Propulsion Laboratory

Robert S. Carmichael
University of Iowa

Habte Giorgis Churnet
*University of Tennessee at
Chattanooga*

Richard A. Crooker
Kutztown University

William A. Dando
Indiana State University

Larry E. Davis
College of St. Benedict

Ronald W. Davis
Western Michigan University

Cyrus B. Dawsey
Auburn University

Frank Day
Clemson University

M. Casey Diana
*University of Illinois at Urbana-
Champaign*

Stephen B. Dobrow
Farleigh Dickinson University

Steven L. Driever
*University of Missouri,
Kansas City*

Sherry L. Eaton
San Diego City College

Femi Ferreira
Hutchinson Community College

Helen Finken
Iowa City High School

Eric J. Fournier
Samford University

Anne Galantowicz
El Camino College

Hari P. Garbharran
*Middle Tennessee State
University*

Keith Garebian
Ontario, Canada

Laurie A. B. Garo
*University of North Carolina,
Charlotte*

Jay D. Gatrell
Indiana State University

Carol Ann Gillespie
Grove City College

Nancy M. Gordon
Amherst, Massachusetts

Noreen A. Grice
Boston Museum of Science

Johnpeter Horst Grill
Mississippi State University

Charles F. Gritzner
South Dakota State University

C. James Haug
Mississippi State University

Douglas Heffington
Middle Tennessee State University

Thomas E. Hemmerly
Middle Tennessee State University

Jane F. Hill
Bethesda, Maryland

Carl W. Hoagstrom
Ohio Northern University

Catherine A. Hooey
Pittsburg State University

Robert M. Hordon
Rutgers University

Kelly Howard
La Jolla, California

Paul F. Hudson
University of Texas at Austin

Huia Richard Hutton
*University of Hawaii/Kapiolani
Community College*

Raymond Pierre Hylton
Virginia Union University

Solomon A. Isiorho
*Indiana University/Purdue
University at Fort Wayne*

Ronald A. Janke
Valparaiso University

Albert C. Jensen
*Central Florida Community
College*

Jeffry Jensen
Altadena, California

Bruce E. Johansen
University of Nebraska at Omaha

Kenneth A. Johnson
*State University of New York,
Oneonta*

Walter B. Jung
University of Central Oklahoma

James R. Keese
*California Polytechnic State
University, San Luis Obispo*

Leigh Husband Kimmel
Indianapolis, Indiana

Denise Knotwell
Wayne, Nebraska

James Knotwell
Wayne State College

Grove Koger
Boise Idaho Public Library

Alvin S. Konigsberg
*State University of New York at
New Paltz*

Steven Lehman
John Abbott College

Denyse Lemaire
Rowan University

Dale R. Lightfoot
Oklahoma State University

Jose Javier Lopez
Minnesota State University

James D. Lowry, Jr.
East Central University

Jinshuang Ma
*Arnold Arboretum of Harvard
University Herbaria*

Dana P. McDermott
Chicago, Illinois

Thomas R. MacDonald
University of San Francisco

Robert R. McKay
Clarion University of Pennsylvania

Nancy Farm Männikkö
L'Anse, Michigan

Carl Henry Marcoux
University of California, Riverside

Christopher Marshall
Unity College

Rubén A. Mazariegos-Alfaro
University of Texas/Pan American

Christopher D. Merrett
Western Illinois University

John A. Milbauer
Northeastern State University

Randall L. Milstein
Oregon State University

Judith Mimbs
Loftis Middle School

Karen A. Mulcahy
East Carolina University

xvi

B. Keith Murphy
Fort Valley State University

M. Mustoe
Omak, Washington

Bryan Ness
Pacific Union College

Kikombo Ilunga Ngoy
Vassar College

Joseph R. Oppong
University of North Texas

Richard L. Orndorff
University of Nevada, Las Vegas

Bimal K. Paul
Kansas State University

Nis Petersen
New Jersey City University

Mark Anthony Phelps
Ozarks Technical Community College

John R. Phillips
Purdue University, Calumet

Alison Philpotts
Shippensburg University

Julio César Pino
Kent State University

Timothy C. Pitts
Morehead State University

Carolyn V. Prorok
Slippery Rock University

P. S. Ramsey
Highland Michigan

Robert M. Rauber
University of Illinois at Urbana-Champaign

Ronald J. Raven
State University of New York at Buffalo

Neil Reid
University of Toledo

Susan Pommering Reynolds
Southern Oregon University

Nathaniel Richmond
Utica College

Edward A. Riedinger
Ohio State University Libraries

Mika Roinila
West Virginia University

Thomas E. Rotnem
Brenau University

Joyce Sakkal-Gastinel
Marseille, France

Helen Salmon
University of Guelph

Elizabeth D. Schafer
Loachapoka, Alabama

Kathleen Valimont Schreiber
Millersville University of Pennsylvania

Ralph C. Scott
Towson University

Guofan Shao
Purdue University

Wendy Shaw
Southern Illinois University, Edwardsville

R. Baird Shuman
University of Illinois, Champaign-Urbana

Sherman E. Silverman
Prince George's Community College

Roger Smith
Portland, Oregon

Robert J. Stewart
California Maritime Academy

Toby R. Stewart
Alamosa, Colorado

Ray Sumner
Long Beach City College

Paul Charles Sutton
University of Denver

Glenn L. Swygart
Tennessee Temple University

Sue Tarjan
Santa Cruz, California

Robert J. Tata
Florida Atlantic University

John M. Theilmann
Converse College

Virginia Thompson
Towson University

Norman J. W. Thrower
University of California, Los Angeles

Paul B. Trescott
Southern Illinois University

Robert D. Ubriaco, Jr.
Illinois Wesleyan University

Mark M. Van Steeter
Western Oregon University

Johan C. Varekamp
Wesleyan University

Anthony J. Vega
Clarion University

Kristopher D. White
University of Connecticut

Kay R. S. Williams
Shippensburg University

William T. Walker
Chestnut Hill College

P. Gary White
Western Carolina University

Lisa A. Wroble
*Redford Township District
Library*

William D. Walters, Jr.
Illinois State University

Thomas A. Wikle
Oklahoma State University

Bin Zhou
*Southern Illinois University,
Edwardsville*

Linda Qingling Wang
University of South Carolina, Aiken

Rowena Wildin
Pasadena, California

Annita Marie Ward
Salem-Teikyo University

Donald Andrew Wiley
Anne Arundel Community College

THE NATURE OF
GEOGRAPHY

The History of Geography

The moment that early humans first looked around their world with inquiring minds was the moment that geography was born. The history of geography is the history of human effort to understand the nature of the world. Through the centuries, people have asked of geography three basic questions: What is the earth like? Where are things located? How can one explain these observations?

GEOGRAPHY IN THE ANCIENT WORLD. In the Western world, the Greeks and the Romans were among the first to write about and study geography. Eratosthenes, a Greek scholar who lived in the third century B.C.E., is often called the "father of geography and is credited with first using the word geography (from the Greek words *ge*, which means "earth," and *graphe*, which means "to describe"). The ancient Greeks had contact with many older civilizations and began to gather together information about the known world. Some, such as Hecataeus, described the multitude of places and peoples with which the Greeks had contact and wrote of the adventures of mythical characters in strange and exotic lands. However, the ancient Greek scholars went beyond just describing the world. They used their knowledge of mathematics to measure and locate. The Greek scholars also used their philosophical nature to theorize about Earth's place in the universe.

One Greek scholar who used mathematics in the study of geography was Anaximander, who lived from 610 to 547 B.C.E. Anaximander is credited with being the first person to draw a map of the world to scale, and he also invented a sundial that could be used to calculate time and direction, and to distinguish the seasons. Eratosthenes is also famous for his mathematical calculations, in particular of the circumference of the earth, using observations of the Sun. Hipparchus, who lived around 140 B.C.E., used his mathematical skills to solve geographic problems and was the first person to introduce the idea of a latitude and longitude grid system to locate places.

Such early Greek philosophers as Plato and Aristotle were also concerned with ge-

CURIOSITY: THE ROOT OF GEOGRAPHY

The earliest human beings, as they hunted and gathered food and used primitive tools in order to survive, must have had detailed knowledge of the geography of their part of the world. The environment could be a hostile place, and knowledge of the world meant the difference between life and death. Human curiosity took them one step further. As they lived in an ancient world of ice and fire, human beings looked to the horizon for new worlds, crossing continents and spreading out to all areas of the globe. They learned not only to live as a part of their environment, but also to understand it, predict it, and change it to their needs.

Plato. (Library of Congress)

ography. They discussed such issues as whether the earth was flat or spherical and if it was the center of the universe, and debated the nature of the earth as the home of humankind.

Whereas the Greeks were great thinkers and introduced many new ideas into geography, the Roman contribution was to compile and gather available knowledge. Although this did not add much that was new to geography, it meant that the knowledge of the ancient world was available as a base to work from and was passed down across the centuries. Geography in the ancient world is often said to have ended with the great work of Ptolemy (Claudius Ptolemaus), who lived from 90

to 168 C.E. Ptolemy is best known for his eight-volume *Guide to Geography*, which included a gazetteer of places located by latitude and longitude, and his world map.

The study of geography also was important in ancient China. Chinese scholars described their resources, climate, transportation routes, and travels, and were mapping their known world at the same time as were the great Western civilizations.

GEOGRAPHY IN THE MIDDLE AGES. With the collapse of the Roman Empire in the fifth century C.E., Europe descended into what is commonly known as the Dark or Middle Ages. During this time, which lasted until the fifteenth century, the geographic knowledge of the ancient world was either lost or challenged as being counter to Christian teachings. For example, the early Greeks had theorized that the earth was a sphere, but this was rejected during the Middle Ages. Scholars of the Middle Ages believed that the world was said to be a flat disk, with the holy city of Jerusalem at its center.

The knowledge and ideas of the ancient world might have been lost if they had not been preserved by Muslim scholars. In the Islamic countries of North Africa and the Middle East, some of the scholarship of the ancient world was sheltered in libraries and universities. This knowledge was extensively added to as Muslims traveled and traded across the known world, gathering their own information.

Among the most famous Muslim geographers were Ibn Battutah, al-Idrisi, and Ibn Khaldun. Ibn Battutah traveled east to India and China in the fourteenth century. Al-Idrisi, at the command of King Roger II of Sicily, wrote *Roger's Book*, which systematically described the world. Information from *Roger's Book* was engraved on a huge planisphere (disk), crafted in silver; this once was considered a wonder of the world, but it is thought to have been destroyed. Ibn Khaldun (1332-1406) is best known for his written world history, but he also was a pioneer in focusing on the relationship of human beings to their environment.

THE AGE OF EUROPEAN EXPLORA-TION. Beginning in the fifteenth century, the isolation of Europe came to an end, and Europeans turned their attention to exploration. The two major goals of this

Jerusalem's Church of the Holy Sepulchre, which contains a column marking the earth's center. After Mark Twain visited the spot in 1867, he wrote about his excitement in finding "the exact centre of the earth." However, his tone made it clear he took that claim no more seriously than the claim that a nearby tomb contained the remains of the biblical Adam. (Mark Twain, The Innocents Abroad, 1869)

sudden surge in exploration were to spread the Christian faith and to obtain needed resources. In 1418 Prince Henry the Navigator established a school for navigators and began to gather the tools and knowledge needed for exploration. He was the first of many Europeans who broke out of the darkness of the Middle Ages, traveling beyond the limits of the known world, mapping, describing, and cataloging all that they saw.

The great wave of European exploration brought new interest in geography, and the monumental works of the Greeks and Romans—so carefully preserved by Muslim scholars—were rediscovered and translated into Latin. The maps produced in the Middle Ages were of little use to the explorers who were traveling to, and beyond, the limits of the known world. Christopher Columbus, for example, relied on Ptolemy's work during his voyages to the Americas, but soon newer, more accurate maps were drawn and, for the first time, globes were made. A particularly famous map, which is still used as a base map, is the Mercator projection. On the world map produced by Gerardus Mercator (Gerhard Kremer) in 1569, compass directions appear as straight lines, which was a great benefit on navigational charts.

When the age of European exploration began, the best world maps crudely depicted a few limited areas of the world. Explorers quickly began to gather huge quantities of information, making detailed charts of coastlines, discovering new continents, and eventually filling in the maps of those continents with information about both the natural and human features they encountered. This age of exploration is often said to have ended when Roald Amundsen planted the Norwegian flag at the South Pole in 1911. At that time, the world map became complete, and human beings had mapped and explored every corner of the globe. However, the beginning of modern geography is usually associated with the work of two nineteenth century German geographers: Alexander von Humboldt and Carl Ritter.

THE BEGINNING OF MODERN GEOGRAPHY. The writings of Alexander von Humboldt and Karl Ritter mark a leap into modern geography, because these writers took an important step beyond the work of previous scholars. The explorers of the previous centuries had focused on gathering information, describing the world, and filling in the world map with as much detail as possible. Humboldt and Ritter took a more scientific and systematic approach to geography. They began not only to compile descriptive information, but also to ask why: Humboldt spent his lifetime looking for relationships among such things as climate and topography (landscape), while Ritter was intrigued by the multitude of connections and relationships he observed within human geographic patterns. Both Humboldt and Ritter died in 1859, ending a period when information-gathering had been paramount. They brought geography into a

Alexander von Humboldt. (Library of Congress)

new age in which synthesis, analysis, and theory-building became central.

EUROPEAN GEOGRAPHY. After the work of Humboldt and Ritter, geography became an accepted academic discipline in Europe, particularly in Germany, France, and Great Britain. Each of these countries emphasized different aspects of geographic study. German geographers continued the tradition of the scientific view, using observable data to answer geographic questions. They also introduced the concept that geography could take a chorological view, studying all aspects, physical and human, of a region and of the interrelationships involved.

The chorological view came to dominate French geography. Paul Vidal de la Blache (1845-1918) was the most prominent French geographer. He advocated the study of small, distinct areas, and French geographers set about identifying the many regions of France. They described and analyzed the unique physical and human geographic complex that was to be found in each region. An important concept that emerged from French geography was "possibilism." German geographers had introduced the notion of environmental determinism—that human beings were largely shaped and controlled by their environments. Possibilism rejected the concept of environmental determinism, asserting that the relationship between human beings and the environment works in two directions: The environment creates both limits and opportunities for people, but people can react in different ways to a given environment, so they are not controlled by it.

British geographers, influenced by the French approach, conducted regional surveys. British regional studies were unique in their emphasis on planning and geography as an applied science. From this work came the concept of a functional region—an area that works together as a unit based on interaction and interdependence.

AMERICAN GEOGRAPHY. Prior to World War II, only a small group of people in the United States called themselves geographers. They were mostly influenced by German ideas, but the nature of geography was hotly debated. Two schools of geographers were philosophical adversaries. The Midwestern School, led by Richard Hartshorne, believed that description of unique regions was the central task of geography.

The Western (or Berkeley) School of geography, led by Carl Sauer, agreed that regional study was important, but believed it was crucial to go beyond description. Sauer and his followers included genesis and process as important elements in any study. To understand a region and to know where it is going, they argued, one must look at its past and how it got to its present state.

In the 1930's, environmental determinism was introduced to U.S. geography but ultimately was rejected. Although geography in both Europe and the United States was essentially an all-male discipline, the United States produced the first famous woman geographer, Ellen Churchill Semple (1863-1932).

World War II illustrated the importance of geographic knowledge, and after the war came to an end in 1945, geographers began to blossom in the United States. From the end of World War II to the early 1960's, U.S. geographers produced many descriptive regional studies.

In the early 1960's, what is often called the quantitative revolution occurred. The

---

THE NATIONAL GEOGRAPHIC SOCIETY AND GEOGRAPHIC RESEARCH

In 1888 the National Geographic Society was founded to support the "increase and diffusion of geographic knowledge" of the world. In its first 110 years, the society funded more than five thousand expeditions and research projects with more than sixty-five hundred grants. By the 1990's it was the largest such foundation in the world, and the results of its funded projects are found on television programs, video discs, video cassettes, and books, as well as in the *National Geographic* magazine, established in 1888. Its productions are cutting-edge resources for information about archaeology, ethnology, biology, and both cultural and physical geography.

---

development of computers allowed complex mathematical analysis to be performed on all kinds of geographic data, and geographers began to analyze a wide range of problems using statistics. There was great enthusiasm for this new approach to geography at first, but beginning in the 1970's, many people considered a purely mathematical approach to be somewhat sterile and thought it left out a valuable human element.

In the 1980's and 1990's, many new ways to look at geographic issues and problems were developed, including humanism, behaviorism, Marxism, feminism, realism, structuration, phenomenology, and postmodernism, all of which bring human beings back into focus within geographical studies.

GEOGRAPHY FOR A NEW MILLENNIUM. Geography increasingly uses technology to analyze global space and answer a wide range of questions. The Geographic Information System (GIS), in particular, provides a powerful way for people trained in geography to understand geographic issues, solve geographic problems, and display geographic information. Geographers continue to adopt a wide variety of philosophies, approaches, and methods in

INFORMATION ON THE WORLD WIDE WEB

A good starting point for Internet research on the history of geography is the About.com geography site, which contains links to pages on physical, regional, and human geography. (geography.about.com/education/geography/)

The Web site of the Environmental Systems Research Institute, an early leader in mapping products using GIS technology, allows viewers to create maps using GIS data (www.esri.com/data/online/index.html) and download ArcExplorer, a free GIS data viewer (www.esri.com/software/arcexplorer/index.html).

their quest to answer questions concerning all things spatial.

Wendy Shaw

FOR FURTHER STUDY

Abler, Ronald F., Melvin G. Marcus, and Judy M. Olson, eds. *Geography's Inner Worlds.* New Brunswick, N.J.: Rutgers University Press, 1992.

Agnew, John, David N. Livingstone, and Alisdair Rogers, eds. *Human Geography: An Essential Anthology.* Cambridge, Mass.: Blackwell, 1996.

Flowers, Sarah. *The Age of Exploration.* San Diego, Calif.: Lucent Books, 1999.

Gould, Peter. *Becoming a Geographer (Space, Place, and Society).* Syracuse, N.Y.: Syracuse University Press, 1999.

Holt-Jensen, Arild. *Geography: History and Concepts. A Student Guide.* 3d ed. Walnut Creek, Calif.: AltaMira Press, 1999.

Martin, Geoffrey J., and Preston E. James. *All Possible Worlds. A History of Geographical Ideas.* 3d ed. New York: John Wiley & Sons, 1993.

MAPMAKING IN HISTORY

Cartography is the science or art of making maps. Although workers in many fields have a concern with cartography and its history, it is most often associated with geography.

MAPS OF PRELITERATE PEOPLES. The history of cartography predates the written record, and most cultures show evidence of mapping skills. The earliest surviving maps are those carved in stone or painted on the walls of caves, but modern preliterate peoples still use a variety of materials to express themselves cartographically. For example, the Marshall Islanders use palm fronds, fiber from coconut husks (coir), and shells to make sea charts for their inter-island navigation. The Inuit use animal skins and driftwood, sometimes painted, in mapping. There is a growing interest in the cartography of early and preliterate peoples, but some of their maps do not fit readily into a more traditional concept of cartography.

MAPPING IN ANTIQUITY. Early literate peoples, such as those of Egypt and Mesopotamia, displayed considerable variety in their maps and charts, as shown by the few maps from these civilizations that still exist. The early Egyptians painted maps on wooden coffin bases to assist the departed in finding their way in the afterlife; they also made practical route maps for their mining operations. It is thought that geometry developed from the Egyptians' riverine surveys. The Babylonians made maps of different scales, using clay tablets with cuneiform characters and stylized symbols, to create city plans, regional maps, and "world" maps. They also di-

vided the circle in the sexigesimal system, an idea they may have obtained from India and that is commonly used in cartography to this day.

The Greeks inherited ideas from both the Egyptians and the Mesopotamians and made signal contributions to cartography themselves. No direct evidence of early Greek maps exists, but indirect evidence in texts provides information about their cosmological ideas, culminating in the concept of a perfectly spherical earth. This they attempted to measure and divide mathematically. The idea of climatic zones was proposed and possibly mapped, and the large known landmasses were divided into first two continents, then three.

Perhaps the greatest accomplishment of the early Greeks was the remarkably accurate measurement of the circumference of the earth by Eratosthenes (276-196 B.C.E.). Serious study of map projections began at about this time. The gnomonic, orthographic, and stereographic projections were invented before the Christian era, but their use was confined to astronomy in this period. With the possible single exception of Aristarchus of Samos, the Greeks believed in a geocentric universe. They made globes (now lost) and regional maps on metal; a few map coins from this era have survived.

Later Greeks carried on these traditions and expanded upon them. Claudius Ptolemy invented two projections for his world maps in the second century C.E. These were enormously important in the European Renaissance as they were modified in the light of new overseas discover-

*Map
Pages 32-33*

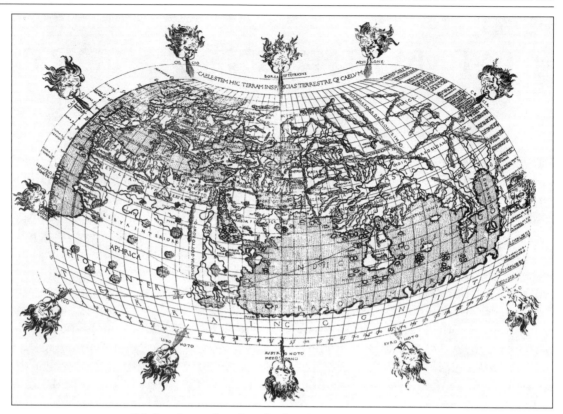

Medieval map based on Ptolemy's early work. (Corbis)

ies. Ptolemy's work is known mainly through later translations and reconstructions, but he compiled maps from Greek and Phoenician travel accounts and proposed sectional maps of different scales in his *Geographia*. Ptolemy's prime meridian (0 degrees longitude) in the Canary Islands was generally accepted for a millennium and a half after his death.

Roman cartography was greatly influenced by later Greeks such as Ptolemy, but the Romans themselves improved upon route mapping and surveying. Much of the Roman Empire was subdivided by instruments into hundredths, of which there is a cartographic record in the form of marble tablets. In Rome, a small-scale map of the world known to the Romans was made on metal by Marcus Vipsanius Agrippa, the son-in-law of Augustus Caesar, and displayed publicly. This map no longer exists, however.

CARTOGRAPHY IN EARLY EAST ASIA. As these developments were taking place in the West, a rich cartographic tradition developed in Asia, particularly China. The earliest survey of China (Yu Kung) is approximately contemporaneous with the oldest reported mapmaking activity of the Greeks. Later, maps, charts, and plans accompanied Chinese texts on various geographical themes. Early rulers of China had a high regard for cartography—the science of princes. A rectangular grid was introduced by Chang Heng, a contemporary of Ptolemy, and the south-pointing needle was used for mapmaking in China from an early date.

These traditions culminated in Chinese cartographic primacy in several areas: the earliest printed maps (about 1155 C.E.), early printed atlases, and terrestrial globes (now lost). Chinese cartography greatly influenced that in other parts of

Asia, particularly Korea and Japan, which fostered innovations of their own. It was

Fanciful medieval engraving of Ptolemy. (Library of Congress)

only after the introduction of ideas from the West, in the Renaissance and later, that Asian cartographic advances were superseded.

ISLAMIC CARTOGRAPHY. A link between China and the West was provided by the Arabs, particularly after the establishment of Islam. It was probably the Arabs who brought the magnetized needle to the Mediterranean, where it was developed into the magnetic compass.

Some scholars have argued that the Arabs were better astronomers than cartographers, but did make several clear advances in mapmaking. Both fields of study were important in Muslim science, and the astrolabe, invented by the Greeks in antiquity but developed by the Arabs, was used in both their astronomical and terrestrial surveys. They made and used many maps, as indicated by the output of their most famous cartographer, al-Idrisi (who lived about 1100-1165). Some of his work still exists, including a zonal world map

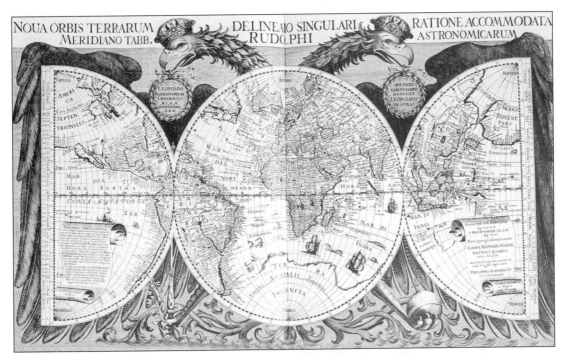

World map prepared by the German astronomer Johannes Kepler and published in 1630, the year of his death. (Corbis)

and detailed charts of the Mediterranean islands.

At about the same time, the magnetic compass was invented in the coastal cities of Italy, which gave rise to advanced navigational charts, including information on ports. These remarkably accurate charts were used for navigating in the Mediterranean Sea. They were superior to the European maps of the Middle Ages, which often were concerned with religious iconography, pilgrimage, and crusade. The scene was now set for the great overseas discoveries of the Europeans, which were initiated in Portugal and Spain in the fifteenth century.

Mercator projection Page 168

In the next four centuries, most of the coasts of the world were visited and mapped. The early, projectionless navigational charts were no longer adequate, so new projections were invented to map the enlarged world as revealed by the European overseas explorations. The culmination of this activity was the development of the projection, in 1569, of Gerardus

Gerardus Mercator. (Library of Congress)

Mercator, which bears his name and is of special value in navigation.

EARLY MODERN MAPMAKING. Europeans began mapping their own countries with greater accuracy. New surveying instruments were invented for this purpose, and a great land-mapping activity was undertaken to match the worldwide coastal surveys. For about a century, the Low Countries of Belgium, Luxembourg, and the Netherlands dominated the map and chart trades, producing beautiful hand-colored engraved sheet wall maps and atlases.

France and England established new national observatories, and by the middle of the seventeenth century, the Low Countries had been eclipsed by France in surveying and making maps and charts. The French adopted the method of triangulation of Mercator's teacher, Gemma Friisius. Under four generations of the Cassini family, a topographic survey of France more comprehensive than any previous survey was completed. Rigorous coastal surveys were undertaken, as well as the precise measurement of latitude (parallels).

The invention of the marine chronometer by John Harrison made it possible for ships at sea to determine longitude. This led to the production of charts of all the oceans, with England's Greenwich eventually being adopted as the international prime meridian.

Quantitative, thematic mapping was advanced by astronomer Edmond Halley (1656-1742) who produced a map of the trade winds; the first published magnetic variation chart, using isolines; tidal charts; and the earliest map of an eclipse. The Venetian Vincenzo Coronelli made globes of greater beauty and accuracy than any previous ones. In the German lands, the study of map projections was vigorously pursued. Johann H. Lambert and others invented a number of equal-area projections that were still in use in the twentieth century.

Ideas developed in Europe were transmitted to colonial areas, and to countries such as China and Russia, where they were grafted onto existing cartographic traditions and methods. The oceanographic explorations of the British and the French built on the earlier charting of the Pacific Ocean and its islands by native navigators and the Iberians.

NINETEENTH CENTURY CARTOGRAPHY. Cartography was greatly diversified and developed in the nineteenth century. Quantitative, thematic mapping was expanded to include the social as well as the physical sciences. Alexander von Humboldt used isolines to show mean air temperature, a method that later was applied to other phenomena. Contour lines gradually replaced less quantitative methods of representing terrain on topographic maps. Such maps were made of many areas, for example India, which previously had been poorly mapped.

Extraterrestrial (especially lunar) mapping, had begun seriously in the preceding two centuries with the invention of the telescope. It was expanded in the nineteenth century. In the same period, regular national censuses provided a large body of data that could be mapped. Ingenious methods were created to express the distribution of population, diseases, social problems, and other data quantitatively, using uniform symbols.

The development of steamships during the nineteenth century revolutionized travel. In 1867 Mark Twain sailed to Europe on the steamship Quaker City *(pictured) in what was regarded as the first-ever trans-Atlantic cruise undertaken purely for pleasure. His published account of that cruise,* The Innocents Abroad *(1869) launched his career as an author of books and became the best-selling travel book of the nineteenth century. Mark Twain himself eventually crossed the Atlantic twenty-five times—a feat that would have been many times more difficult in wind-powered ships.* (Mark Twain, *The Innocents Abroad,* 1869)

Geological mapping began in the nineteenth century with the work of William Smith in England, but soon was adopted worldwide and systematized, notably in the United States. The same is true of transportation maps, as the steamship and the railway increased mobility for many people. Faster land travel in an east-west direction, as in the United States, led to the official adoption of Greenwich as the international prime meridian at a conference held in Washington, D.C., in 1884. Time zone maps were soon published and became a feature of the many world atlases then being published for use in schools, offices, and homes.

A remarkable development in cartography in the nineteenth century was the surveying of areas newly occupied by Europeans. This occurred in such places as the

Satellite image Page 31

South American republics, Australia, and Canada, but was most evident in the United States. The U.S. Public Land Survey covered all areas not previously subdivided for settlement. Property maps arising from surveys were widely available, and in many cases, the information was contained in county and township atlases and maps.

MODERN MAPPING AND IMAGING. Cartography was revolutionized in the twentieth century by aerial photography, sonic sounding, satellite imaging, and the computer. Before those developments, however, Albrecht Penck proposed an ambitious undertaking—an International Map of the World (IMW). Cartography historically had been a nationalistic enterprise, but Penck suggested a map of the world in multiple sheets produced cooperatively by all nations at the scale of 1:1,000,000 with uniform symbols. This was started in the first half of the twentieth century but was not completed, and was superseded by the World Aeronautical Chart (WAC) project, at the same scale, during and after World War II.

The WAC project owed its existence to flight information made available following the invention of the airplane. Both photography and balloons were developed before the twentieth century, but the new, heavier-than-air craft permitted overlapping aerial photographs to be taken, which greatly facilitated the mapping process. Aerial photography revolutionized land surveys—maps could be made at less cost, in less time, and with greater accuracy than by previous methods. Similarly, marine surveying was revolutionized by the advent of sonic sounding in the second half of the twentieth century. This enabled mapping of the floor of the oceans, essentially unknown before this time.

Satellite imaging, especially continuous surveillance by Landsat since 1972, allows temporal monitoring of the earth. The computer, through Geographical Information Systems (GIS) and other technologies, has greatly simplified and speeded up the mapping process. During the twentieth century, the most widely available cartographic product was the road map for travel by automobile.

Spatial information now comes through television and computer screens as well as by more traditional cartographic means. The new media also facilitate animated presentations of geographical and extraterrestrial distributions. Cartographers in the twentieth century generally have been responsive to the opportunities provided by new technologies, materials, and ideas.

Norman J. W. Thrower

FOR FURTHER STUDY

Brown, Lloyd A. *The Story of Maps.* New York: Little, Brown, 1947.

Crone, Gerald R. *Maps and Their Makers.* Hamden, Conn.: Anchor Books, 1978.

Harley, J. B., and David Woodward, eds. *The History of Cartography.* Chicago: University of Chicago Press, 1987.

Skelton, R. A. *History of Cartography.* Cambridge. Mass: Harvard University Press, 1964.

Thrower, Norman J. W. *Maps and Civilization.* Chicago: University of Chicago Press, 1999.

MAPMAKING AND NEW TECHNOLOGIES

The field of geography is concerned primarily with the study of the curved surface of the earth. The earth is huge, however, with an equatorial radius of 3,963 miles (6,378 km.). How can one examine anything more than the small patch of earth that can be experienced at one time? Geographers do what scientists do all of the time: create models. The most common model of the earth is a globe—a spherical map that is usually about the size of a basketball.

A globe can show physical features such as rivers, oceans, the continents, and even the ocean floor. Political globes show the division of the earth into countries and states. Globes can even present views of the distant past of the earth, when the continents and oceans were very different than they are today. Globes are excellent for learning about the distributions, shapes, sizes, and relationships of features of the earth. However, there are limits to the use of globes.

How can the distribution of people over the entire world be described at one glance? On a globe, the human eye can see only half of the earth at one time. What if a city planner needs to map every street, building, fire hydrant, and streetlight in a town? To fit this much detail on a globe, the globe might have to be bigger than the town being mapped. Globes like these would be impossible to create and to carry around. Instead of having to hire a fleet of flatbed trucks to haul oversized globes, the curved surface of the globe can be transformed to a flat plane.

The method used to change from a curved globe surface to a flat map surface is called a map projection. There are hundreds of projections, from simple to extremely complex and dating from about two thousand years ago to projections being invented today. One of the oldest is the gnomonic projection. Imagine a clear globe with a light inside. Now imagine holding a piece of paper against the surface of the globe. The coastlines and parallels of latitude and meridians of longitude would show through the globe and be visible on the paper. Computers can do the same thing because there are mathematical formulas for nearly all map projections.

GEOMETRIC MODELS FOR MAP PROJECTIONS. One way to organize map projections is to imagine what kind of geometric shape might be used to create a map. Like the paper (a plane surface) against the globe described above, other useful geometric shapes include a cone and a cylinder. When the rounded surface of any object, including the earth, is flattened there must be some stretching, or tearing. Map projections help to control the amount and kinds of distortion in maps. There are always a few exceptions that cannot be described in this way, but using geometric shapes helps to classify projections into groups and to organize the hundreds of projections.

*Maps
Pages 226-
227*

15

Another way to describe a map projection is to consider what it might be good for. Some map projections show all of the continents and oceans at their proper sizes relative to one another. Another type of projection can show correct distances between certain points.

MAP PROJECTION PROPERTIES. When areas are retained in the proper size relationships to one another, the map is called an equal-area map, and the map projection is called an equal-area projection. Equal-area (also called equivalent or homolographic) maps are used to measure areas or view densities such as a population density.

If true angles are retained, the shapes of islands, continents, and oceans look more correct. Maps made in this way are called conformal maps or conformal map projections. They are used for navigation, topographic mapping, or in other cases when it is important to view features with a good representation of shape. It is impossible for a map to be both equal-area and conformal at the same time. One or the other must be selected based on the needs of the map user or map maker.

One special property—distance—can only be true on a few parts of a map at one time. To see how far it is between places hundreds or thousands of miles apart, an equidistant projection should be used. There will be several lines along which distance is true. The azimuthal equidistant projection shows true distances from the center of the map outward. Some map projections do not retain any of these properties but are useful for showing compromise views of the world.

MODERN MAPMAKING. Modern mapmaking is assisted from beginning to end by digital technologies. In the past, the paper map was both the primary means for communicating information about the world and the database used to store information. At the start of the twenty-first century, the database is a digital database stored in computers, and cartographic visualizations have taken the place of the paper map. Visualizations may still take the form of paper maps, but they also can appear as flashes on computer screens, animations on local television news programs, and even on screens within vehicles to help drivers navigate. Communication of information is one of the primary purposes of making maps. Mapping helps people to explore and analyze the world.

Making maps has become much easier and the capability available to many people. Desktop mapping software and Internet mapping sites can make anyone with a computer an instant cartographer. The maps, or cartographic visualizations, might be quite basic but they are easy to make. The procedures that trained cartographers use to make map products vary in the choice of data, software, and hardware, but several basic design steps should always take place.

First, the purpose and audience for whom the map is being made must be clear. Is this to be a general reference map or a thematic map? What image should be created in the mind of the map reader? Who will use the map? Will it be used to teach young children the shapes of the continents and oceans, or to show scientists the results of advanced research? What form will the cartographic visualization take? Will it be a paper map, a graphic file posted to the Internet, or a video?

The answers to these questions will guide the cartographer in the design process. The design process can be broken down into stages. In the first stage of map design, imagination rules. What map type, size and shape, basic layout, and data will be used? The second stage is more practical and consists of making a specific plan.

Based on the decisions made in the first stage, the symbols, line weights, colors, and text for the map are chosen. By the end of this stage, there should be a fairly clear plan for the map. During the third stage, details and specifications are finalized to account for the production method to be used. The actual software, hardware, and methods to be used must all be taken into consideration.

What makes a good map? Working in the modern digital environment, the mapmaker can change and test various designs easily. The map is a good one when it communicates the intended information, is pleasing to look at, and encourages map readers to ask thoughtful questions.

NEW TECHNOLOGIES. Mapping technology has gone from manual to magnetic, then to mechanical, optical, photochemical, and electronic methods. All of these methods have overlapped one another and each may still be used in some map-making processes. There have been recent advances in magnetic, optical, and most of all, electronic technologies.

All components of mapping systems—data collection, hardware, software, data storage, analysis, and graphical output tools—have been changing rapidly. Collecting location data, like mapping in general, has been more accessible to more people. The development of the Global Positioning System (GPS), an array of satellites orbiting the earth, gives anyone with a GPS receiver access to location information, day or night, anywhere in the world. GPS receivers are also found in planes, passenger cars, and even in the backpacks of hikers.

Satellites also have helped people to collect data about the world from space. Orbiting satellites collect images using visible light, infrared energy, and other parts of the electromagnetic spectrum. Active sensing systems send out radar signals and create images based on the return of the signal. The entire world can be seen easily with weather satellites, and other specialized satellite imagery can be used to count the trees in a yard.

Satellite image Page 31

These great resources of data are all stored and maintained as binary, computer-readable information. Developments in laser technology provide large amounts of storage space on media such as optical disks and compact disks. Advances in magnetic technology also provide massive storage capability in the form of tape storage, hard drives, and floppy drives. This is especially important for saving the large databases used for mapping.

SLIDING ROCKS GET DIGITAL TREATMENT

Dr. Paula Messina studied the trails of rocks that slide across the surface of a flat playa in Death Valley, California. The sliding rocks have been studied in the past, but no one had been able to say for certain how or when the rocks moved. It was unclear whether the rocks were caught in ice floats during the winter, were blown by strong winds coming through the nearby mountains, or were moved by some other method.

Messina gave the mystery a totally digital treatment. She mapped the locations of the rocks and the rock trails using the global positioning system (GPS) and entered her rock trail data into a geographic information system (GIS) for analysis. She was able to determine that ice was not the moving agent by studying the pattern of the trails. She also used digital elevation models (DEM) and remotely sensed imagery to model the environment of the playa. She reported her results in the form of maps using GIS's cartographic output capabilities. While she did not solve completely the mystery of the sliding rocks, she was able to disprove that winter ice caused the rocks to slide along together in rafts and that there are wind gusts strong enough to move the biggest rock on the playa.

Computer hardware and software continue to become more powerful and less expensive. At home or school, personal computers in the year 2000 were more powerful than the mainframe computers at research universities had been ten years earlier. Software continues to be developed to serve the specialized needs that mapping requires. Just as word processing software can format a paper, check spelling and grammar, draw pictures and shapes, import tables and graphics, and perform dozens of other functions, specialized software executes maps. The most common software used for mapping is called Geographic Information System software. These systems provide tools for data input and for analysis and modeling of real-world spatial data, and provide cartographic tools for designing and producing maps.

Karen A. Mulcahy

FOR FURTHER STUDY

Campbell, John. *Map Use and Analysis.* 3d ed. Boston: WCB, McGraw-Hill, 1998.

Clarke, Keith C. *Getting Started with Geographic Information Systems.* Upper Saddle River, N.J.: Prentice-Hall, 1997.

Messina, Paula, Phil Stoffer, and Keith C. Clarke. "From the XY Files: Mapping Death Valley's Wandering Rocks." *GPS World* 8, no. 4 (April, 1997): 34-44.

Robinson, Arthur H., et al. *Elements of Cartography.* 6th ed. New York: John Wiley & Sons, 1995.

Snyder, John P., and Philip M. Voxland. *An Album of Map Projections.* U.S. Geological Survey Professional Paper 1395. Washington, D.C.: U.S. Government Printing Office, 1989.

Van Burgh, Dana. *How to Teach with Topographical Maps.* Washington, D.C.: International Science Teachers Association, 1994.

INFORMATION ON THE WORLD WIDE WEB

In addition to using map projections to make whole maps, one can also divide up, or tessellate, the surface of the earth into various geometric shapes. The Web site of the National Geographic Data Center at the National Oceanic and Atmospheric Administration (NOAA) displays a Surface of the Earth Icosahedron Globe. (www.ngdc.noaa.gov/mgg/announcements/announce_icosahedron.html)

The Modified Collignon is also called Clarke's Butterfly because of its shape when flattened. The completed form is an octahedron composed of eight triangular sides. A version of this projection can be downloaded for free from the site listed below. (geography.hunter.cuny.edu/mp/gif/Butterfly.gif)

THEMES AND STANDARDS IN GEOGRAPHY EDUCATION

Many people believe that the study of geography consists of little more than knowing the locations of places. Indeed, in the past, whole generations of students grew up memorizing states, capitals, rivers, seas, mountains, and countries. Most students found that approach boring and irrelevant. During the 1990's, however, geography education in the United States underwent a remarkable transformation.

While it remains important to know the locations of places, geography educators know that place name recognition is just the beginning of geographic understanding. Geography classes now place greater emphasis on understanding the characteristics of and the connections between places. Three things have led to the renewal of geography education: the five themes of geography, the national geography standards, and the establishment of a network of geographic alliances.

THE FIVE THEMES OF GEOGRAPHY. One of the first efforts to move geography education beyond simple memorization was the National Geographic Society's publication of five themes of geography in 1984: location, place, human-environment interactions, movement, and regions. Not intended to be a checklist or recipe for understanding the world, these themes merely provided a framework for teachers—many of whom did not have a background in the subject—to incorporate geography throughout a social studies curriculum. The five themes were promoted widely by the National Geographic Society and are still used by some teachers to organize their classes.

Location is about knowing where things are. Both the absolute location (where a place is on earth's surface) and relative location (the connections between places) are important. The concept of place involves the physical and human characteristics that distinguish one place from another. The theme of human/environment interaction recognizes that people have relationships within defined places and are influenced by their surroundings. For example, many different types of housing have been created as adaptations to the world's diverse climates. The theme of movement involves the flow of people, goods, and ideas around the world. Finally, regions are human creations to help organize and understand the earth, and geography studies how they form and change.

THE NATIONAL GEOGRAPHY STANDARDS. Geography was one of six subjects identified by President George H. W. Bush and the governors of the U.S. states when they formulated the National Education Goals in 1989. While the goals themselves foundered amid the political debate that

GEOGRAPHY STANDARDS

The geographically informed person knows and understands the following:

- how to use maps and other geographic representations, tools, and technologies to acquire, process, and report information from a spatial perspective;

- how to use mental maps to organize information about people, places, and environments in a spatial context;

- how to analyze the spatial organization of people, places, and environments on Earth's surface;

- the physical and human characteristics of places;

- that people create regions to interpret Earth's complexity;

- how culture and experience influence people's perceptions of places and regions;

- the physical processes that shape the patterns of Earth's surface;

- the characteristics and spatial distribution of ecosystems on Earth's surface;

- the characteristics, distribution, and migration of human populations on Earth's surface;

- the characteristics, distribution, and complexity of Earth's cultural mosaics;

- the patterns and networks of economic interdependence on Earth's surface;

- the processes, patterns, and functions of human settlement;

- how the forces of cooperation and conflict among people influence the division and control of Earth's surface;

- how human actions modify the physical environment;

- how physical systems affect human systems;

- the changes that occur in the meaning, use, distribution, and importance of resources;

- how to apply geography to interpret the past;

- how to apply geography to interpret the present and plan for the future.

Source: National Geography Standards Project. *Geography for Life: National Geography Standards, 1994.* Washington, D.C.: National Geographics Research and Exploration, 1994.

followed their adoption, one tangible result of the initiative was the creation of Geography for Life: The National Geography Standards. More than one thousand teachers, professors, business people, and government officials were involved in the writing of Geography for Life. The project wassupported by four geography organizations: the American Geographical Society, the Association of American Geographers, the National Council for Geographic Education, and the National Geographic Society. The resulting book defines what every U.S. student should know and be able to accomplish in geography.

Each of the eighteen standards is designed to develop students' geographic skills, including asking geographic questions; acquiring, organizing, and analyzing geographic information; and answering the questions. Each standard features explanations, examples, and specific requirements for students in grades four, eight, and twelve.

GEOGRAPHY ALLIANCES AND THE FUTURE OF GEOGRAPHY EDUCATION. To publicize efforts in geography education, a network of geography alliances was established between 1986 and 1993. Each U.S. state has a geography alliance that links university professors, practicing teachers, and organizations such as the National Geographic Society and the National Council for Geographic Education. The alliances sponsor summer workshops, teacher training sessions, field experiences, and other ways of sharing the best in geographic teaching and learning.

In 2000 the future for geography education in the United States appeared to be bright. The geography alliances created a network of motivated teachers eager to share their excitement about the world. Enrollment in geography classes had risen at all levels, an advanced placement course in geography had been approved, and new learning materials guided by the national standards were being developed for students at all levels.

Eric J. Fournier

FOR FURTHER STUDY

Bednarz, Robert S., and James F. Peterson, eds. *A Decade of Reform in Geographic Education: Inventory and Prospect.* Indiana, Pa.: National Council for Geographic Education, 1994.

Geography Education Standards Project. *Geography for Life: National Geography Standards.* Washington, D.C.: National Geographic Society, 1994.

Hill, A. David. "Geography and Education: North America." *Progress in Human Geography* 16, no. 2 (May, 1994): 232-242.

Maps, the Landscape, and Fundamental Themes in Geography. Washington, D.C.: National Geographic Society, 1986.

NAEP Geography Consensus Project. *Geography Framework for the 1994 Assessment of Educational Progress.* Washington, D.C.: National Assessment Governing Board, 1992.

Vining, James W. *The National Council for Geographic Education: The First Seventy-five Years and Beyond.* Indiana, Pa.: The National Council for Geographic Education, 1990.

INFORMATION ON THE WORLD WIDE WEB

National Council for Geographic Education maintains a Web site that offers publications and activities for teachers and students. (www.ncge.org)

The National Geographic Society's Education site features sections on on-line adventures, maps and geography, lesson plans, and teacher support. (magma.nationalgeographic.com/education/index.cfm)

PHYSICAL GEOGRAPHY

THE EARTH IN SPACE

THE SOLAR SYSTEM

Earth's solar system comprises the Sun and its planets, as well as all the natural satellites, asteroids, meteors, and comets that are captive around it. The solar system formed from an interstellar cloud of dust and gas, or nebula, about 4.6 billion years ago. Gravity drew most of the dust and gas together to make the Sun, a medium-size star with an estimated life span of ten billion years. Its system is located in the Orion arm of the Milky Way galaxy, about two-thirds of the way out from the center.

During the Sun's first 100 million years, the remaining rock and ice smashed together into increasingly larger chunks, or planetesimals, until the planets, moons, asteroids, and comets reached their present state. The resulting disk-shaped solar system can be divided into four regions—terrestrial planets, giant planets, the Kuiper Belt, and the Oort Cloud—each containing its own types of bodies.

TERRESTRIAL PLANETS. In the first region are the terrestrial (Earth-like) planets Mercury, Venus, Earth, and Mars. Mercury, the nearest to the Sun, orbits at an average distance of 36 million miles (58 million km.) and Mars, the farthest, at 142 million miles (228 million km.). Astronomers call the distance from the Sun to Earth (93 million miles/150 million km.) an astronomical unit (AU) and use it to measure planetary distances.

Terrestrial planets are rocky and warm and have cores of dense metal. All four planets have volcanoes, which long ago spewed out gases that created atmospheres on all but Mercury, which is too close to the Sun to hold onto an atmosphere. Mercury is heavily cratered, like the earth's moon.

Venus has a permanent thick cloud cover and a surface temperature hot enough to melt lead. The air on Mars is very thin and usually cold, made mostly of carbon dioxide. Its dry, rock-strewn surface has many craters. It also has the largest known volcano in the solar system, Olympus Mons, which is 16 miles (25 km.) high.

Average temperatures and air pressures on Earth

*Planets
Page 34*

OTHER EARTHS

By the year 2000 astronomers had detected twenty-eight planets circling stars in the Sun's neighborhood of the galaxy. Planets, they think, are common. Those found were all gas giants the size of Saturn or larger. Earth-size planets are much too small to spot at such great distances. Where there are gas giants, there also may be terrestrial dwarfs, as in Earth's solar system. Where there are terrestrial planets, there may be liquid water and, possibly, life.

FORMATION OF THE SOLAR SYSTEM

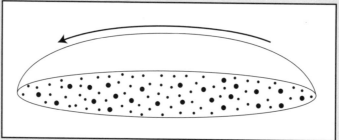

1. The solar system began as a cloud of rotating interstellar gas and dust.

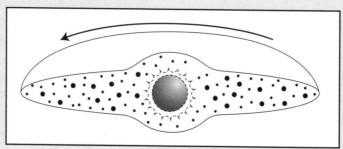

2. Gravity pulled some gases toward the center.

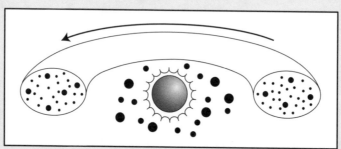

3. Rotation accelerated, and centrifugal force pushed icy, rocky material away from the proto-Sun. Small planetesimals rotate around the Sun in interior orbits.

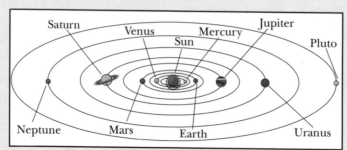

4. The interior, rocky material formed Mercury, Venus, Earth, and Mars. The outer, gaseous material formed Jupiter, Saturn, Uranus, Neptune, and Pluto.

Artist's rendition of the formation of the solar system. Scientists theorize that the system began as a nebula, a spinning cloud of gas and dust that collapsed under its own weight, forming the Sun at the center. The planets are believed to have formed from dustballs that were then melted into rocky spheres by bolts of lightning. (Painting by Don Dixon, NASA)

allow liquid water to collect on the surface, a unique feature among planets within the solar system. Meanwhile, Earth's atmosphere—mostly nitrogen and oxygen— and a strong magnetic field protect the surface from harmful solar radiation. These are the conditions that nurture life, according to scientists. Mars also might have had such conditions long ago. Space probes have photographed features there that look like river channels and lake beds, and scientists think the Martian atmosphere was much thicker at one time. Like Earth, Mars has polar ice caps, although those on Mars are made up mostly of carbon dioxide ice (dry ice), while those on Earth are made up of water ice.

A single natural satellite, the Moon, orbits Earth, probably created by a collision with a huge planetesimal more than four billion years ago. Mars has two tiny moons that may have drifted to it from the aster-

Among the Solar System's other planets, Mars is the one that most resembles Earth. (PhotoDisc)

The Moon Page 35

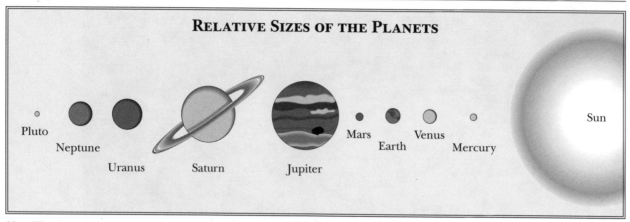

RELATIVE SIZES OF THE PLANETS

Pluto · Neptune · Uranus · Saturn · Jupiter · Mars · Earth · Venus · Mercury · Sun

Note: The size of the Sun and distances between planets are not to scale.

Jupiter moon Page 40

Jupiter Page 31

oid belt. A broad ring from 2 to 3.3 AU from the Sun, this belt is composed of space rocks as small as dust grains and as large as 600 miles (1,000 km.) in diameter. Asteroids are made of mineral compounds, especially those containing iron, carbon, and silicon. Although the asteroid belt contains enough material for a planet, one did not form there because Jupiter's gravity prevented the asteroids from crashing together. The belt separates the first region of the solar system from the second.

THE GIANT PLANETS. The second region belongs to the gas giants Jupiter, Saturn, Uranus, and Neptune. The closest, Jupiter, is 5.2 AU from the Sun, and the most distant, Neptune, is 30.11 AU. Jupiter is the largest planet in the solar system, its diameter 109 times larger than Earth's. The giant planets have solid cores, but most of their immense size is taken up by hydrogen, helium, and methane gases that grow thicker and thicker until they are like sludge near the core. On Jupiter, Saturn, and Uranus, the gases form wide bands over the surface. The bands sometimes have immense circular storms like hurricanes, but hundreds of times larger. The Great Red Spot of Jupiter is an example. It has winds of up to 250 miles (400 km.) per hour, and is at least a century old.

These planets have such strong gravity that each has attracted many moons to orbit it. In fact, they are like miniature solar systems. Jupiter has the most moons—eighteen—and Neptune has the fewest—eight—but Neptune's moon Triton is the largest of all. Most moons are balls of ice and rock, but Jupiter's Europa and Saturn's Titan may have liquid water below ice-bound surfaces. Several moons appear to have volcanoes, and a wispy atmosphere covers Titan. Additionally, the giant planets have rings of broken rock and ice around them, no more than 330 feet (100 meters) thick. Saturn's hundreds of rings are the brightest and most famous.

THE KUIPER BELT. The third region of the solar system, the Kuiper Belt, contains the ninth planet from the Sun, Pluto. Pluto has a single moon, Charon. It does not orbit on the same plane, called the ecliptic, as the rest of the planets do. Instead, its orbit diverges more than seventeen degrees above and below the ecliptic. Its orbit's oval shape brings Pluto within the orbit of Neptune for a large percentage of its long year, which is equal to 248 Earth years. Two-thirds the size of the earth's moon, Pluto has a thin, frigid methane atmosphere. Charon is half Pluto's size and orbits less than 32,000 miles (20,000 km.) from Pluto's surface.

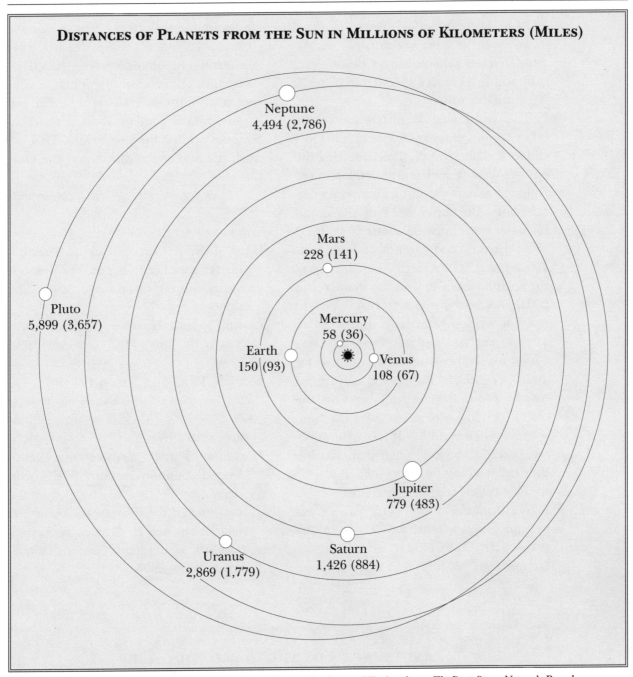

DISTANCES OF PLANETS FROM THE SUN IN MILLIONS OF KILOMETERS (MILES)

Source: Data are from Jet Propulsion Laboratory, California Institute of Technology. *The Deep Space Network.* Pasadena, Calif.: JPL, 1988, p. 17.

Some astronomers consider Pluto and Charon to be a double planet.

Some astronomers long regarded Pluto as not a true planet at all. They instead thought it was merely the largest of several dozen icy bodies discovered in the Kuiper Belt. The Kuiper Belt holds asteroids and the "short-period" comets that pass by Earth in orbits of twenty to two hundred years. These bodies are the remains of planet formation and did not collect into planets because distances between them are too great for many collisions to occur. Most of them are loosely compacted bodies of ice and mineral—"dirty snowballs," in the words of a famous astronomer. An estimated 200 million Kuiper Belt objects orbit within a band of space from 30 to 50 AU from the Sun.

THE OORT CLOUD. In contrast to the other regions of the solar system, the Oort Cloud is a spherical shell surrounding the entire solar system. It is also a collection of comets—as many as two trillion, scientists calculate. The inner edge of the cloud forms at a distance of about 20,000 AU from the Sun and extends as far out as 100,000 AU. The Oort Cloud thus gives the solar system a theoretical diameter of 200,000 AU—a distance so vast that light needs more than three years to cross it. No astronomer has yet detected an Oort Cloud object, because the cloud is so far away. Occasionally, however, gravity from a nearby star dislodges an object in the cloud, causing it to fall toward the Sun. When observers on Earth see such an object sweep by in a long, cigar-shaped orbit, they call it a long-period comet.

The outer edge of the Oort Cloud marks the farthest reach of the Sun's gravitational power to bind bodies to it. In one respect, the Oort Cloud is part of interstellar space.

In addition to light, the Sun sends out a constant stream of charged particles—atoms and subatomic particles—called the solar wind. The solar wind shields the solar system from the interstellar medium, but it only does so out to about 100 AU, a boundary called the heliopause. That is a small fraction of the distance to the Oort Cloud.

Roger Smith

FOR FURTHER STUDY

Beatty, J. Kelly, Carolyn Collins Petersen, and Andrew Chaikin, eds. *The New Solar System.* 4th ed. Cambridge, Mass.: Sky, 1999.

Booth, Nicholas. *Exploring the Solar System.* New York: Cambridge University Press, 1995.

Gribbin, John, and Simon Goodwin. *Empire of the Sun: Planets and Moons of the Solar System.* New York: New York University Press, 1998.

Hartmann, William K. *Moons and Planets.* 4th ed. Belmont, Calif.: Wadsworth, 1999.

Taylor, Stuart Ross. *Destiny or Chance: Our Solar System and Its Place in the Cosmos.* Cambridge, England: Cambridge University Press, 1998.

INFORMATION ON THE WORLD WIDE WEB

The Web site of the Lunar and Planetary Institute, a NASA-funded institute in Houston, Texas, devoted to the study of the solar system, has current data and photos, many from space probes. (cass.jsc.nasa.gov/lpi.html)

The Planetary Society, a nonprofit, nongovernmental organization founded in 1980 by Carl Sagan, Bruce Murray, and Louis Friedman, encourages the exploration of the solar system and the search for extraterrestrial life. The society's Web site features recent planetary news stories, an interactive learning center, and links to other space exploration sites. (planetary.org)

Views of the Solar System is an educational, interactive Web site sponsored by the Hawaiian Astronomical Society. (www.solarviews.com/homepage.htm)

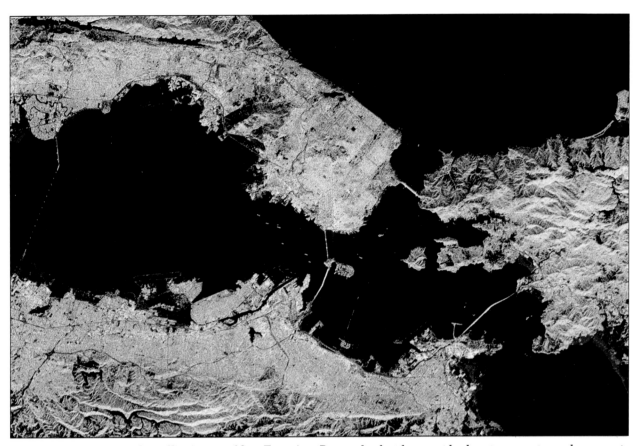

Features in this infrared satellite image of San Francisco Bay make the photograph almost as easy to read as a map. San Francisco (at the top) is linked to Marin County (right) by Golden Gate Bridge. The picture was taken from the space shuttle Discovery *in 1991.* (PhotoDisc)

A gas giant, Jupiter is the largest planet in the solar system. Although it has a solid core, most of its immense size is taken up by hydrogen, helium, and methane gases that grow thicker and thicker until they are like sludge near the core. Jupiter's famous Great Red Spot is actually an immense, centuries-old circular storm in a band of surface gases, with winds of up to 250 miles (400 km) per hour. (PhotoDisc)

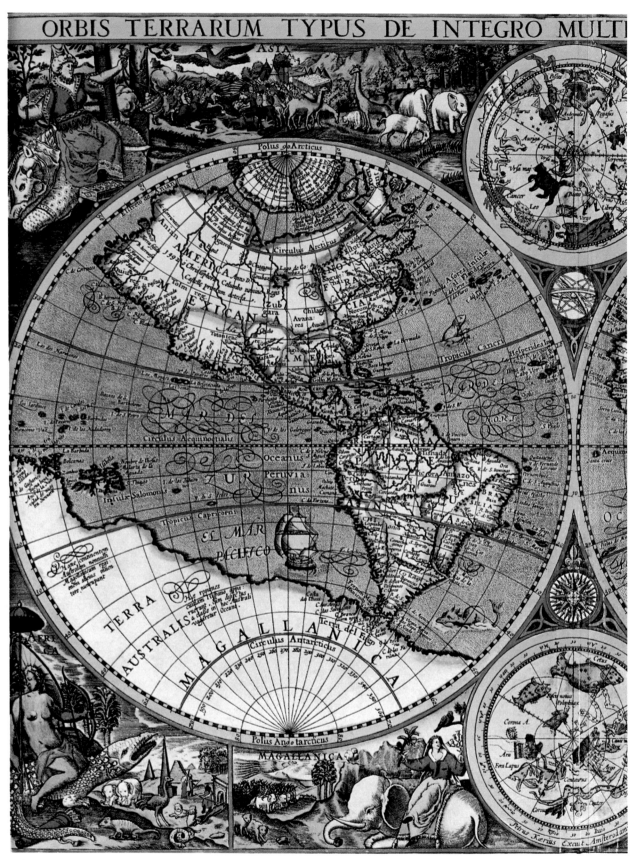

This map of the world was published in 1607, when European geographers had only the vaguest ideas of what lay west of the Americas and the great southern continent, "Terra Australis," existed only in theory. (Corbis)

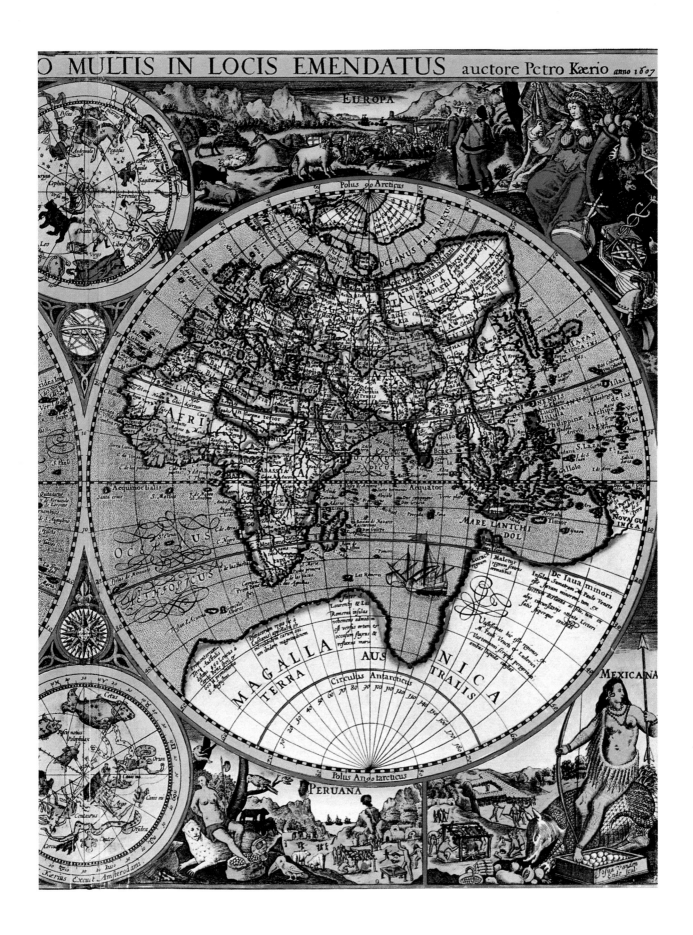

Composite picture (not to scale) with photographs of the Solar System's planets, showing their relative positions, from closest to most distant from the Sun. Earth (with its Moon to the right) is the third from the top. (PhotoDisc)

The Hubble Telescope, which was placed in orbit in 1990, has made possible revolutionary advances in telescopic space exploration. (Cuba is visible to the left.) (PhotoDisc)

Earthrise seen from the surface of the Moon. (PhotoDisc)

Earth's Moon, showing impact craters. (PhotoDisc)

The Moon, viewed over the shoulder of Washington's Mount Rainier in the Cascade Range. (PhotoDisc)

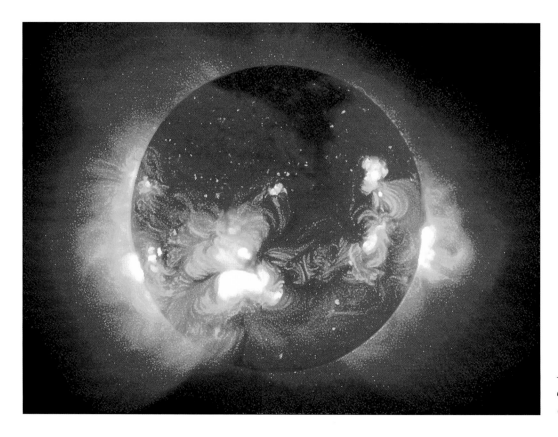

X-ray image of the Sun. (PhotoDisc)

Hailstones originate in cumulonimbus clouds as small ice pellets that grow larger as they collect water droplets within the clouds. When they become too heavy to stay aloft, they fall. Hailstones can weigh a pound or more. (PhotoDisc)

Typical winter landscape in a boreal forest. (PhotoDisc)

Fall landscape in Kent, England. (PhotoDisc)

Yellowstone is one of at least one hundred hot spots whose heat sources, known as mantle plumes, have remained stationary while the tectonic plates have moved over them, producing volcanoes, geysers, and hot springs. (Digital Stock)

MAJOR TECTONIC PLATES AND MID-OCEAN RIDGES

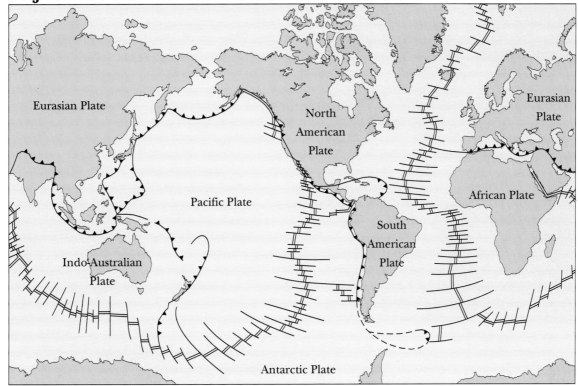

Eurasian Plate

North American Plate

Eurasian Plate

Pacific Plate

African Plate

South American Plate

Indo-Australian Plate

Antarctic Plate

Types of Boundaries: Divergent   Convergent   Transform

The caldera of Oregon's Crater Lake owes its origins to a large pumice eruption that occurred about 76,000 years ago, when the large volume of ejected matter caused the volcano to empty itself and collapse into the hole. (PhotoDisc)

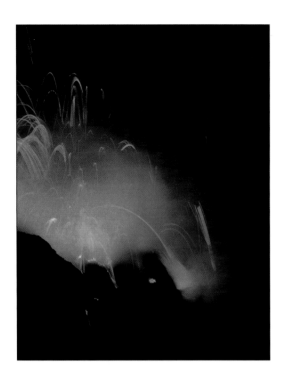

Erupting volcano with lava, or fire, fountains, which occur when large lava flows are blown by strong winds. (PhotoDisc)

The fact that Earth is not the only body in the system with active volcanoes is proven in this photograph of an volcanic eruption on Jupiter's moon Io. (PhotoDisc)

Destruction caused by the eruption of Mount St. Helens in 1980. The eruption knocked down thousands of acres of trees and spewed more than three cubic miles of material into the atmosphere. (Corbis)

Forest flattened by a volcanic eruption. (PhotoDisc)

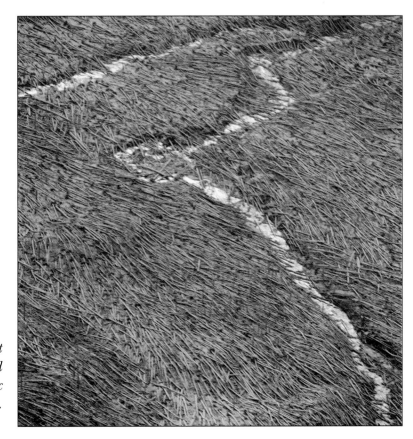

SOME VOLCANIC HOT SPOTS AROUND THE WORLD

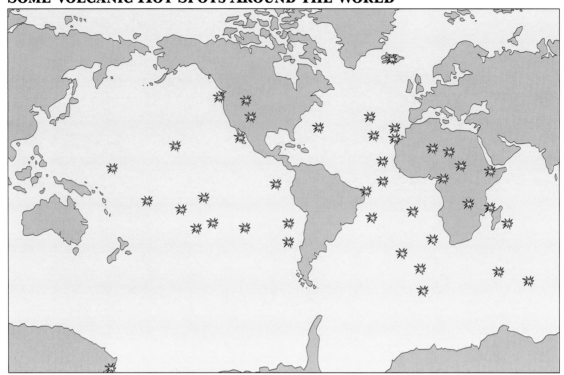

CALIFORNIA'S SAN ANDREAS FAULT

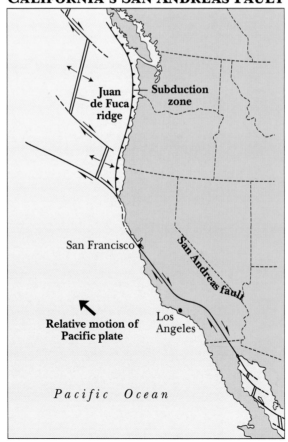

Juan
de Fuca
ridge

Subduction
zone

San Francisco

San Andreas fault

Los
Angeles

Relative motion of
Pacific plate

Pacific Ocean

The stratigraphic layers of sandstone, mudstone, coal and shale found in the Drumheller Badlands of Alberta, Canada, are a rich source of dinosaur fossils and skeletons, and they help scientists attach dates to geologic events. (Digital Stock)

Zion National Park. The walls of Zion Canyon, cut by the Virgin River, reveal fossils and other traces of the geologic past, going back as early as the Mesozoic Era. (PhotoDisc)

In late 1999, the remote northern Peruvian village of La Pucara was partly buried by a debris avalanche triggered by heavy rains merging with underground thermal springs. (AP/Wide World Photos)

In climates with fluctuating temperatures, unusual rock formations such as these granite piles in Zimbabwe's Matopo Hills, are typically the products of mechanical weathering, which causes solid rocks to break into pieces as they expand and contract. (R. Kent Rasmussen)

Utah's Bryce Canyon National Park was created in 1928 to preserve its oddly shaped and multicolored cliffs and towers, which were created by erosion in finely jointed rock. (Corbis)

Bryce Canyon in winter. (PhotoDisc)

Glacier National Park has many examples of glacial troughs, valleys carved by the movement of glaciers pressing rocks against valley walls. (PhotoDisc)

Alaska's Kennicott Glacier shows many landforms associated with glaciers, such as cirques, arêtes, and moraines. (PhotoDisc)

EARTH'S MOON

The fourth largest natural satellite in the solar system, Earth's moon has a diameter of 2,160 miles (3,476 km.)—less than one-quarter the diameter of Earth. The Moon's mass is less than one-eightieth that of Earth.

The Moon orbits Earth in an elliptical path. When it is at perigee (when it is closest to Earth), it is 221,473 miles (356,410 km.) distant. When it is at apogee (farthest from Earth), it is 252,722 miles (406,697 km.) distant.

The Moon completes one orbit around Earth every 27.3 Earth days. Because it rotates at about the same rate that it orbits the earth, observers on Earth only see one side of the Moon. The changing angles between Earth, the Sun, and the Moon determine how much of the Moon's illuminated surface can be seen from Earth and cause the Moon's changing phases.

The Moon Pages 35, 36

ECLIPSES

The Sun's diameter is four hundred times larger than the Moon's; however, the Moon is four hundred times closer to Earth than the Sun, making the two objects appear nearly the same size in the sky to observers on Earth. As the Moon orbits Earth, it crosses the plane of the Earth-Sun orbit twice each month. If one of the orbit-crossing points (called nodes) occurs during a new or full moon phase, a solar or lunar eclipse can occur.

A solar eclipse occurs when the Moon and the Sun appear to be in the exact same place in the sky during a new moon phase. When that happens, the Moon blocks the light of the Sun for up to seven minutes. Because solar eclipses can be seen only from certain places on Earth, some people travel around the world—sometimes to remote places—to view them.

A lunar eclipse occurs when Earth is positioned between the Sun and the Moon and casts its shadow on the Moon. In contrast to solar eclipses, lunar eclipses are visible from every place on Earth from which the Moon can be seen.

THE PHASES OF THE MOON

New Moon   Waxing Quarter

Full Moon   Waning Quarter

VOLCANISM. Naked-eye observations of the Moon from Earth reveal dark areas called *maria*, the plural form of the Latin word *mare* for sea. The maria are the remains of ancient lava flows from inside gigantic impact craters; the last eruptions were more than three billion years ago. The lava consists of basalt, similar in com-

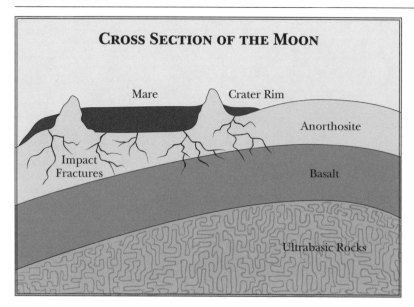

CROSS SECTION OF THE MOON

Mare

Crater Rim

Anorthosite

Impact Fractures

Basalt

Ultrabasic Rocks

position to Earth's oceanic crust and many volcanoes. The maria have names such as Mare Serenitatis (15 to 40 degrees north latitude, longitude 5 to 20 degrees east) and Mare Tranquillitatis (0 to 20 degrees north latitude, longitude 15 to 45 degrees east). Some of the smaller dark areas on the Moon also have names that are water-related: lacus (lake), sinus (bay), and palus (marsh).

IMPACT CRATERS. Observing the Moon with an optical aid, such as a telescope or a pair of binoculars, provides a closer view of impact craters. Impact craters of various sizes cover 83 percent of the Moon's surface. More than 33,000 craters have been counted on the Moon.

One of the easiest craters to observe from the Earth is Tycho. Located at 43.3 degrees south latitude, longitude 11.2 degrees west, it is about 50 miles (85 km.) wide. Surrounding Tycho are rays of dusty material, known as ejecta, that appear to radiate from the crater. When an object from space, such as a meteoroid, slams into the Moon's surface, it is vaporized upon impact. The dust and debris from the interior of the crater fall back onto the lunar surface in a pattern of rays. Because

the ejecta is disrupted by subsequent impacts, only the youngest craters still have rays. Sometimes, pieces of the ejecta fall back and create smaller craters called secondary craters. The ejecta rays of Tycho extend to almost 1,865 miles (3,000 km.) beyond the crater's edge.

OTHER LUNAR FEATURES. Near the crater called Archimedes is the Apennines mountain range, which has peaks nearly 20,000 feet (60,000 meters) high—altitudes comparable to South America's Andes.

The Moon also has valleys. Two of the most well known are the Alpine Valley, which is about 115 miles (185 km.) long; and the Rheita Valley, located about 155 miles (250 km.) from the Stevinus crater, which is 238 miles (383 km.) long, 15.5 miles (25 km.) wide, and 2,000 feet (609 meters) deep.

Smaller than valleys and resembling cracks in the lunar surface are features called rilles, which are thought to be places of ancient lava flow. Many rilles can be seen near the Aristarchus crater. Rilles are often up to 3 miles (5 km.) wide and can stretch for more than 104 miles (167 km.).

A wrinkle in the lunar surface is called a ridge. Many ridges are found around the boundaries of the maria. The Serpentine Ridge cuts through Mare Serenitatis.

EXPLORATION OF THE MOON. Robotic spacecraft were the first visitors to explore the Moon. The Russian spacecraft Luna 1 made the first flyby of the Moon in January, 1959. Eight months later, Luna 2 made the first impact on the Moon's surface. In October, 1959, Luna 3 was the first spacecraft to photograph the side of the Moon not visible from Earth. In 1994 the United States' *Clementine* spacecraft was

Apollo 17 astronaut collecting soil samples on the moon in December, 1972. (Corbis)

the first probe to map the Moon's composition and topography globally.

The first humans to land on the Moon were the U.S. astronauts Neil Armstrong and Edwin "Buzz" Aldrin. On July 20, 1969, they landed in the *Eagle* lunar module, during the Apollo 11 mission. Armstrong's famous statement, "That's one small step for man, one giant leap for mankind," was heard around the world by millions of people who watched the first humans set foot on the lunar surface, at the Sea of Tranquillity. The last twentieth century human mission to reach the lunar surface, Apollo 17, landed there in December, 1972. Astronauts Gene Cernan and geologist Jack Schmitt landed in the Taurus-Littrow Valley (20 degrees north latitude, longitude 31 degrees east).

Noreen A. Grice

FOR FURTHER STUDY

Alter, Dinsmore. *Pictorial Guide to the Moon.* New York: Thomas Crowell, 1967.

Cherrington, Ernest H. *Exploring the Moon Through Binoculars.* New York: McGraw-Hill, 1969.

Fraknoi, Andrew, David Morrison, and

INFORMATION ON THE WORLD WIDE WEB

NASA's Spacelink education Web site, featuring educational services, instructional materials, and news about NASA projects, is a good starting place from which to find information about the Moon.

(spacelink.nasa.gov/index.html)

Sidney Wolff. *Voyages Through the Universe*. Philadelphia: W. B. Saunders, 1997.

_____. *Voyages to the Planets*. 2d ed. Philadelphia: W. B. Saunders, 2000.

Spudis, Paul D. *The Once and Future Moon*. Washington, D.C.: Smithsonian Institution Press, 1996.

THE SUN AND THE EARTH

The Sun Page 36

Of all the astronomical phenomena that one can consider, few are more important to the survival of life on Earth than the relationship between Earth and the Sun. With the exception of small amounts of residual (endogenic) energy that have remained inside the earth from the time of its formation some 4.5 billion years ago and which sustain some specialized forms of life along some oceanic rift systems, almost all other forms of life, including human, depend on the exogenic light and energy that the earth receives directly from the Sun.

The enormous variety of ecosystems on Earth are highly dependent on the angles at which the Sun's rays strike Earth's spherical surface. These angles, which vary greatly with latitude and time of year, determine many commonly observed phenomena, such as the height of the Sun above the horizon, the changing lengths of day and night throughout the year, and the rhythm of the seasons. Daily and seasonal changes have profound effects on the many climatic regions and life cycles found on earth.

THE SUN. The center of Earth's solar system, the Sun is but one ordinary star among some 100 billion stars in an ordinary cluster of stars called the Milky Way galaxy. There are at least ten billion galaxies in the universe, each with billions of stars. Statistically, the chances are good that many of these stars have their own solar systems. Late twentieth century astronomical observations discovered the presence of what appear to be planets, large ones similar in size to Jupiter, orbiting other stars.

Earth's Sun is an average star in terms of its physical characteristics. It is a large sphere of incandescent gas that has a diameter more than 100 times that of Earth, a mass more than 300,000 times that of Earth, and a volume 1.3 million times that of Earth. The Sun's surface gravity is thirty-four times that of Earth.

The conversion of hydrogen into helium in the Sun's interior, a process known as nuclear fusion, is the source of the Sun's energy. The amount of mass that is lost in the fusion process is miniscule, as evidenced by the fact that it will take perhaps 15 million years for the Sun to lose one-millionth of its total mass. The Sun is expected to continue shining through another several billion years.

EARTH REVOLUTION. The earth moves about the Sun in a slightly elliptical orbit called a revolution. It takes one year for the earth to make one revolution at an average orbital velocity of about 29.6 kilometers per second (18.5 miles per second). Earth-sun relationships are described by a tropical year, which is defined as the period of time (365.25 average solar days) from one vernal equinox to another. To balance the tropical year with the calendar year, a whole day (February 29) is added every fourth year (leap year). Other minor adjustments are necessary so as to balance the system.

PERIHELION AND APHELION. The average distance between Earth and the Sun is approximately 93 million miles (150 million km.). At that distance, sunlight, which travels at the speed of light (186,000 miles/300,000 kilometers per second), takes about 8.3 minutes to reach the earth. Since the earth's orbit is an ellipse rather than a circle, the earth is closest to the Sun on about January 3—a distance of 91.5 million miles (147 million km.). This position in space is called perihelion, which comes from the Greek *peri*, meaning "around" or "near," and *helios*, meaning the Sun. Earth is farthest from the Sun on about July 4 at aphelion (Greek *ap*, "away from," and *helios*), with a distance of 152 million kilometers (94.5 million miles).

AXIAL INCLINATION. Astronomers call the imaginary surface on which Earth orbits around the Sun the plane of the ecliptic. The earth's axis is inclined 66.5 degrees to the plane of the ecliptic (or 23.5 degrees from the perpendicular to the plane of the ecliptic), and it maintains this orientation with respect to the stars. Thus, the North Pole points in the same direction to Polaris, the North Star, as it revolves about the Sun. Consequently, the Northern Hemisphere tilts away from the Sun during one-half of Earth's orbit and toward the Sun through the other half.

Winter solstice occurs on December 21 or 22, when the tilt of the Northern Hemisphere away from the Sun is at its maximum. The opposite condition occurs during summer solstice on June 21 or 22, when the Northern Hemisphere reaches its maximum tilt toward the Sun. The equinoxes occur midway between the solstices when neither the Southern nor the Northern Hemisphere is tilted toward the Sun. The vernal and autumnal equinoxes occur on March 20 or 21 and September 22 or 23, respectively.

The axial inclination of 66.6 degrees (or 23.5 degrees from the perpendicular) explains the significance of certain parallels on the earth. The noon sun shines directly overhead on the earth at varying latitudes on different days—between 23.5 degrees south latitude and 23.5 degrees north latitude. The parallels at 23.5 de-

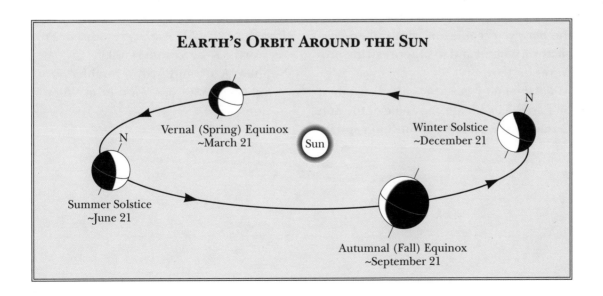

EARTH'S ORBIT AROUND THE SUN

Vernal (Spring) Equinox
~March 21

Winter Solstice
~December 21

Summer Solstice
~June 21

Sun

Autumnal (Fall) Equinox
~September 21

grees south latitude and 23.5 degrees north latitude are called the Tropics of Capricorn and Cancer, respectively.

During the winter and summer solstices, the area on the earth between the Arctic Circle (at 66.5 degrees north latitude) and the North Pole has twenty-four hours of darkness and daylight, respectively. The same phenomena occurs for the area between the Antarctic Circle (at 66.5 degrees south latitude) and the South Pole, except that the seasons are reversed in the Southern Hemisphere. At the poles, the Sun is below the horizon for six months of the year.

For those living outside the Tropics (poleward of 23.5 degrees north and south latitude), the noon sun will never shine directly overhead. Hours of daylight will also vary greatly during the year. For example, daylight will range from approximately nine hours during the winter solstice to fifteen hours during the summer solstice for persons living near 40 degrees north latitude, such as in Philadelphia, Denver, Madrid, and Beijing.

SOLAR RADIATION. Given the size of the earth and its distance from the Sun, it is estimated that this planet receives only about one two-billionth part of the total energy released by the Sun. However, this seemingly small amount is enough to drive the massive oceanic and atmospheric circulation systems and to support all life processes on Earth.

Solar energy is not evenly distributed on Earth. The higher the angle of the Sun in the sky, the greater the duration and intensity of the insolation. To illustrate this, note how easy it is look at the Sun when it is very low on the horizon—near dawn and sunset. At those times, the Sun's rays have to penetrate much more of the atmosphere, so more of the sunlight is absorbed. When the Sun's rays are coming in at a low angle, the same solar energy is spread over a larger area, thereby leading to less insolation per unit of area. Thus, the equatorial region receives much more solar energy than the polar region. This radiation imbalance would make the earth decidedly less habitable were it not for the atmospheric and oceanic circulation systems (such as the warm Gulf Stream) that move the excess heat from the Tropics to the middle and high latitudes.

Robert M. Hordon

FOR FURTHER STUDY

Henbest, Nigel. *The New Astronomy.* 2d ed. New York: Cambridge University Press, 1996.

Jones, A. W. *Innovations in Astronomy.* Santa Barbara, Calif.: ABC-CLIO Information Services, 1999.

McKnight, Tom L. *Physical Geography.* 6th ed. Upper Saddle River, N.J.: Prentice Hall, 1999.

North, Gerald. *Astronomy Explained.* New York: Springer, 1997.

Pasachoff, Jay M. *Astronomy.* 4th ed. Fort Worth, Tex.: Saunders, 1995.

Strahler, Alan, and Arthur Strahler. *Physical Geography.* New York: John Wiley & Sons, 1997.

THE SEASONS

Earth's 365-day year is divided into seasons. In most parts of the world, there are four seasons—winter, spring, summer, and fall (also called autumn). In some tropical regions—those close to the equator—there are only two seasons. In areas close to the equator, temperatures change little throughout the year; however, amounts of rainfall vary greatly, resulting in distinct wet and dry seasons. The polar regions of the Arctic and Antarctic also have little variation in temperature, remaining cold throughout the year. Their seasons are light and dark, because the Sun shines almost constantly in the summer and hardly at all in the winter.

The four seasons that occur throughout the northern and southern temperate zones—between the Tropics and the polar regions—are climatic seasons, based on temperature and weather changes. Winter is the coldest season; it is the time when days are short and few crops can be grown. It is followed by spring, when the days lengthen and the earth warms; this is the time when planting typically begins, and animals that hibernate (from the French word for winter) during the winter leave their dens.

Summer is the hottest time of the year. In many areas, summer is marked by drought, but other regions experience frequent thunderstorms and humid air. In the fall, the days again become shorter and cooler. This is the time when many crops are harvested. In ancient cultures, the turning of the seasons was marked by festivals, acknowledging the importance of seasonal changes to the community's survival.

Each season is defined as lasting three months. Winter begins at the winter solstice, which is the time when the Sun is farthest from the equator. In the Northern Hemisphere, this occurs on December 21 or 22, when the Sun is directly over the tropic of Capricorn. Summer begins at the other solstice, June 20 or 21 in the Northern Hemisphere, when the Sun is directly over the tropic of Cancer. The winter solstice is the shortest day of the year; the summer solstice is the longest.

Spring and fall begin on the two equinoxes. At an equinox, the Sun is directly above the earth's equator and the lengths of day and night are approximately equal everywhere on Earth. In the Northern Hemisphere, the vernal (spring) equinox occurs on March 21 or 22, and the autumnal equinox occurs on September 22 or 23.

SEASONS AND THE HEMISPHERES. The relationship of the seasons to the calendar is opposite in the Northern and Southern Hemispheres. On the day that a summer solstice occurs in the Northern Hemisphere, the winter solstice occurs in the Southern Hemisphere. Thus, when it is summer in the Southern Hemisphere, it is winter in the Northern Hemisphere, and vice versa.

THE SUN AND THE SEASONS. The reason why summers and winters differ in the temperate zones is often misunderstood. Many people think that winter happens when the Sun is more distant from the earth than it is in summer. What causes Earth's seasons is not the changing distances between the earth and the Sun, but the tilt of the earth's axis. A line drawn from the North Pole to the South Pole through the center of the earth (the

Fall landscape Page 38

earth's axis) is not perpendicular to the plane of the earth's orbit (the ecliptic). The earth's axis and the perpendicular to the ecliptic make an angle of 23.5 degrees. This tilts the Northern Hemisphere toward the Sun when the earth is on one side of its orbit around the Sun, and tilts the Southern Hemisphere toward the Sun when the earth moves around to the Sun's opposite side. When the Sun appears to be at its highest in the sky, and its rays are most direct, summer occurs. When the Sun appears to be at its lowest, and its rays are indirect, there is winter.

LOCAL PHENOMENA. Local conditions can have important effects on seasonal weather. At locations near oceans, sea breezes develop during the day, and evenings are characterized by land breezes. Sea breezes bring cooler ocean air in toward land. This results in temperatures at the shore often being 5 to 11 degrees Fahrenheit (3 to 6 degrees Celsius) lower than temperatures a few miles inland.

At night, when land temperatures are lower than ocean temperatures, land breezes move air from the land toward the water. As a result, coastal regions have less seasonal temperature variations than inland areas do. For example, coastal areas seldom become cold enough to have snow in the winter, even though inland areas at the same latitude do.

HAILSTORMS. Hail usually occurs during the summer, and is associated with towering thunderstorm clouds, called cumulonimbus. Hail is occasionally confused with sleet. Sleet is a wintertime event, and occurs when warmer layers of air sit above freezing layers near the ground. Rain that forms in the warmer, upper layer solidifies into tiny ice pellets in the lower, subfreezing layer before hitting the ground.

Hail is an entirely different phenome-

Hail
Page 37

non. When cold air plows into warmer, moist air—called a cold front boundary—powerful updrafts of rising air can be created. The warm, moist air propelled upward by the heavier cold air can reach velocities approaching 100 miles (160 kilometers) per hour. Ice crystals form above the freezing level in the cumulonimbus clouds and fall into lower, warmer parts of the clouds, where they become coated with water. Picked up by an updraft, the coated ice crystals are carried back to a higher, colder levels where their water coatings freeze. This cycle can repeat many times, producing hailstones that have multiple, concentric layers of ice.

Hailstorms can be very damaging. Hail can ruin crops, dent car bodies, crack windshields, and injure people. The Midwest of the United States is particularly susceptible to hailstorms. There, warm, moist air from the Gulf of Mexico often meets much colder, drier air originating in Canada. This combination produces the extreme atmospheric instability necessary for that kind of weather.

Alvin S. Konigsberg

FOR FURTHER STUDY

Barry, Roger Graham. *Mountain Weather and Climate.* London: Methuen Publishers, 1981.

Gokhale, Narayan R. *Hailstorms and Hailstone Growth.* Albany: State University of New York Press, 1975.

Lockwood, John G. *Causes of Climate.* New York: John Wiley & Sons, 1979.

Simpson, John E. *Sea Breeze and Local Winds.* Cambridge, England: Cambridge University Press, 1994.

Stover, Dawn. "Heat Islands." *Popular Science* 255, no. 6 (December, 1999).

EARTH'S INTERIOR

EARTH'S INTERNAL STRUCTURE

Earth is one of the nine known planets in the Sun's solar system that formed from a giant cloud of cosmic dust called a nebula. This event is thought to have happened between 4.44 billion years ago (based on the age of the oldest-known Moon rock) and 4.56 billion years ago (the age of meteorite bombardment). After Earth's formation, heat released by colliding particles combined with the heat energy released by the decay of radioactive elements to cause some or all of Earth's interior to melt. This melting began the process of differentiation, which allowed the heavier elements, mainly iron and nickel, to sink toward Earth's center while the lighter, rocky components moved upward, as a result of the contrast in density of the earth's forming elements.

This process of differentiation was probably the most important event of Earth's early history. It changed the planet from a homogeneous mixture with neither continents nor oceans to a planet with three layers: a dense core beginning at 1,800 miles (2,900 km.) deep and ending at Earth's center, 3,977 miles (6,400 km.) below the surface; a mantle beginning between 3 and 44 miles (5-70 km.) deep and ending at

Earth's core; and a crust going from Earth's surface to about 3-6 miles (5-10 km.) deep for oceanic crust and 22-44 miles (35-70 km.) deep for continental crust.

LAYERING OF THE EARTH. Earth's layers can be classified either by their composition (the traditional method) or by their mechanical behavior (strength). Compositional classification identifies several distinct concentric layers, each with its own properties. The outermost layer of

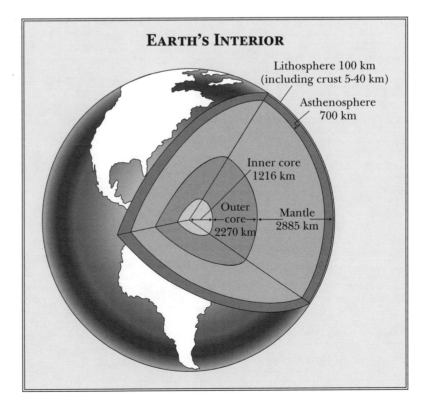

EARTH'S INTERIOR

Lithosphere 100 km
(including crust 5-40 km)

Asthenosphere
700 km

Inner core
1216 km

Outer
core
2270 km

Mantle
2885 km

*Tectonic
plates
Page 39*

Earth is the crust or skin. This is divided into continental and oceanic crusts. The continental crust varies in thickness between 22 and 25 miles (35 and 40 km.) under flat continental regions and up to 44 miles (70 km.) under high mountains. The oceanic crust is made up of igneous rocks rich in iron and magnesium, such as basalt and peridotite. The upper continental crust is composed mainly of alumino-silicates. The oldest continental crustal rock exceeds 3.8 billion years, while oceanic crustal rocks are not older than 180 million years. The oceanic crust is heavier than the continental crust.

Earth's next layer is the mantle, which is made up mostly of ferro-magnesium silicates. It is about 1,800 miles (2,900 km.) thick and is separated into the upper and lower mantle. Most of Earth's internal heat is contained within the mantle. Large convective cells in the mantle circulate heat and may drive plate-tectonic processes.

The last layer is the core, which is separated into the liquid outer core and the solid inner core. The outer core is 1,429 miles (2,300 km.) thick, twice as thick as the inner core. The outer core is mainly composed of a nickel-iron alloy, while the

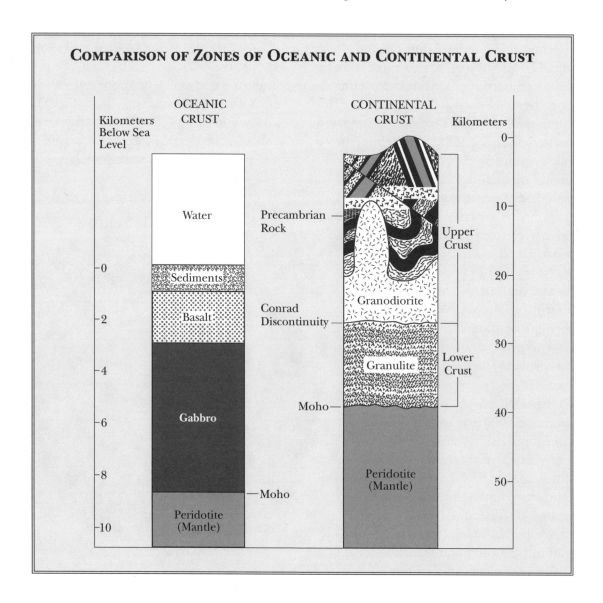

COMPARISON OF ZONES OF OCEANIC AND CONTINENTAL CRUST

inner core is almost entirely composed of iron. Earth's magnetic field is believed to be controlled by the liquid outer core.

In the mechanical layering classification of the earth's interior, the layers are separated based on mechanical properties or strength (resistance to flowing or deformation) in addition to composition. The uppermost layer is the lithosphere (sphere of rock), which comprises the crust and a solid portion of the upper mantle. The lithosphere is divided into many plates that move in relation to each other due to tectonic forces. The solid lithosphere floats atop a semiliquid layer known as the asthenosphere (weak sphere), which enables the lithosphere to move around.

EXPLORING EARTH'S INTERIOR. Volcanic activity provides natural samples of the outer 124 miles (200 km.) of Earth's interior. Meteorites—samples of the solar system that have collided with Earth—also provide clues about Earth's composition and early history. The most ambitious human effort to penetrate Earth's interior was made by the former Soviet Union, which drilled a super-deep research well, named the Kola Well, near Murmansk, Russia. This was an attempt to penetrate the crust and reach the upper mantle. The reported depth of the Kola Well is a little more than 7.5 miles (12 km.). Although impressive, the drilled depth represents less than 0.2 percent of the distance from the earth's surface to its center.

A great deal of knowledge about Earth's composition and structure has been obtained through computer modeling, high-pressure laboratory experiments, and meteorites, but most of what is known about Earth's interior has been ac-

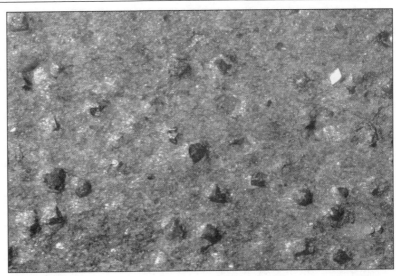

Peridotite, a primary mantle rock, is a heavy, dark-green igneous rock, consisting of the elements magnesium, oxygen, and silicon. This sample is composed of pyroxene omphacite with small garnets. (William E. Ferguson)

quired by studying seismic waves generated by earthquakes and nuclear explosions. As seismic waves are transmitted, reflected, and refracted through the earth, they carry information to the surface about the materials through which they have traveled. Seismic waves are recorded at receiver stations (seismographic stations) and processed to provide a picturelike image of Earth's interior.

Changes in P- and S-wave velocities within Earth reveal the sequence of layers that make up Earth's interior. P-wave velocity depends on the elasticity, rigidity, and density of the material. By contrast, S-wave velocity depends only on the rigidity and density of the material. There are sharp variations in velocity at different depths, which correspond to boundaries between the different layers of Earth. P-wave velocity within crustal rocks ranges from 3.6-4.2 miles} (6-7 km.) per second.

The boundary between the crust and the mantle is called the Mohorovičić discontinuity or Moho. At Moho, P-wave velocity increases from 4.2-4.8 miles (7-8

PROPERTIES OF SEISMIC WAVES

Seismologists use two types of body waves—primary (P-waves) and secondary (S-waves) waves—to estimate seismic velocities of the different layers within the earth. In most rock types P-waves travel between 1.7 and 1.8 times more quickly than S-waves; therefore, P-waves always arrive first at seismographic stations. P-waves travel by a series of compressions and expansions of the material through which they travel. P-waves can travel through solids, liquids, or gases. When P-waves travel in air, they are called sound waves.

The slower S-waves, also called shear waves, move like a wave in a rope. This movement makes the S-wave more destructive to structures like buildings and highway overpasses during earthquakes. Because S-waves can travel only through solids and cannot travel through Earth's

outer core, seismologists concluded that Earth's outer core must be liquid or at least must have the properties of a fluid.

MOVEMENT OF SEISMIC WAVES

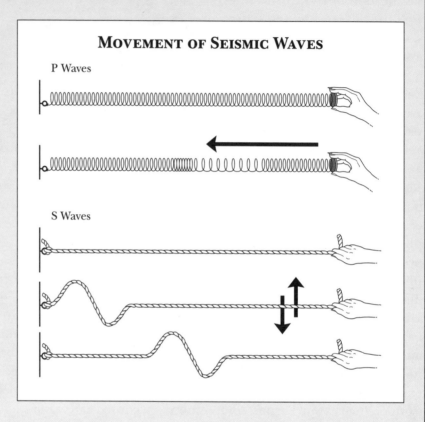

P Waves

S Waves

km.) per second. Beyond the crust-mantle boundary, P-wave velocity increases gradually up to about 8.1 miles (13.5 km.) per second at the core-mantle boundary. At this depth, S-waves are not transmitted and P-wave velocity, decreases from 8.1 to 4.8 miles (13.5 to 8 km.) per second, which strongly supports the concept that the outer core is liquid, since S-waves cannot travel through liquids. As P-waves enter the inner core, their velocity again increases, to about 6.8 miles (11.3 km.) per second.

Earth's interior seems to be characterized by a gradual increase with depth in temperature, pressure, and density. Exten-

sive experimental and modeling work indicates that the temperature at 62 miles (100 km.) is between 1,200 and 1,400 degrees Celsius (2,192 to 2,552 degrees Fahrenheit). The temperature at the core-mantle boundary—about 1,802 miles (2,900 km.) deep—is calculated to be about 8,130 degrees Fahrenheit (4,500 degrees Celsius). At Earth's center the temperature may exceed 12,092 degrees Fahrenheit (6,700 degrees Celsius). Although at Earth's surface, heat energy is slowly but continuously lost as a result of outgassing, such as from volcanic eruptions, its interior remains hot.

SEISMIC TOMOGRAPHY AND FUTURE EXPLORATION. Seismic tomography is one of the newest tools that earth scientists are using to develop three-dimensional velocity images of Earth's interior. In seismic tomography, several crossing seismic waves from different sources (earthquakes and nuclear explosions) are analyzed in much the same way that computerized axial tomography (CAT) scanners are used in medicine to obtain images of human organs. Seismic tomography is providing two- and three-dimensional images from the crust to the core-mantle boundary. Fast P-wave velocities have been correlated to cool material—for example, a piece of sinking lithosphere (cool rigid layer) such as in regions underneath the Andes Mountains (subduction zone); slow P-wave velocities have been correlated with hot materials—for example, rising mantle plumes of hot spots such as the one responsible for volcanic activity in the Hawaiian Islands.

Rubén A. Mazariegos-Alfaro

FOR FURTHER STUDY

Bolt, Bruce A. *Earthquakes.* Rev. ed. New York: W. H. Freeman, 1993.

Brown, G. C. *The Inaccessible Earth.* London: George Allen & Unwin, 1981.

Brumbaugh, David S. *Earthquakes: Science and Society.* Upper Saddle River, N.J.: Prentice Hall, 1999.

Fowler, C. M. R. *The Solid Earth.* Cambridge, England: Cambridge University Press, 1990.

Levy, Matthys, and Mario Salvadori. *Why the Earth Quakes: The Story of Earthquakes and Volcanoes.* New York: W. W. Norton, 1995.

McKenzie, D. P. "The Earth's Mantle." *Scientific American* 249, no. 3 (1983): 66-78.

Officer, C., and J. Page. *Tales of the Earth.* New York: Oxford University Press, 1993.

Siever, R. "The Dynamic Earth." *Scientific American* 249, no. 3 (1983): 46-55.

Yeats, Robert S., Kerry Sieh, and Clarence R. Allen. *The Geology of Earthquakes.* New York: Oxford University Press, 1997.

Yellowstone Page 38

INFORMATION ON THE WORLD WIDE WEB

The AskGeoMan site at the University of Oregon has answers to many typical questions about Earth's interior and gives users the opportunity to formulate questions about general topics in geoscience. (jersey.uoregon.edu/~mstrick/AskGeoMan)

Hot spots Page 42

PLATE TECTONICS

The theory of plate tectonics provides an explanation for the present-day structure of the large landforms that constitute the outer part of the earth. The theory accounts for the global distribution of continents, mountains, hills, valleys, plains, earthquake activity, and volcanism, as well as various associations of igneous, meta-

Map Page 39

morphic, and sedimentary rocks, the formation and location of mineral resources, and the geology of ocean basins. Everything about the earth is related either directly or indirectly to plate tectonics.

BASIC THEORY. Plate-tectonic theory is based on an Earth model in which a rigid, outer shell—the lithosphere—lies above a hotter, weaker, partially molten part of the mantle called the asthenosphere. The lithosphere varies in thickness between 6 and 90 miles (10 and 150 km.), and comprises the crust and the underlying, upper mantle. The asthenosphere extends from the base of the lithosphere to a depth of about 420 miles (700 km.). The brittle lithosphere is broken into a pattern of in-

ternally rigid plates that move horizontally relative to each other across the earth's surface.

More than a dozen plates have been distinguished, some extending more than 2,500 miles (4,000 km.) across. Exhibiting independent motion, the plates grind and scrape against each other, similar to chunks of ice in water, or like giant rafts cruising slowly on the asthenosphere. Most of the earth's dynamic activity, including earthquakes and volcanism, occurs along plate boundaries. The global distribution of these tectonic phenomena delineates the boundaries of the plates.

Geological observations, geophysical data, and theoretical models support the

THE SUPERCONTINENTS

The theory of plate tectonics explains the present-day distribution of major landforms, seismic and volcanic activity, and physiographic features of ocean basins. Many scientists also use the the-

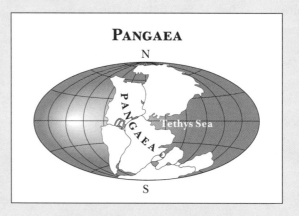

PANGAEA

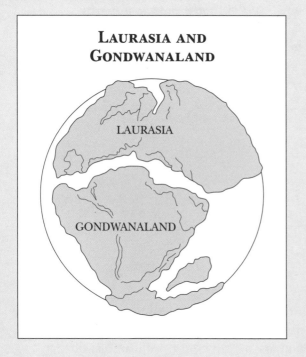

LAURASIA AND GONDWANALAND

ory to explain the history of Earth's surface. Evidence indicates that the modern continents once formed a single landmass called Pangaea, meaning "all lands." According to the theory of plate tectonics, approximately 200 million years ago Pangaea began to split into two supercontinents, Laurasia and Gondwanaland. Eventually, as a result of tectonic forces, Laurasia split into North America, Europe, and most of Asia. Gondwanaland broke up into India, South America, Africa, Australia, and Antarctica.

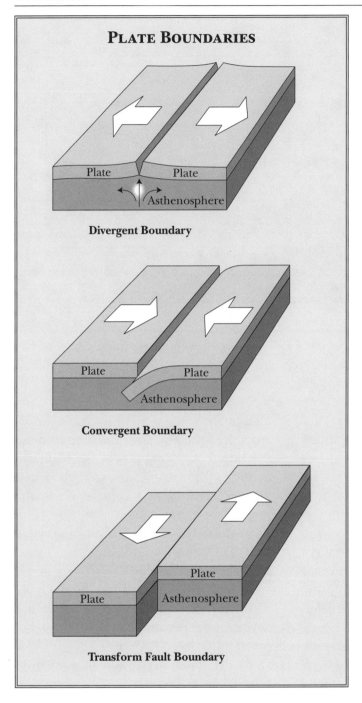

PLATE BOUNDARIES

Plate   Plate
Asthenosphere

Divergent Boundary

Plate   Plate
Asthenosphere

Convergent Boundary

Plate
Plate   Asthenosphere

Transform Fault Boundary

The continents were formed by the movement at plate boundaries, and continental landforms were generated by volcanic eruptions and continental plates colliding with each other. The velocity of plate movement varies from plate to plate and even within portions of the same plate, ranging from 0.8 to 8 inches (2 to 20 centimeters) per year. The rates are calculated from the distance to the midoceanic ridge crests, along with the age of the sea floor as determined by radioactive dating methods.

Convection currents that are driven by heat from radioactive decay in the mantle are important mechanisms involved in moving the huge plates. Convection currents in the earth's mantle carry magma (molten rock) up from the asthenosphere. Some of this magma escapes to form new lithosphere, but the rest spreads out sideways beneath the lithosphere, slowly cooling in the process. Assisted by gravity, the magma flows outward, dragging the overlying lithosphere with it, thus continuing to open the ridges. When the flowing hot rock cools, it becomes dense enough to sink back into the mantle at convergent boundaries.

A second plate-driving mechanism is the pull of dense, cold, down-flowing lithosphere in a subduction zone on the rest of the trailing plate, further opening up the spreading centers so magma can move upward.

existence of three types of plate boundaries. Divergent boundaries occur where adjacent plates move away from each other. Convergent boundaries occur where adjacent plates move toward each other. Transform boundaries occur where plates slip past one another in directions parallel to their common boundaries.

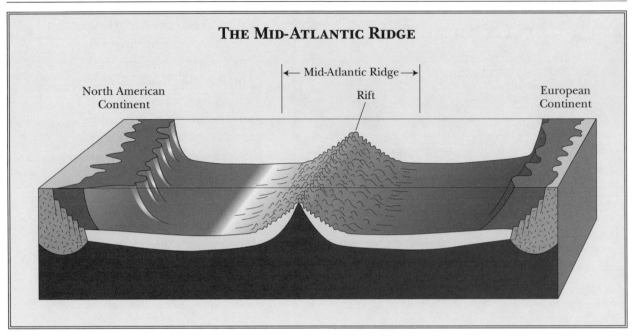

THE MID-ATLANTIC RIDGE

Mid-Atlantic Ridge

North American Continent

Rift

European Continent

The Mid-Atlantic Ridge is a major site of seafloor spreading, where the North American and European plates pull apart.

DIVERGENT PLATE BOUNDARIES. During the 1950's and 1960's, oceanographic studies revealed that Earth's seafloors were marked by a nearly continuous system of submarine ridges, more than 40,000 miles (64,000 km.) in length. Detailed investigations revealed that the midoceanic ridge system has a central rift valley that runs along its length and that the ridge system is associated with volcanic and earthquake activity. The earthquakes are frequent, shallow, and mild.

Magnetic studies of the seafloor indicate that the oceanic lithosphere has been segmented into a series of long magnetic strips that run parallel to the axis of the midoceanic ridges. On either side of the ridge, the ocean floor consists of alternating bands of rock, magnetized either parallel to or exactly opposite of the present-day direction of the earth's magnetic field.

Midoceanic ridges, or divergent plate boundaries, are tensional features representing zones of weakness within the earth's crust, where new seafloor is cre-

ated by the welling up of mantle material from the asthenosphere into cracks along the ridges. As rifting proceeds, magma ascends to fill in the fissures, creating new oceanic crust. Iron minerals within the magma become aligned to the existing Earth polarity as the rock cools and crystallizes. The oceanic floor slowly moves away from the oceanic ridge toward deep ocean trenches, where it descends into the mantle to be melted and recycled to the earth's surface to generate new rocks and landforms.

As the seafloor spreads outward from the rift center, about half of the material is carried to either side of the rift, which is later filled by another influx of molten basalt. When the polarity of the earth changes, the subsequent molten basalt is magnetized in the opposite polarity. The continuation of this process over geologic time leads to the young geologic age of the seafloor and the magnetic symmetry around the midoceanic ridges.

Not all spreading centers are underneath the oceans. An example of continental rifting in its embryonic stage can be observed in the Red Sea, where the Arabian plate has separated from the African plate, creating a new oceanic ridge. Another modern-day example of continental divergent activity is East Africa's Great Rift Valley system. If this rifting continues, it will eventually fragment Africa, producing an ocean that will separate the resulting pieces. Through divergence, large plates are made into smaller ones.

CONVERGENT PLATE BOUNDARIES. Because Earth's volume is not changing, the increase in lithosphere created along divergent boundaries must be compensated for by the destruction of lithosphere elsewhere. Otherwise, the radius of Earth would change. The compensation occurs at convergent plate boundaries, where plates are moving together. Three scenarios are possible along convergent boundaries, depending on whether the crust involved is oceanic or continental.

If both converging plates are made of oceanic crust, one will inevitably be older, cooler, and denser than the other. The denser plate eventually subducts beneath the less-dense plate and descends into the asthenosphere. The boundary along the two interacting plates, called a subduction zone, forms a trench. Some trenches are more than 620 miles (1,000 km.) long, 62 miles (100 km.) wide, and 6.8 miles (11 km.) deep. Heated by the hot asthenosphere beneath, the subducted plate becomes hot enough to melt.

Because of buoyancy, some of the melted material rises through fissures and cracks to generate volcanoes along the overlying plate. Over time, other parts of

the melted material eventually migrate to a divergent boundary and rise again in cyclic fashion to generate new seafloor. The volcanoes generated along the overriding plate often form a string of islands called island arcs. Japan, the Philippines, the Aleutians, and the Mariannas are good examples of island arcs resulting from subduction of two plates consisting of oceanic lithosphere. Intense earthquakes often occur along subduction zones.

If the leading edge of one of the two convergent plates is oceanic crust and the other is continental crust, the oceanic plate is always the one subducted, because it is always denser. A classic example of this case is the western boundary of South America. On the oceanic side of the boundary, a trench was formed where the oceanic plate plunged underneath the continental plate. On the continental side, a fold mountain belt—the Andes—was formed as the oceanic lithosphere pushed against the continental lithosphere.

When the oceanic plate descends into

*Red Sea
Page 238*

California's Big Sur coast. The state's coastline lies at a plate boundary where the Pacific plate is subducting under the North American plate. (William E. Ferguson)

*Earthquake
damage
Page 238*

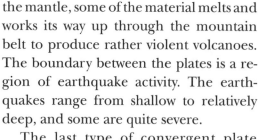

*Tectonic
plates
Page 39*

*Yellowstone
Page 38*

*San
Andreas
fault
Page 42*

the mantle, some of the material melts and works its way up through the mountain belt to produce rather violent volcanoes. The boundary between the plates is a region of earthquake activity. The earthquakes range from shallow to relatively deep, and some are quite severe.

The last type of convergent plate boundary involves the collision of two continental masses of lithosphere, which can result in folding, faulting, metamorphism, and volcanic activity. When the plates collide, neither is dense enough to be forced into the asthenosphere. The collision compresses and thickens the continental edges, twisting and deforming the rocks and uplifting the land to form unusually high fold mountain belts. The prototype example is the collision of India with Asia, resulting in the formation of the Himalayas. In this case, the earthquakes are typically shallow, but frequent and severe.

TRANSFORM PLATE BOUNDARIES. The actual structure of a seafloor spreading ridge is more complex than a single, straight crack. Instead, ridges comprise many short segments slightly offset from one another. The offsets are a special kind of fault, or break in the lithosphere, known as a transform fault, and their function is to connect segments of a spreading ridge. The opposite sides of a transform fault belong to two different plates that are grinding against each other in opposite directions.

Transform faults form the boundaries that allow the plates to move relative to each another. The classic case of a transform boundary is the San Andreas Fault. It slices off a small piece of western California, which rides on the Pacific plate, from the rest of the state, which resides on the North American plate. As the two plates scrape past each other, stress builds up, eventually being released in earthquakes that can be quite violent.

MANTLE PLUMES AND HOT SPOTS. Most plate tectonic features are near plate boundaries, but the Hawaiian Islands are not. In the late twentieth century, the only active volcanoes in the Hawaiian Islands were on the island of Hawaii, at the southeast end of the chain. Radiometric dating and examination of states of erosion show that, when proceeding along the chain to the northwest, successive islands are progressively older.

Evidently, the same heat source produced all the volcanoes in the Hawaiian chain. Known as a mantle plume, it has remained stationary while the Pacific plate rides over it, producing a volcanic trail from which absolute motion of the plate can be determined. Since mantle plumes do not move with the plates, the plumes must originate beneath the lithosphere, probably far below it. Resulting volcanoes are called hot spots to distinguish them from subduction-zone volcanoes. Iceland is a good example of a hot spot, as is Yellowstone. At least one hundred hot spots are distributed around Earth.

Alvin K. Benson

FOR FURTHER STUDY

Brown, Geoff, and Alan Mussett. *The Inaccessible Earth.* 2d ed. London, England: Chapman and Hall, 1993.

Cox, Allan, and Robert Brian Hart. *Plate Tectonics: How It Works.* London, England: Blackwell Scientific Publications, 1986.

Fowler, C. M. R. *The Solid Earth: An Introduction to Global Geophysics.* Cambridge, England: Cambridge University Press, 1992.

Hamblin, W. Kenneth, and Eric H. Christiansen. *Earth's Dynamic Systems.* 8th ed. Upper Saddle River, N.J.: Prentice Hall, 1998.

Keller, Edward A., and Nicholas Pinter. *Active Tectonics.* Upper Saddle River, N.J.:

Prentice Hall, 1996.

Lillie, Robert J. *Whole Earth Geophysics.* Upper Saddle River, N.J.: Prentice Hall, 1999.

Montgomery, Carla W. *Fundamentals of Geology.* 3d ed. Dubuque, Iowa: William C. Brown, 1997.

Sager, Robert J., David M. Helgren, and Saul Israel. *World Geography Today.* 7th ed. New York: Holt, Rinehart and Winston, 1989.

Strahler, Arthur N. *Plate Tectonics.* Cambridge, Mass.: GeoBooks, 1998.

VOLCANOES

Volcanoes form mountains both on land and in the sea and either do it on a grand scale or merely create minute bumps on the seafloor. Volcanoes do not occur in a random pattern, but are found in distinct zones that are related to plate dynamics. Each of the three types of volcanism on Earth is characterized by specific types of eruptions and magma compositions. Molten magma is the rock material below the earth's crust that forms igneous rock as it cools.

TYPES OF VOLCANOES. Abundant mid-ocean ridge basalt (MORB) volcanism occurs at divergent plate margins, where new ocean floor is created. The mid-Atlantic ridge is a submarine chain of such volcanoes, which emerges above the sea surface in Iceland.

The second type is the hot-spot or plume volcano, which is associated with mantle upwellings from great depth. When the plumes appear below an oceanic plate, large basaltic volcanoes (shield volcanoes) form, such as those on Hawaii and the Galapagos Islands. When the plume occurs below a continent, wholesale melting of the crust may take place, creating a large volcanic area such as Yellowstone National Park in the United States.

Arc volcanoes are found near subduction zones, in which oceanic plates are subducted below other oceanic plates (for example, the Aleutian Arc) or beneath continents, such as the Andes volcanoes. Some of the world's classical examples of cone-shaped stratovolcanoes, such as Mount Fuji in Japan, Mayon in the Philippines, and several Cascade Range volcanoes in Oregon and Washington, are arc volcanoes. Some of the highest volcanoes on Earth are of the arc type, notably Nevado Ojos del Salado (22,600 feet/6,885 meters) in the Chilean Andes.

A cross-section of the earth shows a subduction zone with associated arc volcanism and illustrates the trench and the volcanic arc. MORB volcanoes are shown where two plates are drifting apart, and plume volcanoes form tracks where the plume "burned through" the overriding plate.

VOLCANIC COMPOSITION. Volcanoes in the midocean ridges and plume environments draw most of their magmas from the earth's mantle and produce mainly dark, magnesium-rich basaltic magmas.

Yellowstone Page 38

Cascade Range Page 36

Erupting volcanoes Page 40

When basaltic magmas accumulate in the continental crust (for example, at Yellowstone), the large-scale crustal melting leads to rhyolitic volcanism, the volcanic equivalent of granites. Arc magmas cover a wider range of magmatic compositions, ranging from arc basalt to light-colored, silica-rich rhyolites; the latter are commonly erupted in the form of the silica-rich volcanic rock known as pumice, or the black volcanic glass known as obsidian. Andesites, named after the Andes Mountains, are a common volcanic rock in arc volcanoes, intermediate in composition between basalt and rhyolite.

Magmas form from several processes that lead to partial melting of a solid rock. The simplest is adding heat—for example, plumes carrying heat from deep levels in the mantle to shallower levels, where melting occurs. Decompressional (lowering the pressure) melting of the mantle occurs where the ocean floor is thinned or carried away by seafloor spreading in mid-ocean ridge environments.

Fire fountain Page 40

GENESIS OF MAGMA. Adding a "flux" to a solid mineral mixture may lower the substance's melting point. The most common theory about arc magma genesis invokes the addition of a low-melting-point substance to the arc mantle, a layer of mantle material at about 60 to 90 miles (100 to 150 km.) below the volcanic arc. The relatively dry arc mantle would usually start to melt at about 2,100 to 2,300 degrees Fahrenheit (1,200 to 1,300 degrees Celsius). However, the addition of water and other gases can lower the melting point of the mixture. The water and its dissolved chemicals are supposedly derived from the subducted slab, the former ocean floor that is pushed back into the earth.

Mt. St. Helens Page 41

The sequence of events is as follows: New basaltic ocean floor forms at mid-ocean ridge volcanoes. The new hot magma interacts with seawater, leading to vents at the seafloor with their mineralized deposits. The seafloor becomes hydrated, and sulfur and chlorine from seawater are locked up in newly formed minerals. During subduction, this altered seafloor with slivers of sediment, including limestone, is gradually warmed up and starts to decompose, adding a flux to the surrounding mantle rocks. The mantle rocks then start to melt, and these magmas with minor inherited oceanic materials start to rise and pond at the bottom of the crust. There the magmas sit and wait for an opportunity to erupt, while cooling and crystallizing. Thus, arc magmas bear a chemical signature of subducted oceanic components while their chemical compositions range from basalt to rhyolite.

VOLCANIC ERUPTIONS. Volcanic eruptions occur as a result of the rise of magma into the volcano (from depths as great as several miles) and then into the throat of the volcano. In basaltic volcanoes, the magmas have relatively little gas, and the magma simply overflows and forms large lava flows, sometimes associated with fire fountains. Arc volcanoes can erupt regularly with small explosions or catastrophically after long periods of dormancy. Mount Stromboli, a volcano in Italy, erupts every twenty minutes, with an explosion that creates a column 650 to 980 feet (200 to 300 meters) high. Mount St. Helens in the U.S. state of Washington had a catastrophic eruption in 1980 after about two hundred years of dormancy. It emitted an ash plume that reached more than 12 miles (20 km.) into the atmosphere.

After long magma storage periods in the crust, crystallization and melting of crustal material can lead to silica-rich magmas. These are viscous and can have high dissolved water contents—up to 4 to 6 percent by weight. When these magmas break out, the eruption can be violent and form

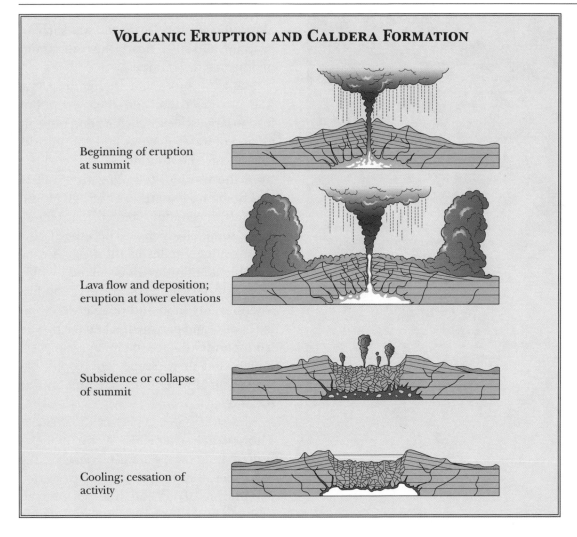

VOLCANIC ERUPTION AND CALDERA FORMATION

Beginning of eruption
at summit

Lava flow and deposition;
eruption at lower elevations

Subsidence or collapse
of summit

Cooling; cessation of
activity

an eruption column 12 to 35 miles (20 to 55 km.) high. Many cubic miles of magma can be ejected. This leads to so-called plinian ash falls, with showers of pumice and ash over thousands of square miles, with the ash commonly carried around the globe by the high-level winds known as jet streams.

If the volume of ejected magma is large, the volcano empties itself and collapses into the hole, leading to a caldera—a volcanic collapse structure. The caldera at Crater Lake in Oregon is related to a large pumice eruption about 76,000 years ago. Basaltic volcanoes can also form collapse calderas when large volumes of lava have been extruded in a short time. Examples

of famous basaltic calderas can be found in Hawaii's Mount Kilauea and the Galapagos Islands.

VOLCANIC PLUMES. The dynamics of volcanic plumes has been studied from eruption photographs, experiments, and theoretical work. The rapidly expanding hot gases force the viscous magma out of the throat of the volcano, where it freezes into pumice. The kinetic energy of the ejected mass carries it 2 to 2.5 miles (3-4 km.) above the volcano. During this phase, air is entrained in the column, diluting the concentration of ash and pumice particles. The hot particles heat the entrained air, the mixture of hot air and solids becomes less dense than the sur-

*Crater
Lake
Page 39*

Eruption of Alaska's Crater Peak volcano in 1992.

rounding atmosphere, and a buoyant column rises high into the sky.

The height of an eruption column is not directly proportional to the force of the eruption but is strongly dependent on the rate of heat release of the volcano. If little of the entrained air is heated up, the column will collapse back to the ground and an ash flow forms, which may deposit ash around the volcano. These types of eruptions are among the most devastating, creating glowing ash clouds traveling at speeds up to 60 miles (100 km.) per hour, burning everything in their path. The 1902 eruption of Mount Pelée on

Martinique in the Caribbean was such an eruption and killed nearly thirty thousand people in a few minutes.

Many volcanoes that are high in elevation are glaciated, and their eruptions lead to large-scale ice melting and possibly mixing of water, magma, and volcanic debris. Massive hot mudflows can race down from the volcano, following river valleys and filling up low areas. The 1980 Mount St. Helens eruption created many mudflows, some of which reached the Pacific Ocean, ninety miles to the west. A catastrophic mudflow event occurred in 1984 at Nevado del Ruiz, a volcano in Colombia, where twenty thousand people were buried in mud and perished. When magma intrudes under the ice, meltwater can accumulate and then escape catastrophically, but such meltwater bursts are rare outside Iceland.

MINERALS AND GASES IN ERUPTIONS. The gas-rich character of arc magmas leads to fluid escape at various levels in the volcanoes, and these fluids tend to be rich in chlorine. They can transport metals such as copper, lead, zinc, and gold at high concentrations, and lead to the enrichment of these metals in the fractured volcanic rocks. Many of the world's largest copper ore deposits are associated with older arc volcanism, where erosion has removed most of the volcanic structure and laid the volcano innards bare. Many active volcanoes have modern hydrothermal (hot-water) systems, leading to acid hot springs and crater lakes and the potential to harness geothermal energy. Some areas in Japan, New Zealand, and Central America have an abundance of geothermal energy resources, which are gradually being developed.

Apart from the dangers of eruptions, continuous emissions of large amounts of sulfur dioxide, hydrochloric acid, and hydrofluoric acid present a danger of air pollution and acid rain. Incidences of emphysema and other irritations of the respiratory system are common in people living on the slopes of active volcanoes. The large lava emissions in Iceland in the eighteenth century led to acid fogs all over Europe. Many cattle died in Iceland during this period from the hydrofluoric acid vapors. High levels of fluorine in drinking water can lead to fluorosis, a disease that attacks the bone structure. The discharge of highly acidic fluids from hot springs and crater lakes can cause widespread environmental contamination, which can present a danger for crops gathered from fields irrigated with these waters and for local ecosystems in general.

Johan C. Varekamp

FURTHER READING

Bardintzeff, J. M., and A. R. McBirney. *Volcanology.* Sudbury, Mass.: Jones and Bartlett, 2000.

Cook, Diane. *Hot Spots: America's Volcanic Landscapes.* Boston: Little, Brown, 1996.

Decker, Robert W., and Barbara B. Decker. *Mountains of Fire.* New York: Cambridge University Press, 1991.

_____. *Volcanoes.* 3d ed. New York: Freeman, 1998.

Erickson, Jon. *Volcanoes and Earthquakes.* Blue Ridge Summit, Pa.: TAB Books, 1987.

Francis, P. *Volcanoes: A Planetary Perspective.* Oxford, England: Oxford University Press, 1995.

Harris, Stephen L. *Agents of Chaos: Earthquakes, Volcanoes, and Other Natural Disasters.* Missoula, Mont.: Mountain Press, 1990.

MacDonald, G. A. *Volcanoes.* Englewood Cliffs, N.J.: Prentice Hall, 1971.

Sigurdsson, Haraldur. *Encyclopedia of Volcanoes.* San Diego, Calif.: Academic Press, 2000.

GEOLOGIC TIME SCALE

A major difference between the geosciences (earth sciences) and other sciences is the great enormity of their time scale. One might compare the magnitude of geologic time for geoscientists to the vastness of space for astronomers. Every geological process, such as the movement of crustal plates (plate tectonics), the formation of mountains, and the advance and retreat of glaciers, must be considered within the context of time.

Although certain geologic events, such as floods and earthquakes, seem to occur over short periods of time, the vast majority of observed geological features formed over a great span of time. Consequently, modern geoscientists consider Earth to be exceedingly old. Using radiometric age-dating techniques, they calculate the age of Earth as 4.6 billion years old.

Early miners were probably the first to recognize the need for a scale by which rock and mineral units could be compared over large geographic areas. However, before a time scale—and even geology as a science—could develop, certain principles had to be established. This did not occur until the late eighteenth century when James Hutton, a Scottish naturalist, began his extensive examinations of rock relationships and natural processes at work on the earth. His work was amplified by Charles Lyell in his textbook *Principles of Geology* (1830-1833). After careful observation, Hutton concluded that the natural processes and functions he observed had operated in the same basic manner in the past, and that, in general, natural laws were invariable. That idea became known as the principle of uniformitarianism.

THE BIRTH OF STRATIGRAPHY. In 1669 Nicholas Steno, a Danish physician working in Italy, recognized that horizontal rock layers contained a chronological record of Earth history and formulated three important principles for interpreting that history. The principle of superposition states that in a succession of undeformed strata, the oldest stratum lies at the bottom, with successively younger ones above. The principle of original horizontality states that because sedimentary particles settle from fluids under graviational influence, sedimentary rock layers must be horizontal; if not, they have suffered from subsequent disturbance. The principle of original lateral continuity states that strata originally extended in all directions until they thinned to zero or terminated against the edges of the original area of deposition.

In the late eighteenth century, the English surveyor William Smith recognized the wide geographic uniformity of rock layers and discovered the utility of fossils in correlating these layers. By 1815, Smith had completed a geologic map of England and was able to correlate English rock layers with layers exposed across the English Channel in France.

From the need to classify and organize rock layers into an orderly form arose a

*Rock layers
Page 43*

subdiscipline of modern geology—stratigraphy, the study of rock layers and their age relationships. In 1835 two British geologists, Adam Sedgwick and Roderick Murchison, began organizing rock units into a formal stratigraphic classification. Large divisions, called eras, were based upon well known and characteristic fossils, and included a number of smaller subdivisions, called periods.

The periods are often subdivided into smaller units called epochs. Each period is defined by a representative sequence of rock strata and fossils. For instance, the Devonian period is named for exposures of rock in Devonshire in southern England, while the Jurassic period is defined by strata exposed in the Jura Mountains in northern Switzerland.

Approximately 80 percent of Earth's history is included in the Crypotozoic era (meaning obscure life). Fossils from the Crypotozoic era are rare, and the rock record is very incomplete. After the Crypotozoic era came the Paleozoic (ancient life), Mesozoic (middle life), and Cenozoic (recent life) eras. Most of the life forms that evolved during the Paleozoic and Mesozoic eras are now extinct, whereas 90 percent of the life forms that evolved up to the middle Cenozoic era still exist.

THE GEOLOGIC TIME SCALE. The geologic time scale is continually in revision as new rock formations are discovered and dated. The ages shown in the table below are in millions of years ago (MYA) before the present and represent the beginning of that particular period. It would be impossible to list all the significant events in Earth's history, but one or two are provided for each period. Note that in the United States, the Carboniferous period has been subdivided into the Mississippian period (older) and the Pennsylvanian period (younger).

THE GEOLOGIC TIME SCALE

MYA	Eon	Era	Period	Epoch	Developments
0.01	Phanerozoic Eon (544 mya-present)	Cenozoic (65 mya-today)	Quaternary (1.8 mya-today)	Holocene (11,000 ya-today)	Ice Age ends; humans begin to impact biosphere
1.8				Pleistocene (1.8 mya-11,000 ya)	Glaciation leads to Ice Age; modern humans evolve
5			Tertiary (65-1.8 mya)	Pliocene (5-1.8 mya)	Cooling period leads to Ice Age
23				Miocene (23-5 mya)	Erect-walking human ancestors
38				Oligocene (38-23 mya)	Primate ancestors of humans
54				Eocene (54-38 mya)	Intense mountain building: Alps, Himalaya, Rockies. Modern mammals: rodents, hoofed animals
65				Paleocene (65-54 mya)	Cretaceous-Tertiary event (?) leads to dinosaurs' extinction c. 65 mya
165		Mesozoic (245-65 mya)	Cretaceous (146-65 mya)		Birds arise; breakup of super-continents into present form
208			Jurassic (208-146 mya)		Earliest mammals
245			Triassic (245-208 mya)		Dinosaurs develop
286		Paleozoic (544-245 mya)	Permian (286-245 mya)		Permian extinction
325			Carboniferous (360-286 mya)	Pennsylvanian (325-286 mya)	Supercontinent Pangaea forms
360				Mississippian (360-325 mya)	Reptiles
410			Devonian (410-360 mya)		Amphibians; vascular plants; diverse insects
440			Silurian (440-410 mya)		Early land plants, insects
505			Ordovician (505-440 mya)		Life colonizes land; earliest vertebrates appear in fossil record
544			Cambrian (544-505 mya)	Tommotian (530-527 mya)	Cambrian diversification of life
900	Precambrian Time (4,500-544 mya)	Proterozoic (2500-544 mya)	Neoproterozoic (900-544 mya)	Vendian (650-544 mya)	Earliest invertebrates Marine plants, animals
1600			Mesoproterozoic (1600-900 mya)		
2500			Paleoproterozoic (2500-1600 mya)		Transition from prokaryotic to eukarotic life leads to multi-cellular oganisms, c. 2 bya
3800		Archaean (3800-2500 mya)			Microbial life as early as 3.5 bya
4500		Hadean (4500-3800 mya)			Earth forms 4.5 bya

Notes: mya = millions of years ago; bya = billions of years ago.

Source: Data on time periods in this version of the geologic time scale are based on new findings in the last decade of the twentieth century as presented by the Geologic Society of America, which notably moves the transition between the Precambrian and Cambrian times from 570 mya to 544 mya.

EARTH'S HISTORY COMPRESSED INTO ONE CALENDAR YEAR

One way to visualize events in Earth's history is to compress geologic events into a single calendar year. Earth's birth, 4.6 billion years ago, would occur during the first minute of January 1. The first three-quarters of Earth's history is obscure and would take place from January to mid-October. During this time, Earth gained an oxygenated atmosphere, and the earliest life-forms evolved. The first organisms with hard parts preserved in the fossil record (approximately 570 million years ago) would appear around November 15. The extinction of the dinosaurs (65 million years ago) would occur on Christmas Day. *Homo sapiens* would first appear at approximately 11 P.M. on December 31, and all of recorded human history would occur in the last few seconds of New Year's Eve.

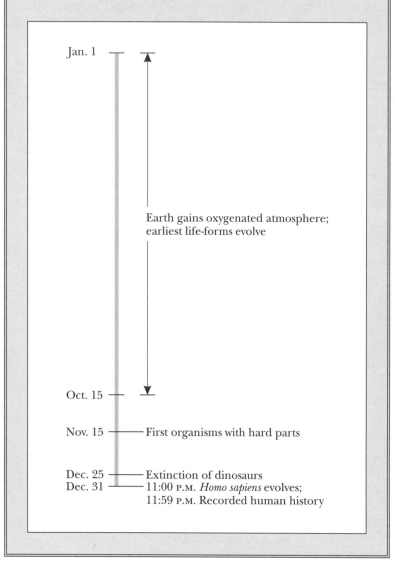

Jan. 1

Earth gains oxygenated atmosphere; earliest life-forms evolve

Oct. 15

Nov. 15 — First organisms with hard parts

Dec. 25 — Extinction of dinosaurs
Dec. 31 — 11:00 P.M. *Homo sapiens* evolves; 11:59 P.M. Recorded human history

THE FOSSIL RECORD. The word "fossil" comes from the Latin *fossilium*, meaning "dug from beneath the surface of the ground." Fossils are defined as any physical evidence of past life. Fossils can include not only shells, bones, and teeth, but also tracks, trails, and burrows. The latter group are referred to as trace fossils. Fossils demonstrate two important truths about life on Earth: First, thousands of species of plants and animals have existed and later became extinct. Second, plants and animals have evolved through time, and the communities of life that have existed on Earth have changed.

Some organisms are slow to evolve and may exist in several geologic time periods, while others evolve quickly and are restricted to small intervals of time within a particular period. The latter, referred to as index fossils, are the most useful to geoscientists for correlating rock layers over wide geographic areas and for recognizing geologic time.

The fossil record is incomplete, because the process of preservation favors organisms with hard parts that are rapidly buried by sediments soon after death. For this reason, the vast majority of fossils are represented by marine invertebrates with exoskeletons, such as clams and snails. Under special circumstances, soft-bodied organism can be preserved, for instance the preservation of insects in amber, made famous by the feature film *Jurassic Park* (1993).

Fossil leaves from sandstone of the Gerome Andesite, Northwest Uranium Mine, Stevens County, Washington. (U.S. Geological Survey)

The processes by which the various rock types change over time are illustrated in the rock cycle.

Larry E. Davis

FOR FURTHER STUDY

Ausich, William I., and N. Gary Lane. *Life of the Past.* Upper Saddle River, N.J.: Prentice Hall, 1999.

Berry, William B. N. *Growth of a Prehistoric Time Scale.* Palo Alto, Calif.: Blackwell, 1987.

Cooper, John D., Richard H. Miller, and Jacqueline Patterson. *A Trip Through Time: Principles of Historical Geology.* Columbus, Ohio: Merrill, 1986.

THE ROCK CYCLE. A rock is a naturally formed aggregate of one or more minerals. Three types of rocks exist in the earth's crust, each reflecting a different origin. Igneous rocks have cooled and solidified from molten material either at or beneath Earth's surface. Sedimentary rocks form when preexisting rocks are weathered and broken down into fragments that accumulate and become compacted or cemented together. Fossils are most commonly found in sedimentary rocks. Metamorphic rocks form when heat, pressure, or chemical reactions in Earth's interior change the mineral or chemical composition and structure of any type of preexisting rock.

Over the huge span of geologic time, rocks of any one of these basic types can change into either of the other types or into a different form of the same type. For this reason, older rocks become increasingly more rare.

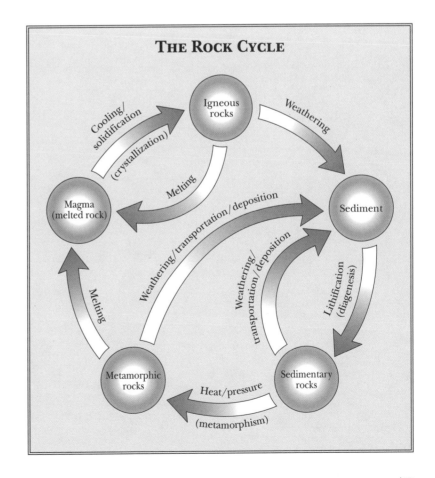

THE ROCK CYCLE

INFORMATION ON THE WORLD WIDE WEB

Geologic Time: Online Edition, a Web site maintained by the U.S. Geological Survey, contains links to a wide range of topics, including the geologic time scale, index fossils, and radiometric dating. (pubs.usgs.gov/gip/geotime/)

The Web site of the Geological Society of America, headquartered in Boulder, Colorado, features information about current issues, publications, and careers in Earth science. (www.geosociety.org)

The Museum of Paleontology at the University of California, Berkeley, maintains a Web site with on-line exhibits on the history of life, geologic time periods, and fossils. (www.ucmp.berkeley.edu/index.html)

Dott, Robert H., and Donald R. Prothero. *Evolution of the Earth.* New York: McGraw-Hill, 1994

Eicher, Don L. *Geologic Time.* Englewood Cliffs, N.J.: Prentice-Hall, 1976.

Hallam, A. *Great Geological Controversies.* Oxford, England: Oxford Press, 1992.

McPhee, John. *Basin and Range.* New York: Farrar, Straus & Giroux, 1981.

Stanley, Steven M. "Extinctions—Or, Which Way Did They Go?" *Earth* (January, 1991): 18-27.

Wicander, Reed, and James S. Monroe. *Historical Geology: Evolution of Earth and Life Through Time.* 3d ed. Pacific Grove, Calif.: Brooks/Cole, 2000.

EARTH'S SURFACE

INTERNAL GEOLOGICAL PROCESSES

The earth is layered into a core, a mantle, and a crust. The topmost mantle and the crust make up the lithosphere. Beneath this is a layer called the asthenosphere, which is composed of moldable and partly liquid materials. Heat transference within the asthenosphere sets up convection cells that diverge from hot regions and converge to cold regions. Consequently, the overlying lithosphere is segmented into ridged plates that are moved by the convection process. The hot asthenosphere does not rise along a line. This causes the development of a structure called a transform plate boundary, which is perpendicular to and offsetting the divergent boundary.

The topographic features at Earth's surface, such as mountains, rift valleys, oceans, islands, and ocean trenches, are produced by extension or compression forces that act along divergent, convergent, or transform plate boundaries. The extension and compression forces at Earth's surface are powered by convection within the asthenosphere.

MOUNTAINS AND DEPRESSIONS IN ZONES OF COMPRESSION. Compression along convergent plate boundaries yields three types of mountain: island arcs that are partly under water; mountains along a continental edge, such as the Andes; and mountains at continental interiors, such as the Alps. At convergent plate bound-aries, the denser of the two colliding plates slides down into the asthenosphere and causes volcanic activity to form on the leading edge of the upper plate. Island arcs such as the Aleutians and the Carib-bean are formed when an oceanic plate descends beneath another oceanic plate.

Volcanic mountain chains such as the Andes of South America are formed when an oceanic plate descends beneath a conti-nental plate. In both the island arc type and Andean type collisions, a deep depres-sion in the oceans, called a trench, marks the place where neighboring plates are colliding and where the denser plates are pulled downward into the asthensophere. If the colliding plates are of similar den-sity, neither plate will go into the astheno-sphere. Instead, the edges of the neigh-boring plates will be folded and faulted and excess material will be pushed upward to form a block mountain, such as the mountain chain that stretches from the Alps through to the Himalayas. This type of mountain chain is not associated with a trench.

The Appalachians of the eastern United States are an example of the alpine type of mountain belt. When the Appala-chians were forming 300 million years ago, rock layers were deformed. The deforma-tion included folding to form ridges and valleys; fracturing along joint sets, with one joint set being parallel to ridges, while the other set is perpendicular; and thrust faulting, in which rock blocks were detached and shoved upward and northwestward.

Millions of years of erosion have reduced the height of the mountains and have produced topographic inversion in the foothills. Topographic inversion occurs because joints create wider fractures at upfolded ridges and narrower fractures at downfolded valleys. Erosion is then accelerated at upfolded ridges, converting ancient ridges into valleys, while ancient valleys stand as ridges. The Valley and Ridge Province of the Appalachians is noted for such topographic inversion.

West of the Valley and Ridge Province of the Appalachians is the Allegheny Plateau, which is bounded by a cliff on its eastern side. In general, plateaus are flat topped because the rock layer that covers the surface is resistant to weathering. The cliff side is formed by erosion along joint or fault surfaces.

The Sierra Nevada range, which formed seventy million years ago, is an example of an Andean type of mountain belt. Millions of years of erosion there has exposed igneous rocks that formed at depth. Over the years, the force of compression that formed the Sierras has evolved to form a zone of extension between the Sierras and the Colorado Plateau.

MOUNTAINS AND DEPRESSIONS IN ZONES OF EXTENSION. Extension is a strain that involves an increase in length and causes crustal thinning and faulting. Extension is associated with convergent boundaries, divergent boundaries, and transform boundaries.

EXTENSION ASSOCIATED WITH A CONVERGENT BOUNDARY. During the formation of the Sierra Nevada, an oceanic plate that was subducted beneath California declined at a shallow angle eastward toward the Colorado Plateau. Later, the subducted plate peeled off and molten asthenosphere took its place. From the asthenosphere, lava ascended through fractures to form volcanic mountains in Arizona and Utah, and lava flowed and volcanic ash fell as far west as California. The lithosphere has been heated up and has become buoyant, so the Colorado Plateau rises to higher elevations, and rock layers slide westward from it in a zone of extension that characterizes the Basin and Range Province.

In the extension zone, the top rock layers move westward on curved displacement planes that are steep at the surface and nearly horizontal at depth. When rock layers move westward over a curved detachment surface, the trailing edge of the rock layers roll over and are tilted toward the east so they do not leave space in buried rocks. On the other hand, a west-facing slope is left behind on a mountain from which the rock layers were detached. Therefore, movement along one curved detachment surface creates a valley, and movement along several such detachment surfaces forms a series of valleys separated by ridges, as in the Basin and Range Province. The amount of the displacement along the curved surfaces is not uniform. For example, more displacement has created wide zones of valleys such as the Las Vegas valley in Nevada, and Death Valley in California.

EXTENSION ASSOCIATED WITH A DIVERGENT BOUNDARY. The longest mountain chain on Earth lies under the Pacific Ocean. It is about 37,500 miles (60,000 km.) long, 31.3 miles (50 km.) wide, and 2 miles (3 km.) high. The central part of this midoceanic ridge is marked by a depression, about 3,000 feet (1,000 meters) deep, and is called a rift valley. A part of the submarine ridge, called the East Pacific Rise, forms the seafloor sector in the Gulf of California and reappears off the coast of northern California, Oregon, and Washington as the Juan de Fuca Ridge. Another part forms the seafloor sector in the Gulf of Aden and Red Sea seafloor, part of which is exposed in the Afar of Ethiopia. From the Afar southward to the southern part of Mozambique is the longest exposed rift valley on land, the East African Great Rift Valley.

A rift valley is the place where old rocks are pushed aside and new rocks are created. Blocks of rock that are detached from the rift walls slide down by a series of normal fault displacements. The ridge adjacent to the central rift is present because hot rocks are less dense and buoyant. If the process of divergences continues from the rifting stage to a drifting stage, as the rocks move farther away from the central rift, the rocks become older, colder, and denser, and push on the underlying asthenosphere to create basins. These basins will be flooded by oceanic water as neighboring continents drift away. However, not all processes of divergence advance from the rifting to the drifting stage.

EXTENSION ASSOCIATED WITH TRANSFORM BOUNDARY. The best-known example of a transform boundary is the San Andreas Fault that offsets the East Pacific Rise from the Juan de Fuca Ridge, and is exposed on land from the Gulf of Califor-

nia to San Francisco. Along transform boundaries, there are pull-apart basins that may be filled to form lakes, such as the Salton Sea in Southern California. Another example is the Aqaba transform of the Middle East, along which the Sea of Galilee and the Dead Sea are located.

H. G. Churnet

FOR FURTHER STUDY

Bolt, B. A. *Inside the Earth: Evidence from Earthquakes.* San Francisco: W. H. Freeman, 1982.

Burns, George. *Exploring the World of Geology.* New York: Franklin Watts, 1995.

Christian, Spencer, and Antonia Felix. *Shake, Rattle, and Roll: The World's Most Amazing Volcanoes, Earthquakes, and Other Forces.* New York: John Wiley & Sons, 1997.

Erickson, Jon. *Rock Formations and Unusual Geologic Structures.* New York: Facts On File, 1993.

Hubler, Clark. *America's Mountains: An Exploration of Their Origins and Influences from the Alaskan Range to the Appalachians.* New York: Facts on File, 1995.

Montgomery, Carla W. *Fundamentals of Geology.* 3d ed. Dubuque, Iowa: William C. Brown, 1997.

Olsen, Kenneth H., ed. *Continental Rifts: Evolution, Structure, Tectonics.* Amsterdam: Elsevier, 1995.

San Andreas Fault Page 42

Gulf of Aden Pages 225, 238

EXTERNAL PROCESSES

Continuous processes are at work shaping the earth's surface. These include breaking down rocks, moving the pieces, and depositing the pieces in new locations.

Weathering breaks down rocks through atmospheric agents. The process of moving weathered pieces of rock by wind, water, ice, or gravity is called erosion. The materi-

Weathered rocks Page 44

als that are deposited by erosion are called sediment.

Mechanical weathering occurs when a rock is broken into smaller pieces but its chemical makeup is not changed. If the rock is broken down by a change in its chemical composition, the process is called chemical weathering.

MECHANICAL WEATHERING. Different types of mechanical weathering occur, depending on climatic conditions. In areas with moist climates and fluctuating temperatures, rocks can be broken apart by frost wedging. Water fills in cracks in rocks, then freezes during cold nights. As the ice expands and pushes out on the crack walls, the crack enlarges. During the warm days, the water thaws and flows deeper into the enlarged crack. Over time, the crack grows until the rock is broken apart. This process is active in mountains, producing a pile of rock pieces at the mountain base called talus.

Salt weathering occurs in areas where much salt is available or there is a high evaporation rate, such as along the seashore. Salt crystals form when salty moisture enters rock cracks. Growing crystals settle in the bottom of the crack and apply pressure on the crack walls, enlarging the crack.

Thermal expansion and contraction occur in climates with fluctuating temperatures, such as deserts. All minerals expand during hot days and contract during cold nights, and some minerals expand and contract more than others. This process continues until the rock loosens up and breaks into pieces.

Mechanical exfoliation can happen to a rock body overlain by a thick rock or sediment layer. If the heavy overlying layer over a portion of the rock body is removed, pressure is relieved and the exposed rock surface will expand in response. This expanding surface will break off into sheets parallel to the surface, but the remaining rock body remains under pressure and unchanged.

When plant roots grow into cracks in rocks, they enlarge the cracks and break up the rocks. Finally, abrasion can occur to rock fragments during transport. Either the fragments collide, breaking apart, or fragments are scraped against rocks, breaking off pieces.

CHEMICAL WEATHERING. Water and oxygen create two common causes of chemical weathering. For example, dissolution occurs when water or another solution dissolves minerals within a rock and carries them away. Hydrolysis can occur when water flows through earth materials. The hydrogen ions or the hydroxide ions of the water may react with minerals in the rocks. When this occurs, the chemical composition of the mineral is changed, and a new mineral is formed. Hydrolysis often produces clay minerals.

Some elements in minerals combine with oxygen from the atmosphere, creating a new mineral. This process is called oxidation. Some of these oxidation minerals are commonly referred to as rust.

MASS MOVEMENT. Weathered rock pieces (sediments) are transported (eroded) by one or more of four transport processes: water (streams and oceans), wind, ice (glaciers), or gravity. Mass movement transports earth materials down slopes by the pull of gravity. Gravity, constantly working to pull surface materials down, parallel to the slope, is the most important factor affecting mass movement. There is also a force involved perpendicular to the slope that contributes to the effects of friction.

Friction, the second factor, is determined by the earth material type involved. For example, weathering may create cracks in rocks, which form planes of weakness on which the mass movement

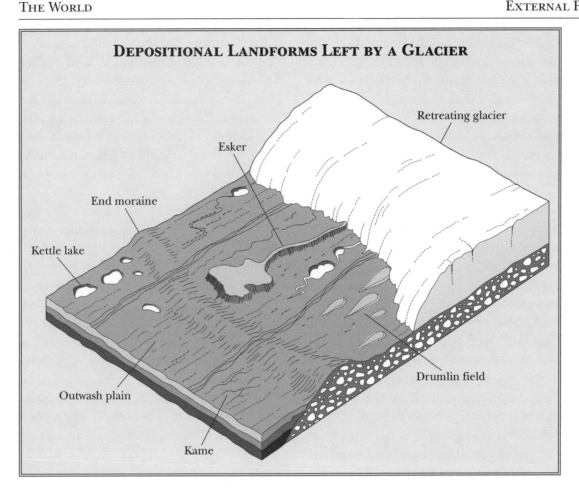

DEPOSITIONAL LANDFORMS LEFT BY A GLACIER

Retreating glacier

Esker

End moraine

Kettle lake

Drumlin field

Outwash plain

Kame

can occur. Loose sediments always tend to roll downhill.

The third factor is the slope angle. Each earth material has its own angle of repose, which is the steepest slope angle on which the materials remain stable. Beyond this slope angle, earth materials will move downslope.

Water, the fourth factor, affects the stability of the earth material in the slope. Friction is weakened by water between the mineral grains in the rock. For example, water can make clay quite slippery, causing the mass movement.

The rooting system of vegetation, the fifth factor, helps make the surficial materials of the slope stable by binding the loose materials together.

Mass movements can be classified by their speed of movement. Creep and solifluction are the two types of slow mass movement, which are measured in fractions of inches per year. Creep is the slowest mass movement process, where unconsolidated materials at the surface of a slope move slowly downslope. The materials move slightly faster at the surface than below, so evidence of creep commonly can be seen by slanted telephone poles. During solifluction, the warm sun of the brief summer season in cold regions thaws the upper few feet of the earth. This waterlogged soil flows downslope over the underlying permafrost.

Rapid mass movement processes occur at feet per second or miles per hour. Falls occur when loose rock or sediment is dislodged and drops from a steep slope, such as along sea cliffs where waves erode the cliff base. Topples occur when there is an overturning movement of the mass. A topple can turn into a fall or a slide. A slide is a mass of rock or sediment that becomes dislodged and moves along a plane of weakness, such as a fracture. A slump is a slide that separates along a concave surface. Lateral spreads occur when a fractured earth mass spreads out at the sides.

A flow occurs when a mass of wet or dry rock fragments or sediment moves downslope as a highly viscous fluid. There are several different flow types. A debris flow is a mass of relatively dry, broken pieces of earth material that suddenly has water added. The debris flow occurs on steeper slopes and moves at speeds of 1-25 miles (2-40 km.) per hour. A debris avalanche occurs when an entire area of soil and underlying weathered bedrock becomes detached from the underlying bedrock and moves quickly down the slope. This flow type is often triggered by heavy rains in areas where vegetation has been removed. An earthflow is a dry mass of clayey or silty material that moves relatively slowly down the slope. A mudflow is a mass of earth material mixed with water that moves quickly down the slope.

A quick clay can occur when partially saturated, solid, clayey sediments are subjected to an earthquake, explosion, or loud noise and become liquid instantly.

Sherry L. Eaton

Debris flow
Page 44

FOR FURTHER STUDY

Burns, George. *Exploring the World of Geology.* New York: Franklin Watts, 1995.

Christian, Spencer, and Antonia Felix. *Shake, Rattle, and Roll: The World's Most Amazing Volcanoes, Earthquakes, and Other Forces.* Spencer Christian's World of Wonder Series. New York: John Wiley & Sons, 1997.

Goodwin, Peter H. *Landslides, Slumps, and Creeps.* New York: Franklin Watts, 1998.

Redfern, Martin. *The Kingfisher Young People's Book of Planet Earth.* New York: KingFisher Books, 1999.

Sharpe, C. F. Stewart. *Landslides and Related Phenomena, A Study of Mass Movements of Soil and Rock.* New York: Columbia University Press, 1938.

Stotsky, Sandra. *Geology: The Active Earth (Ranger Rick's Naturescope).* New York: Chelsea House, 1998.

INFORMATION ON THE WORLD WIDE WEB

Information on North American geology can be found through the American Geological Institute. (www.agiweb .org)

FLUVIAL AND KARST PROCESSES

Earth's landscape has been sculptured into an almost infinite variety of forms. The earth's surface has been modified by various processes for thousands, even hundreds of millions, of years to arrive at the modern configuration of landscapes.

Each process that transforms the surface is classified as either endogenic or exogenic. Endogenic processes are driven by the earth's internal heat and energy and are responsible for major crustal deformation. Endogenic processes are considered constructional, because they build up the earth's surface and create new landforms, such as mountain systems. Conversely, exogenic processes are considered destructional because they result in the wearing away of landforms created by endogenic processes. Exogenic processes are driven by solar energy putting into motion the earth's atmosphere and water, resulting in the lowering of features originally created by endogenic processes.

The most effective exogenic processes for wearing away the landscape are those that involve the action of flowing water, commonly referred to as fluvial processes. Water flows over the surface as runoff, after it evaporates into the atmosphere and infiltrates into the soil. The water that is left over flows down under the influence of gravity and has tremendous energy for sculpturing the earth's surface. Although flowing water is the most effective agent for modifying the landscape, it represents less than 0.01 percent of all the water on Earth's surface. By comparison, nearly 75 percent of the earth's surface water is stored within glaciers.

DRAINAGE BASINS. Fluvial processes can be considered from a variety of spatial scales. The largest scale is the drainage basin. A drainage basin is the area defined by topographic divides that diverts all water and material within the basin to a single outlet. Every stream of any size has its own drainage basin, and every portion of the earth's land surfaces are located within a drainage basin. Drainage basins vary tremendously in size, depending on the size of the river considered. For example, the largest drainage basin on earth is the Amazon, which drains about 2.25

million square miles (5.83 million sq. km.) of South America.

The Amazon Basin is so large that it could contain nearly the entire continent of Australia. By comparison, the Mississippi River drainage basin, the largest in North America, drains an area of about 1,235,000 square miles (3,200,000 million sq. km.). Smaller rivers have much smaller basins, with many draining only an area roughly the size of a football field. While basins vary tremendously in size, they are spatially organized, with larger basins receiving the drainage from smaller basins, and eventually draining into the ocean. Because drainage basins receive water and material from the landscape within the basin, they are sensitive to environmental change that occurs within the basin. For example, during the twentieth century, the Mississippi River was influenced by many human-imposed changes that occurred either within the basin or directly within the channel, such as agriculture, dams and reservoirs, and levees.

DRAINAGE NETWORKS AND SURFACE EROSION. Drainage basins can be subdivided into drainage networks by the arrangement of their valleys and interfluves. Interfluves are the ridges of higher elevation that separate adjacent valleys. Where an interfluve represents a natural boundary between two or more basins, it is referred to as a drainage divide. Valleys contain the larger rivers and are easily distinguished from interfluves by their relatively low, flat surfaces. Interfluves have relatively steep slopes and, for this reason, are eroded by runoff. The term erosion refers to the transport of material, in this case sediment that is dislodged from the surface.

Runoff starts as a broad sheet of slow-moving water that is not very erosive. As it continues to flow downslope, it speeds up and concentrates into rills, which are narrow, fast-moving lines of water. Because

HOW HYDROLOGY SHAPES GEOGRAPHY

Water and ice sculpt the landscape over time. Fast-flowing rivers erode the soil and rock through which they flow. When rivers slow down in flatter areas, they deposit eroded sediments, creating areas of rich soils and deltas at the mouths of the rivers. Over time this process wears down mountain ranges. The Appalachian Mountain range on the eastern side of the North American continent is hundreds of millions of years older than the Rocky Mountain range on the continent's western side. Although the Appalachians once rivaled the Rockies in size, they have been made smaller by time and erosion.

Canyons are carved by rivers, as the Grand Canyon was carved by the Colorado River, which exposed rocks billions of years old. Ice also changes the landscape. Large ice sheets from past ice ages could have been well over a mile (1,600 meters) thick, and they scoured enormous amounts of soil and rock as they slowly moved over the land surface. Terminal moraines are the enormous mounds of soil pushed directly in front of the ice sheets. Long Island, New York, and Cape Cod, Massachusetts, are two examples of enormous terminal moraines that were left behind when the ice sheets retreated.

ment, particularly intensive agricultural and grazing practices. A change in land use from natural vegetation, such as forests or prairie, can result in a type of land cover that is not suited for preventing erosion. Such land surfaces become susceptible to the formation of gullies during heavy, prolonged rains.

At a smaller scale, fluvial processes can be considered from the perspective of the river channel. River channels are located within the valleys of basins, offering a permanent conduit for drainage. Higher in the basin, river channels and valleys are relatively narrow, but grow larger toward the mouth of the basin as they receive drainage from smaller rivers within the basin. River channels may be categorized by their planform pattern, which refers to their overhead appearance, such as would be viewed from the window of an airplane.

Gully erosion on farmland. (William E. Ferguson)

the runoff is concentrated within rills, the water travels faster and has more energy for erosion. Thus, rills are responsible for transporting sediment from higher points of elevation within the basin to the valleys, which are at a lower elevation. Rills can become powerful enough to scour deeply into the surface, developing into permanent channels called gullies.

The presence of many gullies indicates significant erosion on the landscape and represents an expensive and long-lasting problem if it is not remedied after initial development. The formations of gullies is often associated with human manipulation of the earth. For example, gullies can develop after improper land manage-

The two major types of rivers are meandering and braided. Meandering rivers have a single channel that is sinuous and winding. These rivers are characterized as having orderly and symmetrical bends, causing the river to alternate directions as it flows across its valley. In contrast, braided rivers contain numerous channels divided by small islands, which results in a disorganized pattern. The islands within a braided river channel are not permanent. Instead, they erode and form over the course of a few years, or even during large flood events. Meandering channels usually have narrow and deep channels, but braided river channels are shallow and wide.

SEDIMENT AND FLOODPLAINS. Another distinction between braided and meandering river channels is the types of sediment they transport. Braided rivers transport a great amount of sediment that is deposited into midchannel islands within the river. Also, because braided rivers are frequently located higher in the drainage basin, they may have larger sediments from the erosion of adjacent slopes. In contrast, meandering river channels are located closer to the mouth of the basin and transport fine-grained sediment that is easily stored within point bars, which results in symmetrical bends within the river.

The sediments of both meandering and braided rivers are deposited within the valleys onto floodplains. Floodplains are wide, flat surfaces formed from the accumulation of alluvium, which is a term for sediment that is deposited by water. Floodplain sediments are deposited with seasonal flooding. When a river floods, it transports a large amount of sediment from the channel to the adjacent floodplain. After the water escapes the channel, it loses energy and can no longer transport the sediment. As a result, the sediment falls out of suspension and is de-

posited onto the floodplain. Because flooding occurs seasonally, floodplain deposits are layered and may accumulate into very thick alluvial deposits over thousands of years.

KARST PROCESSES AND LANDFORMS. A specialized type of exogenic process that is also related to the presence of water is karst. Karst processes and topography are characterized by the solution of limestone by acidic groundwater into a number of distinctive landforms. While fluvial processes lower the landscape from the surface, karst processes lower the landscape from beneath the surface. Because limestone is a very permeable sedimentary rock, it allows for a large amount of groundwater flow. The primary areas for solution of the limestone occur along bedding planes and joints. This creates a positive feedback by increasing the amount of water flowing through the rock, thereby further increasing solution of the limestone. The result is a complex maze of underground conduits and caverns, and a surface with few rivers because of the high degree of infiltration.

The surface topography of karst regions often is characterized as undulating. A closer inspection reveals numerous depressions that lack surface outlets. Where this is best developed, it is referred to as cockpit karst. It occurs in areas underlain by extensive limestone and receiving high amounts of precipitation, for example, southern Illinois and Indiana in the midwestern United States, and in Puerto Rico and Jamaica.

Sinkholes are also common to karstic regions. Sinkholes are circular depressions having steep-sided vertical walls. Sinkholes can form either from the sudden collapse of the ceiling of an underground cavern or as a result of the gradual solution and lowering of the surface. Sinkholes can fill with sediments washed in

Bryce Canyon Page 45

from surface runoff. This reduces infiltration and results in the development of small circular lakes, particularly common in central Florida. Over time, erosion causes the vertical walls to retreat, resulting in uvalas, which are much larger flat-floored depressions.

Where there are numerous adjacent sinkholes, the retreat and expansion of the depressions causes them to coalesce, resulting in the formation of poljes. Unlike uvalas, poljes have an irregular shape, and the floor of the basin is not flat because of differences between the coalescing sinkholes.

Caves are among the most characteristic features of karst regions, but can only be seen beneath the surface. Caves can traverse the subsurface for miles, developing into a complex network of interconnected passages. Some caves develop spectacular formations as a result of the high amount of dissolved limestone transported by the groundwater. The evaporation of water results in the accumulation of carbonate deposits, which may grow for thousands of years. Some of the most common deposits are stalactites, which grow downward from the ceiling of the cave, and stalagmites, which grow upward and occasionally connect with stalactites to form large vertical columns.

Paul F. Hudson

FOR FURTHER STUDY

Barry, John M. *Rising Tide: The Great Mississippi Flood of 1927 and How It Changed America.* New York: Simon & Schuster, 1997.

Davies, W. E., and I. M. Morgan. *Geology of Caves.* U.S. Geological Survey, 1991.

Erickson, Jon. *Craters, Caverns, and Canyons: Delving Beneath the Earth's Surface.* Chicago: Facts On File, 1993.

Exley, Sheck. *Caverns Measureless to Man.* St. Louis, Mo.: Cave Books, 1994.

Fincham, Alan G. *Jamaica Underground: The Caves, Sinkholes and Underground Rivers of the Island.* Kingston, Jamaica: University of the West Indies Press, 1998.

Karr, James R., and Ellen W. Chu. *Restoring Life in Running Waters: Better Biological Monitoring.* Washington, D.C.: Island Press, 1999.

Knighton, Andrew. D. *Fluvial Forms and Processes.* New York: John Wiley & Sons 1998.

Middleton, John, and Anthony C. Waltham. *The Underground Atlas: A Gazetteer of the World's Cave Regions.* New York: St. Martin's Press, 1987.

Mount, Jeffrey F., and Janice C. Fong. *California Rivers and Streams.* Berkeley: University of California Press, 1995.

Rapp, Valerie. *What the River Reveals: Understanding and Restoring Healthy Watersheds.* Seattle, Wash.: Mountaineers Books, 1997.

GLACIATION

In areas where more snow accumulates each winter than can thaw in summer, glaciers form. Glacier ice, called firn, looks like rock but is not as strong as most rocks and is subject to intermittent thawing and freezing. Glacier ice can be brittle and fracture readily into crevasses, while other ice behaves as a plastic substance. A glacier is thickest in the area receiving the most snow, called the zone of accumulation. As the thickness piles up, it settles down and squeezes the limit of the ice outward in all directions. Eventually, the ice reaches a climate where the ice begins to melt and evaporate. This is called the zone of ablation.

ALPINE GLACIATION. Varied topographic evidence throughout the alpine environment attests to the sculpturing ability of glacial ice. The world's most spectacular mountain scenery has been produced by alpine glaciation, including the Matterhorn, Yosemite Valley, Glacier National Park, Mount Blanc, the Tetons, and Rocky Mountain National Park, all of which are visited by large numbers of people annually. Although alpine glaciation is still an active process of land sculpture in the high mountain ranges of the world, it

is much less active than it was in the Ice Age of the Pleistocene epoch.

The prerequisites for alpine, or mountain, glaciation to become active are a mountainous terrain with Arctic climatic conditions in the higher elevations, and sufficient moisture to help snow and ice develop into glacial ice. As glaciers move out from their points of origin, they erode into the sides of mountains and increase the local relief in the higher elevations. The erosional features produced by alpine glaciation dominate mountain topography and usually are the most visible features on topographic maps. The eroded material is transported downvalley and deposited in a variety of landforms.

One kind of an erosional feature is a cirque, a hollow bowl-shaped depression. The bowl of the cirque commonly contains a small round lake or tarn. A steep-

*Cirques
Page 46*

Nineteenth century engraving of Mont Blanc's Mer de Glace (Sea of Ice), a major tourist attraction in France both because it is the second-longest glacier in the Alps and because it seems alive, moving down the north slope of Mont Blanc at a rate of about 425 feet (130 meters) a year. (Mark Twain, A Tramp Abroad, 1880)

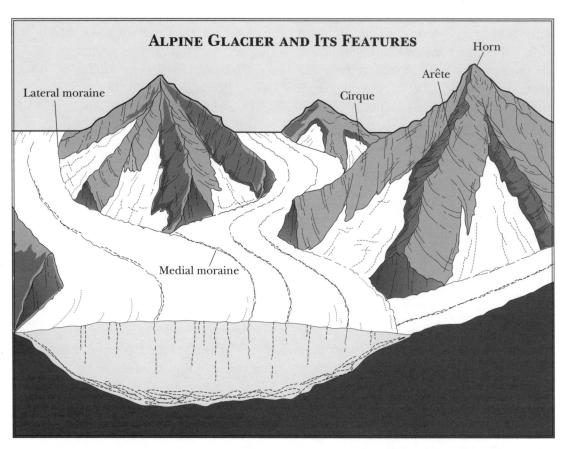

ALPINE GLACIER AND ITS FEATURES

Lateral moraine

Cirque

Arête

Horn

Medial moraine

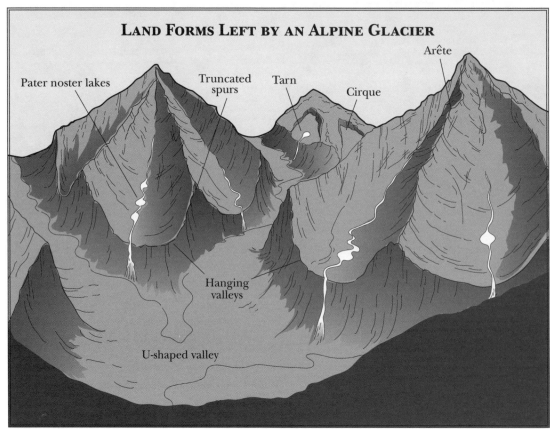

LAND FORMS LEFT BY AN ALPINE GLACIER

Pater noster lakes

Truncated spurs

Tarn

Cirque

Arête

Hanging valleys

U-shaped valley

walled mountain ridge called an arête forms between two cirques. A high pyramidal peak, called horn, is formed by the intersecting walls of three or more cirques.

Erosion is particularly rapid at the head of a glacier. In valleys, moving glaciers press rock fragments against the sides, widening and deepening them by abrasion and forming broad U-shaped valleys. When glaciers recede, tributary streams become higher than the floor of the U-shaped valley and waterfalls occur over these hanging valleys. As the ice continues to melt, residual sediments called moraines may be deposited. Moraines are made up of glacier till, a collection of sediment of all sizes. Bands of sediment along the side of a valley glacier are lateral moraines; those crossing the valley are end or recessional moraines; where two glaciers join, a medial moraine is formed. Meltwater may also sort out the finer materials, transport them downvalley, and deposit them in beds as outwash.

CONTINENTAL GLACIATION. In the modern world, continental glaciation operates on a large scale only in Greenland and Antarctica. However, its existence in previous geologic ages is evidenced by strata of tillite (a compacted rock formed of glacial deposits) or, more frequently, by surficial deposits of glacial materials.

Much of the geomorphology of the northeastern quadrant of North America and the northwestern portion of Europe was formed during the Ice Age. During that time, great masses of ice accumulated on the continents and moved out from centers near the Hudson Bay and the Fenno-Scandian Shield, extending over the continents in great advancing and retreating lobes. In North America, the four major stages of lobe advance were the Wisconsin (the most recent), the Illinoian, the Kansan, and the Nebraskan (the oldest). Between each of these major ad-

Kinnerly Peak, in Montana's Glacier National Park, is an example of a horn—a high pyramidal peak formed by the intersection of several cirques. (U.S. Geological Survey)

vances were pluvial periods in which the ice melted and great quantities of water rushed over or stood on the continents, creating distinctive features which can still be detected today.

The two major functions of gradation are accomplished by the processes of scour (degradation) in the areas close to the centers and deposition (aggradation) adjacent to the terminal or peripheral areas of the lobes. Thus, the overall effect of continental glaciation is to reduce relief—to scour high areas and fill in lower regions—unlike the changes caused by alpine glaciation.

Although continental glaciation usually does not result in the spectacular scenery of alpine glaciation, it was responsible for creating most of the Great Lakes and the

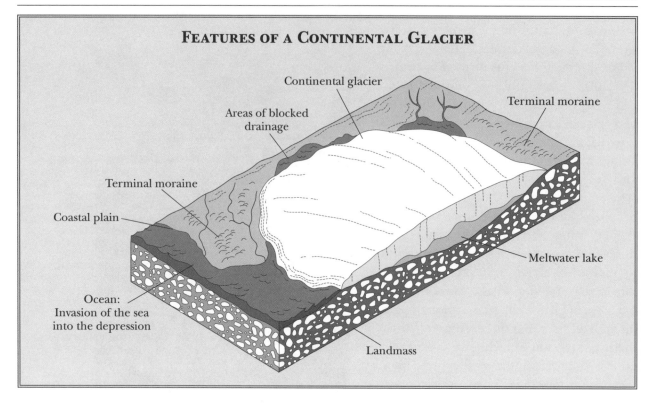

FEATURES OF A CONTINENTAL GLACIER

Continental glacier
Areas of blocked drainage
Terminal moraine
Terminal moraine
Coastal plain
Meltwater lake
Ocean: Invasion of the sea into the depression
Landmass

lakes of Wisconsin, Michigan, Minnesota, Finland, and Canada; for gravel deposits; and for the rich agricultural lands of the Midwest, to mention just a few of its effects.

While glaciers were leveling hilly sections of North America and Europe by scraping them bare of soil and cutting into the ice itself, they acquired a tremendous load of material. As a glacier warms and melts, there is a tremendous outflow of water, and the streams thus formed carry with them the debris of the glacier. The material deposited by glaciers is called drift or outwash. Glaciofluvial drift can be recognized by its separation into layers of finer sands and coarser gravels.

Kettles and kames are the most common features of the end moraines found at the outermost edges of a glacier. A kettle is a depression left when a block of ice, partially or completely buried in deposits of drift, melts away. Most of the lakes in the upper Great Lakes of the United States are kettle lakes. A kame is a round, cone-

A FUTURE ICE AGE

If history is an indicator, some time in the future conditions again will become favorable for the growth of glaciers. As recently as 1300 to 1600 C.E., a cold period known at the Little Ice Age settled over Northern Europe and Eastern North America. Viking colonies perished as agriculture became unfeasible, and previously ice-free rivers in Europe froze over.

Another ice age would probably develop rapidly and be impossible to stop. Active mountain glaciers would bury living forests. Great ice caps would again cover Europe and North America, moving at a rate of 100 feet (30 meters) per day. Major cities and populations would shift to the subtropics and the topics.

shaped hill. Kames are produced by deposition from glacial meltwater. Sometimes, the outwash material poured into a long and deep crevasse, rather than a hole. These tunnels have had their courses choked by debris, revealed today by long, narrow ridges, generally referred to as eskers.

Ron Janke

FOR FURTHER STUDY

Benn, Douglas I., and David J. A. Evans. *Glaciers and Glaciation.* New York: John Wiley & Sons, 1998.

Bennett, Matthew R., and Neil F. Glasser. *Glacial Geology: Ice Sheets and Landforms.* New York: John Wiley & Sons, 1996.

Ferguson, Sue A. *Glaciers of North America: A Field Guide.* Golden, Colo.: Fulcrum Publications, 1992.

Guyton, Bill. *Glaciers of California.* Berkeley: California University Press, 1998.

Hambrey, Michael, and Jurg Alean. *Glaciers.* New York: Cambridge University Press, 1992.

Nesje, Atle, and Svein Olaf Dahl. *Glaciers and Environmental Change.* London: Edward Arnold, 2000.

Owen, Lewis A., ed. *Mountain Glaciation.* New York: John Wiley & Sons, 1998.

Post, Austin, and Edward R. Lachapelle. *Glacier Ice.* Rev. ed. Seattle: University of Washington Press, 2000.

Pyne, Stephen. *The Ice.* Seattle: University of Washington Press, 1998.

DESERT LANDFORMS

Deserts are often striking in color, form, or both. The underlying lack of water in deserts produces unique desert features not found in humid regions. Arid lands cover approximately 30 percent of the earth's land surface, an area of about 15.4 million square miles (40 million sq. km.). Arid lands include deserts and surrounding steppes, semiarid regions that act as transition zones between arid and humid lands.

Many of the world's largest and driest deserts are found between 20 and 40 degrees north and south latitude. These include the Mojave and Sonoran Deserts of the United States, the Sahara in northern Africa, and the Great Sandy Desert in Australia. In these deserts, the subtropical high prevents cloud formation and precipitation while increasing rates of surface evaporation.

Some arid lands, like China's Gobi Desert, form because they are far from oceans that are the dominant source for atmospheric water vapor and precipitation. Others, like California's Death Valley, are arid because mountain ranges block moisture from coming from the sea. The combination of mountain barriers and very low elevations makes Death Valley the hottest, driest desert in North America.

SAND DUNES. Many people envision deserts as vast expanses of blowing sand. Although wind plays a more important role in deserts than it does elsewhere, only

Mojave Desert Page 97

Dunes Page 95

DEATH VALLEY PLAYA

California's Death Valley is the driest desert in the United States, with an average rainfall of only 1.5 inches (38 millimeters) per year at the town of Furnace Creek. It is also consistently one of the hottest places on Earth, with a record high of 134 degrees Fahrenheit (57 degrees Celsius). In the distant past, however, Death Valley held lakes that formed in response to global cooling. Over 120,000 years ago, Death Valley hosted a 295-foot-deep (90 meters) body of water called Lake Manley. Evidence of this lake remains in evaporite deposits that make up the playa in the valley's center, in wave-cut shorelines, and in beach bars.

pended within the moving air mass, while sand grains bounce along the surface. Removal of material often leaves behind depressions called blowouts or deflation hollows. Moving grains abrade cobbles and boulders at the surface, creating uniquely sculpted and smoothed rocks known as ventifacts. Bedrock outcrops can be streamlined as they are blasted by wind-borne grains to form features called yardangs. As these rocks are ground away, they contribute additional sediment to the wind.

Desert sand dunes are not stationary features—instead, they represent accumulations of moving sand. Wind blows sand along the desert floor. Where it collects, it forms dunes. Typically, dunes have relatively shallow windward faces and steeper slip faces. Sand grains bounce up the windward face then eventually cascade down the slip face, the movement of individual grains driving movement of the entire dune in a downwind direction.

Four major dune types are found within arid regions. Barchan dunes are crescent-shaped features, with arms that point downwind. They may occur as isolated structures or within fields. They form where winds blow in a single direction and where the supply of sand is limited. With a larger supply of sand, barchan dunes can join with one another to form a transverse dune field.

There, ridges are perpendicular to the predominant wind direction. With quartering winds (that is, winds that vary in direction throughout a range of about 45 degrees) dune ridges form that are parallel to the average wind direction.

about 25 percent of arid lands are covered by sand. Broad regions that are covered entirely in sand (such as portions of northwestern Africa, Arabia, and Australia) are referred to as sand seas. Why is wind more effective here than elsewhere?

The lack of soil water and vegetation, both of which act to bind grains together, allows enhanced eolian (wind) erosion. Very small particles are picked up and sus-

Desert plants Page 109

Transverse dune field in the Moroccan Sahara. (Corbis)

These so-called longitudinal dunes have no clearly defined windward and slip faces. Where winds blow sand from all directions, star dunes form. Sand collects in the middle of the feature to form a peaked center with arms that spiral outward.

BADLANDS, MESAS, AND BUTTES. As scarce as it may be, water is still the dominant force in shaping desert landscapes. Annual precipitation may be low, but the amount of precipitation in a single storm may be a large fraction of the yearly total. An arid landscape that is underlain by poorly cemented rock or sediment, such as that found in western South Dakota, may form badlands as a result of the erosive ability of storm-water runoff. Overall aridity prevents vegetation from establishing the interconnected root system that holds soil particles together in more humid regions.

Cloudbursts cause rapid erosion that forms numerous gullies, deeply incised washes, and hoodoos. The latter structures are created when rock or sediment that is more resistant protects underlying material from erosion. Over time, protected sections stand as prominent spires while surrounding material is removed. Landscapes like those found in Badlands National Park in South Dakota are devoid of vegetation and erode rapidly during storms.

Arid regions that are underlain by flat-lying rock units can form mesas and buttes. Water follows fractures and other lines of weakness, forming ever-widening canyons. Over time, these grow into broad valleys. In northern Arizona's Monument Valley, remnants of original bedrock stand as isolated, flat-topped structures. Broad mesas are marked by their flat tops (made of a resistant rock

like sandstone or basalt) and steep sides. Buttes are much narrower, with a small resistant cap, but are often as tall and steep as neighboring mesas.

DESERT PAVEMENT AND DESERT VARNISH. Much of the desert floor is covered by desert pavement, an accumulation of gravel and cobbles that forms a surface fabric that can interconnect tightly. Fine material has been removed by wind and water, leaving behind larger fragments that inhibit further erosion. In many areas, desert pavements have been stable for long periods of time, as evidenced by their surface patina of desert varnish. Desert varnish is a thin outer coating of wind-deposited clay mixed with iron and manganese oxides. Varying in color from light brown to black, these coatings are thought to adhere to rocks by the action of single-celled microorganisms. Under a microscope, desert varnish can be seen to be made up of very fine layers. A thick, dark patina means that a rock has been exposed for a long time.

PLAYAS. Where neither dunes nor rocky pavements cover the desert floor,

Badlands Pages 43, 95

Monument Valley Page 96

Desert floors are often covered by desert pavement—a term for tightly interconnected accumulations of gravel and cobbles left behind after finer materials has been removed by wind and water. (William E. Ferguson)

one may find an accumulation of saline minerals. A playa is a flat surface that is often blindingly white in color. Playas are usually found in the centers of desert valleys and contain material that mineralized during the evaporation of a lake. Dry lake beds are a common feature of the Great Basin in the western United States. During glacial stages, the last of which occurred about twenty thousand years ago, lakes grew in what are now arid, closed valleys. As the climate warmed, these lakes shrank, and many dried completely. As a lake evaporates, minerals that were held in solution crystallize, forming salts, including halite (table salt). These salt deposits frequently are mined for useful household and industrial chemicals.

Richard L. Orndorff

FOR FURTHER STUDY

Durham, M. S. *The Desert States.* New York: Stewart, Tabori & Chang, 1990.

Hartman, W. K. *The American Desert.* New York: Crescent Books, 1991.

Lancaster, N. *Geomorphology of Desert Dunes.* London: Routledge, 1995.

Larson, Peggy. *A Sierra Club Naturalist's Guide to the Deserts of the Southwest: The Deserts of the Southwest.* Vol. 1. San Francisco: Sierra Club Books, 1982.

Mabbutt, J. A. *Desert Landforms.* Cambridge, Mass.: MIT Press, 1977.

Mares, M. A., ed. *Encyclopedia of Deserts.* Norman: University of Oklahoma Press, 1999.

Sullivan, C., and S. Weiley, eds. *Travels in the American West.* Washington, D.C.: Smithsonian Institution Press, 1992.

OCEAN MARGINS

Oil spill
Page 97

Ocean margins are the areas where land borders the sea. Although often referred to as coastlines or beaches, ocean margins cover far greater territory than beaches. An ocean margin extends from the coastal plain—the fertile farming belt of land along the seacoast—to the edge of the gently sloping land submerged in water, called the continental shelf.

Ocean margin constitutes 8 percent of the world's surface. It is rich in minerals, both above and below water, and is home to 25 percent of Earth's people, along with 90 percent of the marine life. This fringe of land at the border of the ocean is ever changing. Tides wash sediment in and leave it behind, just below sea level. This process, called deposition, builds up land in some areas of the coastline. At the same time, ocean waves, winds, and storms wear away or erode parts of the shoreline. As land is worn away or built up, the amount of land above sea level changes. Factors such as climate, erosion, deposition, changes in sea level, and the effects of humans constantly change the shape of the ocean margin on Earth.

BEACH DYNAMICS. The two types of coasts or land formations at the ocean margin are primary coasts and secondary coasts. Primary coasts are formed by systems on land, such as the melting of glaciers, wind or water erosion, and sediment deposited by rivers. Deltas and fjords are examples of primary coasts. Secondary coasts are formed by ocean patterns, such as erosion by waves or currents, sediment deposition by waves or currents, or changes by marine plants or animals. Beaches, coral reefs, salt marshes, and mangrove swamps are examples of secondary coasts.

Sediment carried by rivers to the sea is deposited to form deltas at the mouths of the rivers. Some of the sediment can wash out to sea, causing formations to build up at a distance from the shore. These formations eventually become barrier islands, which are often little more than 10 feet (3 km.) above sea level. As a consequence, heavy storms, such as hurricanes, can cause great damage to barrier islands. Barrier islands naturally protect the coastline from erosion, however, especially during heavy coastal storms.

Sea level changes also affect the shape of the coastline. As oceans slowly rise, land is slowly consumed by the ocean. Barrier islands, having low sea levels, may slowly be covered with water. The melting of continental glaciers increased the sea level 0.06 inch (0.15 centimeter) per year during the twentieth century. As ocean waters warm, they expand, eating away at sea levels. Global warming caused by carbon dioxide levels in the atmosphere could cause sea levels to rise as much as 0.24 inch (0.6 centimeter) per year as a result of the warming of the water and glacial melting.

HUMAN INFLUENCE. The shape of the ocean margin also changes radically as a result of human influence. According to the United States Geological Survey, half of the people living in the United States live within fifty miles (80.5 km.) of the coasts. Pollution from toxins, dredging, recreational boating, and waste disposal kills plants and animals along the ocean margin. This changes the coastal shape, as mangrove forests, coral reefs, and other coastal lifeforms die.

Pollution from toxins, dredging, recreational boating, and careless waste disposal kills plants and animals along the ocean margin, making the water unhealthful even to humans. (PhotoDisc)

A greater concern along the coastal fringe, however, is human development. Not only are people drawn to the fertile soil along the coastal zone of the continent, but they also develop islands and coves into resort communities. To protect homes and hotels along the coastal zone from coastal erosion, people build breakwalls, jetties, and sand and stone bars called groins.

These human-made barriers disrupt the natural method by which the ocean carries material along the coast. Longshore drift, a zigzag movement, deposits sediment from one area of the beach farther along the shoreline. Breakwalls, jetties, and groins disrupt this flow. As the ocean smashes against a breakwall, the property behind it may be safe for the present, but the coastline neighboring the breakwall takes a greater beating. The silt and sediment from upshore, which would replace that carried downshore, never arrives. Eventually, the breakwall will break down under the impact of the ocean force.

Hurricane damage Page 105

Global warming Page 233

Polluted beach Page 98

A groin built to protect the coastline of Cape Hatteras, North Carolina, traps sand that normally moves along the shoreline. (U.S. Geological Survey)

ocean margin, it continues to provide a stable supply of resources—fish, seafood, minerals, sponges, and other marine plants and animals. Offshore drilling of oil and natural gas often takes place within 200 miles (322 km.) of shorelines.

Lisa A. Wroble

FOR FURTHER STUDY

Ackerman, Jennifer. "Islands at the Edge." *National Geographic* (August, 1997): 4-31.

Buchanan, Noel, ed. *Discovering the Wonders of Our World.* New York: Reader's Digest, 1993.

Erickson, Jon. *Marine Geology: Undersea Landforms and Life Forms.* New York: Facts on File, 1996.

Kemper, Steve. "This Beach Boy Sings a Song Developers Don't Want to Hear." *Smithsonian* 23, no. 7 (October, 1992): 72-86.

Lye, Keith. *Our World: Coasts.* Englewood Cliffs, N.J.: Silver Burdett Press, 1988.

Miller, Christina, and Louise A. Berry. *Coastal Rescue: Preserving Our Seashores.* New York: Atheneum, 1989.

Wroble, Lisa A. *Endangered Animals and Habitats: The Oceans.* San Diego, Calif.: Lucent Books, 1998.

Areas with breakwalls and jetties often suffer greater damage in coastal storms than areas that remain naturally open to the changing forces of the ocean.

To compensate for the destructive nature of human-made barriers, many recreational beaches replace lost sand with dredgings or deposit truckloads of sand from inland sources. For example, Virginia Beach in the United States spends $800,000 annually to restore beaches for the tourist season in this way.

Despite the changes in the shape of the

Offshore drill Page 172

INFORMATION ON THE WORLD WIDE WEB

Ocean Planet, part of a traveling exhibit prepared by the Smithsonian Institution, features interesting facts on oceans and ocean margins. (seawifs.gsfc.nasa.gov/OCEAN_PLANET/HTML/oceanography_geography.html)

The National Oceanic and Atmospheric Administration (NOAA) home page provides maps, photos, and links to general information about NOAA programs involving the ocean, coastlines, and weather relationships. (www.noaa.gov/)

The Globe Program is a worldwide program in which schools help scientists with data collection as students learn about science in their own regions of the world. Student findings are posted to the Globe Web site and are accessible to computer users. (www.globe.gov/)

Sand dunes such as these conform to popular images of desert landforms; however, only about 25 percent of arid lands are covered by sand. (PhotoDisc)

Badlands National Park in South Dakota is a good example of desert badlands that are largely devoid of vegetation and erode rapidly during storms. (PhotoDisc)

These isolated, flat-topped structures in Arizona's Monument Valley are actually remnants of original bedrock, standing in broad valleys formed over time by water that followed fractures and other lines of weakness and formed ever-widening canyons. (PhotoDisc)

Moonrise over California's Mojave Desert. (PhotoDisc)

Ships discharging wastes or spilling oil can cause catastrophic damage to ocean margins. (PhotoDisc)

Workers cleaning oil spilled from a tanker ship. A disturbing late twentieth century development in commerce was the greatly escalating sizes of tanker ships carrying oil and thereby increasing the risk of catastrophic oil spills. (PhotoDisc)

Ocean margins are often perceived as coastlines or beaches, but they actually extend from coastal plains to the edges of the land sloping into the seawater known as the continental shelf. (PhotoDisc)

The tropical wet climate is an almost seasonless climate, characterized by year-round warm, humid, rainy conditions that allow land areas to support a dense broadleaf forest cover. (PhotoDisc)

In exceptionally cold climates, it is not unusual for the surfaces of streams to freeze. (PhotoDisc)

WORLD CLIMATE REGIONS

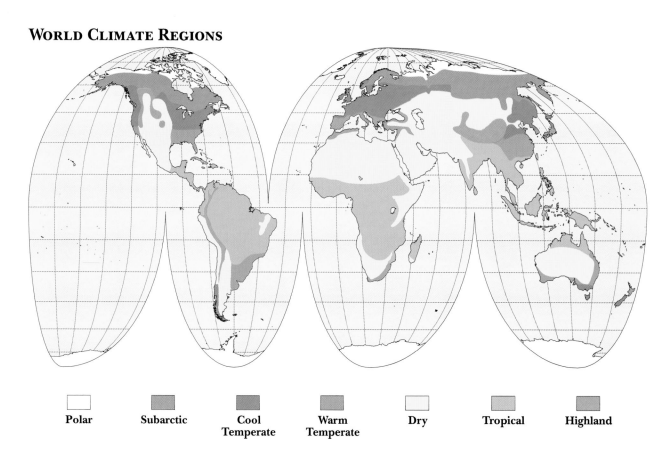

Polar   Subarctic   Cool Temperate   Warm Temperate   Dry   Tropical   Highland

Storm clouds forming over a desert. (PhotoDisc)

The extent of Earth's cloud cover can be seen in this composite satellite photograph of the planet, on which a portion of northern Mexico and the central United States is the only visible landmass. (PhotoDisc)

Rainbow over a Scottish field. Rainbows appear wherever the sun's rays reach raindrops in the atmosphere. They are created by the refraction of light on individual drops of water, and their appearance varies with the position of the viewer. (PhotoDisc)

Clouds forming over low-lying mountains. (PhotoDisc)

Lightning storm over Seattle, Washington. Lightning is the product of positive and negative electrical charges in storm clouds creating giant sparks while attempting to balance out. Lightning that finds its way to the surface heats the air around it to such high temperatures that the air expands explosively, creating the shock waves called thunder. (PhotoDisc)

Satellite view of a thunderstorm. (PhotoDisc)

Because tornado winds cannot be measured directly, tornadoes are ranked according to the damage they cause. (PhotoDisc)

The immense size that a hurricane can reach is dramatically evident in this satellite image of a hurricane off the coasts of Florida and Cuba. (PhotoDisc)

Hurricane-ravaged beach. (PhotoDisc)

The aspen is a deciduous tree that has adapted well to cold winters in boreal forests. (PhotoDisc)

BIOMES OF THE WORLD

Desert	Tropical Rain Forest	Temperate Grassland	Taiga
Monsoon	Savanna	Temperate Forest	Tundra
	Mediterranean	Mountain	Polar

Beaver dams are becoming less common as the wetlands they need are giving way to human settlements. (PhotoDisc)

Wetlands, places where the ground is saturated with water, constitute transition zones between aquatic ecosystems and terrestrial ecosystems. (PhotoDisc)

Washington's Olympic National Park contains a well-known example of a temperate-zone rain forest. (Digital Stock)

Evergreen forest in Glacier National Park. (PhotoDisc)

The great herds of buffalo that once roamed North America's plains played an important role in maintaining the native grasslands. (PhotoDisc)

Desert vegetation adapts to arid conditions and temperature extremes. Many desert plants are short-lived annuals whose life cycles are keyed to rainfall. (PhotoDisc)

Wildflowers in Arizona's Grand Canyon, a desert biome. (PhotoDisc)

Skeleton Coast of northern Namibia in Africa. Namibia's coast Namib Desert—after which the country is named—is an example of a desert that has developed along a warm tropical coast adjacent to a cold ocean current. The air above the ocean currents is cooled and contains little moisture. As this cool, dry air moves inland, it brings moisture in the form of fog which enables plants, animals, and people to live in this harsh environment. (Corbis)

EARTH'S CLIMATES

THE ATMOSPHERE

The thin layer of gases that envelops the earth is the atmosphere. This layer is so thin that if the earth were the size of a desktop globe, more than 99 percent of its atmosphere would be contained within the thickness of an ordinary sheet of paper. Despite its thinness, the atmosphere sustains life on Earth, protecting it from the Sun's searing radiation and regulating the earth's temperature. Storms of the atmosphere carry water to the continents, and weathering by its wind and rain helps shape their form.

COMPOSITION OF THE ATMOSPHERE. The earth's atmosphere consists of gases, microscopic particles called aerosol, and clouds consisting of water droplets and ice particles. Its two principal gases are nitrogen and oxygen. In dry air, nitrogen occupies 78 percent, and oxygen 21 percent, of the atmosphere's volume. Argon, neon, xenon, helium, hydrogen, and other trace gases together equal less than 1 percent of the remaining volume.

These gases are distributed homogeneously in a layer called the homosphere, which occurs between the earth's surface and about 50 miles (80 km.) altitude. Above 50 miles altitude, in the heterosphere, the concentration of heavier gases decreases more rapidly than lighter gases.

The atmosphere has no firm top. It simply thins out until the concentration of its gas molecules approaches that of the gases in outer space. The concentration of nitrogen and oxygen remains essentially constant in the atmosphere because a balance exists between the production and removal of these gases at the earth's surface. Decaying organic matter adds nitrogen to the atmosphere, while soil bacteria remove nitrogen. Oxygen enters the atmosphere primarily through photosynthesis and is removed through animal respiration, combustion, and decay of organic material, and by chemical reactions involving the creation of oxides.

The atmosphere contains many gases that are present in small, variable concentrations. Three gases—water vapor, carbon dioxide and ozone—are vital to life on Earth. Water vapor enters the atmosphere through evaporation, primarily from the oceans, and through transpiration by plants. It condenses to form clouds, which provide the rain and snow that sustain life outside the oceans. The concentration of water vapor varies from about 4 percent by volume in tropical humid climates to a small fraction of a percent in polar dry climates. Water vapor plays an important role in regulating the temperature of the earth's surface and the atmosphere. Clouds reflect some of the incoming solar radiation, while water vapor and clouds both absorb earth's infrared radiation.

Carbon dioxide also absorbs the earth's infrared radiation. The concentration of

Clouds forming Page 102

111

carbon dioxide, about 0.037 percent by volume at the turn of the millennium, has increased about 25 percent since the early nineteenth century. Carbon dioxide enters the atmosphere as the result of decay of organic material, through respiration, during volcanic eruptions, and from the burning of fossil fuels. It is removed dur-

THE GREENHOUSE EFFECT

Clouds and atmospheric gases such as water vapor, carbon dioxide, methane, and nitrous oxide absorb part of the infrared radiation emitted by the earth's surface and reradiate part of it back to the earth. This process effectively reduces the amount of energy escaping to space and is popularly called the "greenhouse effect" because of its role in warming the lower atmosphere. The greenhouse effect has drawn worldwide attention because increasing concentrations of carbon dioxide from the burning of fossil fuels may result in a global warming of the atmosphere.

Scientists know that the greenhouse analogy is incorrect. A greenhouse traps warm air within a glass building where it cannot mix with cooler air outside. In a real greenhouse, the trapping of air is more important in maintaining the temperature than is the trapping of infrared energy. In the atmosphere, air is free to mix and move about.

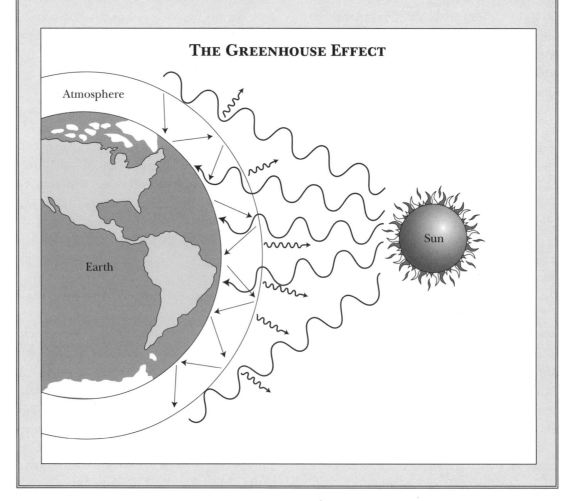

THE GREENHOUSE EFFECT

Atmosphere

Earth

Sun

ing photosynthesis and by dissolving in ocean water, where it is used by organisms and converted to carbonates. The increase in atmospheric carbon dioxide associated with the burning of fossil fuels has raised concerns that the earth's atmosphere may be warming through enhancement of the greenhouse effect.

Ozone, a gas consisting of molecules containing three oxygen atoms, forms in the upper atmosphere when oxygen atoms and oxygen molecules combine. Most ozone exists in the upper atmosphere between 15 and 20 miles (25-35 km.) in altitude, in concentrations of no more than 0.002 percent by volume. This small amount of ozone sustains life outside the oceans by absorbing most of the Sun's ultraviolet radiation, thereby shielding the earth's surface from the radiation's harmful effects on living organisms. Paradoxically, ozone is an irritant near the earth's surface and is the major component of photochemical smog. Other gases that contribute to pollution include methane, nitrous oxide, hydrocarbons, and chlorofluorocarbons.

Aerosols represent another component of atmospheric pollution. Aerosols form in the atmosphere during chemical reactions between gases, through mechanical or chemical interactions between the earth, ocean surface and atmosphere, and during evaporation of droplets containing dissolved or solid material. These microscopic particles are always present in air, with concentrations of about a few hundred per cubic centimeter in clean air to as many as a million per cubic centimeter in polluted air. Aerosols are essential to the formation of rain and snow, because they serve as centers upon which cloud droplets and ice particles form.

ENERGY EXCHANGE IN THE ATMOSPHERE. The Sun is the ultimate source of the energy in Earth's atmosphere. Its radi-

THE OZONE HOLE

Since the 1970's, balloon-borne and satellite measurements of stratospheric ozone have shown rapidly declining stratospheric ozone concentrations over the continent of Antarctica, termed the "ozone hole." The lowest concentrations occur during the Antarctic spring, in September and October. The decrease in ozone has been associated with an increase in the concentration of chlorine, a gas introduced into the stratosphere through chemical reactions involving sunlight and chlorofluorocarbons, synthetic chemicals used primarily as refrigerants. The ozone hole over Antarctica has raised concern about possible worldwide reduction in the concentration of upper atmospheric ozone.

ation, called electromagnetic radiation because it propagates as waves with electric and magnetic properties, travels to the surface of the earth's atmosphere at the speed of light. This energy spans many wavelengths, some of which the human eye perceives as colors. Visible wavelengths make up about 44 percent of the Sun's energy. The remainder of the Sun's radiant energy cannot be seen by human eyes. About 7 percent arrives as ultraviolet radiation, and most of the remaining energy is infrared radiation.

The Sun is not the only source of radiation. All objects emit and absorb radiation to some degree. Cooler objects such as the earth emit nearly all their energy at infrared wavelengths. Objects heat when they absorb radiation and cool when they emit radiation. The radiation emitted by the earth and atmosphere is called terrestrial radiation.

The balance between absorption of solar radiation and emission of terrestrial radiation ultimately determines the average temperature of the earth-atmosphere system. The vertical temperature distribution within the atmosphere also depends on the absorption and emission of radiation within the atmosphere, and the transfer of energy by the processes of conduction, convection, and latent heat exchange. Conduction is the direct transfer of heat from molecule to molecule. This process is most important in transferring heat from the earth's surface to the first few centimeters of the atmosphere. Convection, the transfer of heat by rising or sinking air, transports heat energy vertically through the atmosphere.

Latent heat is the energy required to change the state of a substance, for example, from a liquid to a gas. Energy is transferred from the earth's surface to the atmosphere through latent heat exchange when water evaporates from the oceans and condenses to form rain in the atmosphere.

Only 51 percent of the solar energy reaching the top of the earth's atmosphere is absorbed by the earth's surface. The atmosphere absorbs another 19 percent. The remaining 30 percent is scattered back to space by atmospheric gases, clouds and the earth's surface. To understand the importance of terrestrial radiation and the greenhouse effect in the atmosphere's energy balance, consider the solar radiation arriving at the top of the earth to be 100 energy units, with 51 energy units absorbed by the earth's surface and 19 units by the atmosphere.

The earth's surface actually emits 117 units of energy upward as terrestrial radiation, more than twice as much energy as it receives from the Sun. Only 6 of these units are radiated to space—the atmosphere absorbs the remaining energy. Latent heat exchange, conduction, and convection account for another 30 units of energy transferred from the surface to the atmosphere. The atmosphere, in turn, radiates 96 units of energy back to the earth's surface (the greenhouse effect), and 64 units to space. The earth's and atmosphere's energy budget remains in balance, the atmosphere gaining and losing 160 units of energy, and the earth gaining and losing 147 units of energy.

VERTICAL STRUCTURE OF THE ATMOSPHERE. Temperature decreases rapidly upward away from the earth's surface, to about –58 degrees Farenheit (–50 degrees

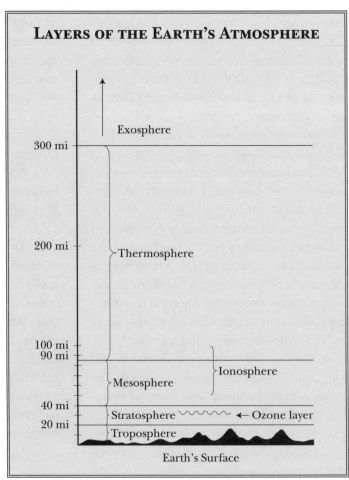

LAYERS OF THE EARTH'S ATMOSPHERE

Exosphere

300 mi

Thermosphere

200 mi

100 mi
90 mi

Mesosphere   Ionosphere

40 mi

Stratosphere   ← Ozone layer
20 mi

Troposphere

Earth's Surface

Celsius) at an altitude of about 7.5 miles (12 km.). Above this altitude, temperature increases with height to about 32 degrees Farenheit (0 degrees Celsius) at an altitude of 31 miles (50 km.). The layer of air in the lower atmosphere where temperature decreases with height is called the troposphere. It contains about 75 percent of the atmosphere's mass. The layer of air above the troposphere, where temperature increases with height, is called the stratosphere. All but 0.1 percent of the remaining mass of the atmosphere resides in the stratosphere.

The stratosphere exists because ozone in the stratosphere absorbs ultraviolet light and converts it to heat. The boundary between the troposphere and stratosphere is called the tropopause. The tropopause is extremely important because it acts as a lid on the earth's weather. Storms can grow vertically in the troposphere, but cannot rise far, if at all, beyond the tropopause. In the polar regions, the tropopause can be as low as 5 miles (8 km.) above the surface, while in the Tropics, the tropopause can be as high as 11 miles (18 km.). For this reason, tropical storms can extend to much higher altitudes than storms in cold regions.

The mesosphere extends from the top of the stratosphere, the stratopause, to an altitude of about 56 miles (90 km.). Temperature decreases with height within the mesosphere. The lowest average temperatures in the atmosphere occur at the mesopause, the top of the mesosphere, where the temperature is about −130 degrees Farenheit (−90 degrees Celsius). Only 0.0005 percent of the atmosphere's mass remains above the mesopause. In this uppermost layer, the thermosphere, there are few atoms and molecules. Oxygen molecules in the thermosphere absorb high-energy solar radiation. In this near vacuum, absorption of even small amounts of

energy causes a large increase in temperature. As a result, temperature increases rapidly with height in the lower thermosphere, reaching about 1,300 degrees Farenheit (700 degrees Celsius) above 155 miles (250 km.) altitude.

The upper mesosphere and thermosphere also contain ions, electrically charged atoms or molecules. Ions are created in the atmosphere when air molecules collide with high-energy particles arriving from space or absorb high-energy solar radiation. Ions cannot exist very long in the lower atmosphere, because collisions between newly formed ions quickly restore ions to their uncharged state. However, above about 37 miles (60 km.) collisions are less frequent and ions can exist for longer times. This region of the atmosphere, called the ionosphere, is particularly important for amplitude-modulated (AM) radio communication because it reflects standard AM radio waves. At night, the lower ionosphere disappears as ions recombine, allowing AM radio waves to travel longer distances when reflected. For this reason, AM radio station signals can sometimes travel great distances at night.

The top of the atmosphere occurs at about 310 miles (500 km.). At this altitude, the distance between individual molecules is so great that energetic molecules can move into free space without colliding with neighbor molecules. In this uppermost layer, called the exosphere, the earth's atmosphere merges into space.

Robert M. Rauber

FOR FURTHER STUDY

Ahrens, C. Donald. *Meteorology Today: An Introduction to Weather, Climate, and the Environment.* 6th ed. Pacific Grove, Calif.: Brooks/Cole, 2000.

Graedel, Thomas E., and Paul J. Crutzen. *Atmosphere, Climate, and Change.* New York: W. H. Freeman, 1995.

Lutgens, Frederick K., and Edward J. Tarbuck. *The Atmosphere: An Introduction to Meteorology.* Englewood Cliffs, N.J.: Prentice-Hall, 1997.

Moran, Joseph M., and Michael D. Morgan: *Meteorology: The Atmosphere and Science of Weather.* Englewood Cliffs, N.J.: Prentice-Hall, 1996.

Schaefer, Vincent, and John A. Day. *The Atmosphere.* Peterson Field Guide Series. Boston: Houghton Mifflin, 1999

INFORMATION ON THE WORLD WIDE WEB

The Weather World 2010 Web site, maintained by the Department of Atmospheric Sciences at the University of Illinois at Urbana-Champaign, features information on many aspects of the atmosphere, including clouds, precipitation, storms, and forecasting. (ww2010.atmos.uiuc.edu)

GLOBAL CLIMATES

*Global climates
Page 100*

*Earth's cloud cover
Page 101*

A region's climate is the sum of its long-term weather conditions. Most descriptions of climate emphasize temperature and precipitation characteristics, because these two climatic elements usually exert more impact on environmental conditions and human activities than do other elements, such as wind, humidity, and cloud cover. Climatic descriptions of a region generally cover both mean conditions and extremes. Climatic means are important because they represent average conditions that are frequently experienced; extreme conditions, such as severe storms, excessive heat and cold, and droughts, are important because of their adverse impact.

IMPORTANT CLIMATE CONTROLS. A region's climate is largely determined by the interaction of six important natural controls: sun angle, elevation, ocean currents, land and water heating and cooling characteristics, air pressure and wind belts, and orographic influence.

Sun angle—the height of the Sun in de-grees above the nearest horizon—largely controls the amount of solar heating that a site on Earth receives. It strongly influences the mean temperatures of most of the earth's surface, because the Sun is the ultimate energy source for nearly all the atmosphere's heat. The higher the angle of the Sun in the sky, the greater the concentration of energy, per unit area, on the earth's surface (assuming clear skies). From a global perspective, the Sun's mean angle is highest, on average, at the equator, and becomes progressively lower poleward. This causes a gradual decrease in mean temperatures with increasing latitude.

Sun angles also vary seasonally and daily. Each hemisphere is inclined toward the Sun during spring and summer, and away from the Sun during fall and winter. This changing inclination causes mean sun angles to be higher, and the length of daylight longer, during the spring and summer. Therefore, most locations, especially those outside the Tropics, have warmer temperatures during these two seasons. The earth's rotation causes sun angles to be higher during midday than in the early morning and late afternoon, resulting in warmer temperatures at midday. Heating and cooling lags cause both seasonal and daily maximum and minimum temperatures typically to occur somewhat after the periods of maximum and minimum solar energy receipt.

Variations in elevation—the distance above sea level—can cause locations at similar latitudes to vary greatly in temperature. Temperatures decrease an average of about 3.5 degrees Fahrenheit per thousand feet (6.4 degrees Celsius per thousand meters). Therefore, high mountain and plateau stations are much colder than low-elevation stations at the same latitude.

Surface ocean currents can transport masses of warm or cold water great distances from their source regions, affecting both temperature and moisture conditions. Warm currents facilitate the evaporation of copious amounts of water into the atmosphere and add buoyancy to the air by heating it from below. This results in a general increase in precipitation totals. Cold currents evaporate water relatively slowly and chill the overlying air, thus stabilizing it and reducing its potential for precipitation.

The influence of ocean currents on land areas is greatest in coastal regions and decreases inland. The west coasts of continents (except for Europe) generally are paralleled by relatively cold currents, and the east coasts by relatively warm currents. For example, the warm Gulf Stream flows northward off the eastern United States, while the West Coast is cooled by the southward-flowing California Current.

Land can change temperature much more readily than water. As a result, the air over continents typically experiences larger annual temperature ranges (that is, larger temperature differences between summer and winter) and shorter heating and cooling lags than does the air over oceans. This same effect causes continental interiors and the leeward (downwind) coasts of continents typically to have larger temperature ranges than do windward (upwind) coasts. Climates that are dominated by air from landmasses are often described as continental climates. Conversely, climates dominated by air from oceans are described as maritime climates.

The seasonal heating and cooling of continents can also produce a monsoon influence, which has to do with annual shifts of wind patterns. Areas influenced by a monsoon, such as Southeast Asia, tend to have a predominantly onshore flow of moist maritime air during the summer. This often produces heavy rains. An offshore flow of dry air predominates in winter, producing fair weather.

Earth's atmosphere displays a banded, or beltlike, pattern of air pressure and wind systems. High pressure is associated with descending air and dry weather; low pressure is associated with rising air, which produces cloudiness and often precipitation. Wind is produced by differences in air pressure. The air blows outward from high-pressure systems and into low-pressure systems in a constant attempt to equalize air pressures.

The direction and speed of movement of weather systems, such as weather fronts and storms, are controlled by wind patterns, especially those several kilometers above the surface. The seasonal shift of global temperatures caused by the movement of the Sun's vertical rays between the Tropics of Cancer and Capricorn produces a latitudinal migration of both air pressure and wind belts. This shift affects the annual temperature and precipitation patterns of many regions.

Four air-pressure belts exist in each hemisphere. The intertropical conver-

California coast
Page 232

gence zone (ITCZ) is a broad belt of low pressure centered within a few degrees of latitude of the equator. The subtropical highs are high-pressure belts centered near 30 degrees north and south latitude, which are responsible for many of the world's deserts. The subpolar lows are low-pressure belts centered about 60 or 65 degrees north and south latitude. Finally, the polar highs are high-pressure centers located near the North and South Poles.

Tropical wet climate Page 99

The air pressure gradient between these belts produces the earth's major wind belts. The regions between the ITCZ and the subtropical highs are dominated by the trade winds, a broad belt in each hemisphere of easterly (that is, moving east to west) winds. The middle latitudes are mostly situated between the subtropical highs and the subpolar lows and are within the westerly wind belt. This wind belt causes winds, and weather systems, to travel generally from west to east in the United States and Canada. Finally, the high-latitude zones between the subpolar lows and polar highs are situated within the polar easterlies.

The final factor affecting climate—orographic influence—is the lifting effect of mountain peaks or ranges on winds that pass over them. As air approaches a mountain barrier, it rises, typically producing clouds and precipitation on the windward (upwind) side of the mountains. After it crosses the crest, it descends the leeward (downwind) side of the mountains, generally producing dry weather. Most of the world's wettest locations are found on the windward sides of high mountain ranges; some deserts, such as those of the western interior United States, owe their aridity to their location on the leeward sides of orographic barriers.

WORLD CLIMATE TYPES. The global distribution of the world climate controls is responsible for the development of four-

teen widely recognized climate types. In this section, the major characteristics of each of these climates will be briefly described. The climates are discussed in a rough poleward sequence.

TROPICAL WET CLIMATE. Sometimes called the tropical rain forest climate, the tropical wet climate exists chiefly in areas lying within 10 degrees of the equator. It is an almost seasonless climate, characterized by year-round warm, humid, rainy conditions that allow land areas to support a dense broadleaf forest cover. The warm temperatures, which for most locations average near 80 degrees Fahrenheit (27 degrees Celsius) throughout the year, result from the constantly high midday sun angles experienced at this low latitude. The heavy precipitation totals result from the heating and subsequent rising of the warm moist air to form frequent showers and thunderstorms, especially during the afternoon hours. The dominance of the ITCZ enhances precipitation totals, helping make this climate type one of the world's rainiest.

TROPICAL MONSOONAL CLIMATE. The tropical monsoonal climate occurs in low-latitude areas, such as Southeast Asia, that have a warm, rainy climate with a short dry season. Temperatures are similar to those of the tropical wet climate, with the warmest weather often occurring during the drier period, when sunshine is more abundant. The heavy rainfalls result from the nearness of the ITCZ for much of the year, as well as the dominance of warm, moist air masses derived from tropical oceans. During the brief dry season, however, the ITCZ has usually shifted into the opposite hemisphere, and windflow patterns often have changed so as to bring in somewhat drier air derived from continental sources.

TROPICAL SAVANNA CLIMATE. The tropical savanna climate, also referred to as the tropical wet and dry climate, occu-

Satellite image of storm patterns over the surface of the earth. (PhotoDisc)

pies a large portion of the Tropics between 5 and 20 degrees latitude in both hemispheres. It experiences a distinctive alternation of wet and dry seasons, caused chiefly by the seasonal shift in latitude of the subtropical highs and ITCZ. Summer is typically the rainy season because of the domination of the ITCZ. In many areas, an onshore windflow associated with the summer monsoon increases rainfalls at this time. In winter, however, the ITCZ shifts into the opposite hemisphere and is replaced by drier and more stable air associated with the subtropical high. In addition, the winter monsoon tendency often produces an outflow of continental air. The long dry season inhibits forest growth, so vegetation usually consists of a cover of drought-resistant shrubs or the tall savanna grasses after which the climate is named.

SUBTROPICAL DESERT CLIMATE. The subtropical desert climate has hot, arid conditions as a result of the year-round dominance of the subtropical highs. Summertime temperatures in this climate soar to the highest readings found anywhere on earth. The world's record high temper-

ature was 136.4 degrees Fahrenheit (58 degrees Celsius), recorded in El Azizia, Libya, in the northern Sahara Desert. Rainfall totals in this type of climate are generally less than 10 inches (25 centimeters) per year. What rainfall does occur often arrives as brief, sometimes violent, afternoon thunderstorms. Although summer temperatures are extremely hot, the dry air enables rapid cooling during the winter, so that temperatures are cool to mild at this time of year.

SUBTROPICAL STEPPE CLIMATE. The subtropical steppe climate is a semiarid climate, found mostly on the margins of the subtropical deserts. Precipitation usually ranges from 10 to 30 inches (25 to 75 centimeters), sufficient for a ground cover of shrubs or short steppe grasses. Areas on the equatorward margins of subtropical deserts typically receive their precipitation during a brief showery period in midsummer, associated with the poleward shift of the ITCZ. Areas on the poleward margins of the subtropical highs receive most of their rainfall during the winter, due to the penetration of cyclonic storms associated with the equatorward shift of the westerly wind belt.

MEDITERRANEAN CLIMATE. The Mediterranean climate, also sometimes referred to as the dry summer subtropics, has a distinctive pattern of dry summers and more humid, moderately wet winters. This pattern is caused by the seasonal shift in latitude of the subtropical high and the westerlies. During the summer, the subtropical high shifts poleward into the Mediterranean climate regions, blanketing them with dry, warm, stable air. As winter approaches, this pressure center retreats equatorward, allowing the westerlies, with their eastward-traveling weather fronts and cyclonic storms, to overspread this region. The Mediterranean climate is found on the windward sides of continents, par-

Desert storm Page 101

ticularly the area surrounding the Mediterranean Sea and much of California. This results in the predominance of maritime air and relatively mild temperatures throughout the year.

HUMID SUBTROPICAL CLIMATE. The humid subtropical climate is found on the eastern, or leeward, sides of continents in the lower middle latitudes. The most extensive land area with this climate is the southeastern United States, but it is also seen in large areas in South America, Asia, and Australia. Temperature ranges are moderately large, with warm to hot summers and cool to mild winters. Mean temperatures for a given location are dictated largely by latitude, elevation, and proximity to the coast. Precipitation is moderate. Winter precipitation is usually associated with weather fronts and cyclonic storms that travel eastward within the westerly wind belt. During summer, most precipitation is in the form of brief, heavy afternoon and evening thunderstorms. Some coastal areas are subject to destructive hurricanes during the late summer and autumn.

MIDLATITUDE DESERT CLIMATE. This type of climate consists of areas within the western United States, southern South America, and Central Asia that have arid conditions resulting from the moisture-blocking influence of mountain barriers. This climate is highly continental, with warm summers and cold winters. When precipitations occurs, it frequently comes in the form of winter snowfalls associated with weather fronts and cyclonic storms. Rainfall in summer typically occurs as afternoon thunderstorms.

MIDLATITUDE STEPPE. The midlatitude steppe climate is located in interior portions of continents in the middle latitudes, particularly in Asia and North America. This climate has semiarid conditions caused by a combination of continentality

*Hurricane
Page 104*

resulting from the large distance from oceanic moisture sources and the presence of mountain barriers. Like the midlatitude desert climate, this climate has large annual temperature ranges, with cold winters and warm summers. It also receives winter rains and snows chiefly from weather fronts and cyclonic storms; summer rains occur largely from afternoon convectional storms. In the Great Plains of the United States, spring can bring very turbulent conditions, with blizzards in early spring and hailstorms and tornadoes in mid to late spring.

MARINE WEST COAST. This type of climate is typically located on the west coasts of continents just poleward of the Mediterranean climate. Its location in the heart of the westerly wind belt on the windward sides of continents produces highly maritime conditions. As a result, cloudy and humid weather is common, along with frequent periods of rainfall from passing weather fronts and cyclonic storms. These storms are often well developed in winter, resulting in extended periods of wet and windy weather. Precipitation amounts are largely controlled by the presence and strength of the orographic effect; mountainous coasts like the northwestern United States and the west coast of Canada are much wetter than are flatter areas like northern Europe. Temperatures are held at moderate levels by the onshore flow of maritime air. As a consequence, winters are relatively mild and summers relatively cool for the latitude.

HUMID CONTINENTAL CLIMATE. The humid continental climate is found in the northern interiors of Eurasia (Europe and Asia) and North America. It does not occur in the Southern Hemisphere because of the absence of large land masses in the upper midlatitudes of that hemisphere. This climate type is characterized by low to moderate precipitation that is largely frontal and cyclonic in nature. Most precipitation occurs in summer, but cold winter temperatures typically cause the surface to be frozen and snow-covered for much of the late fall, winter, and early spring. Temperature ranges in this climate are the largest in the world. In Siberia, for example, mean temperatures in July can average more than 108 degrees Fahrenheit (60 degrees Celsius) warmer than in January. Winter temperatures in parts of both North America and Siberia can fall well below –58 degrees Fahrenheit (–50 degrees Celsius), making these the coldest permanently settled sites in the world.

TUNDRA CLIMATE. The tundra climate is a severely cold climate that exists mostly on the coastal margins of the Arctic Ocean in extreme northern North America and Eurasia, and along the coast of Greenland. The high-latitude location and proximity to icy water cause every month to have average temperatures below 50 degrees Fahrenheit (10 degrees Celsius), although a few months in summer have means above freezing. As a result of the cold temperatures, tundra areas are not forested, but instead typically have a sparse ground cover of grasses, sedges, flowers, and lichens. Even this vegetation is buried by a layer of snow during most of the year. Cold temperatures lower the water vapor holding capacity of the air, causing precipitation totals to be generally light. Most precipitation is associated with weather fronts and cyclonic storms and occurs during the summer half of the year.

ICE CAP CLIMATE. The most poleward and coldest of the world's climates is called the ice cap climate. It is found on the continent of Antarctica, interior Greenland, and some high mountain peaks and plateaus. Because monthly mean temperatures are subfreezing throughout the year, areas with this climate are glaciated and have no permanent human inhabitants.

The coldest temperatures of all occur in interior Antarctica, where a Russian research station named Vostok recorded the world's coldest temperature of –128.6 degrees Fahrenheit (–89.2 degrees Celsius) on July 21, 1983. This climate receives little precipitation because the atmosphere can hold very little water vapor. A major moisture surplus exists, however, because of the lack of snowmelt and evaporation. This causes the build up of a surface snow cover that eventually compacts to form the icecaps that bury the surface. Snowstorms are often accompanied by high winds, producing blizzard conditions.

GLOBAL WARMING. Global temperatures increased significantly during the twentieth century. Recordings taken from both ships and land stations indicate that the global average temperature rose by about 0.5 to 1.1 degrees Fahrenheit (0.3 to 0.6 degrees Celsius) during this period, and much of this increase occurred during the 1990's. It is strongly suspected that human activities that increase the abundance of greenhouse gases (heat-trapping gases) in the atmosphere may play a key role in the temperature rise.

Emissions of carbon dioxide, a gas responsible for nearly two-thirds of the global-warming potential of all human-released gases, rose about 400 percent between 1950 and 2000. Carbon dioxide is released chiefly by the burning of fossil fuels. Atmospheric carbon dioxide concentrations are also increased by deforestation, which is occurring at a rapid rate in several tropical countries. Deforestation causes carbon dioxide levels to rise because trees remove large quantities of this gas from the atmosphere during the process of photosynthesis.

Research indicates that if atmospheric concentrations of greenhouse gases continue to increase at the 1990's pace, global temperatures could rise an additional 1.8

Frozen stream Page 100

Rising sea level Page 233

to 6.3 degrees Fahrenheit (1 to 3.5 degrees Celsius) during the twenty-first century. That level of temperature increase would produce major changes in global climates and plant and animal habitats and would cause sea levels to rise substantially.

Ralph C. Scott

FOR FURTHER STUDY

Aguado, Edward, and James E. Best. *Understanding Weather and Climate.* Englewood Cliffs, N.J.: Prentice Hall, 1999.

Ahrens, C. Donald. *Meteorology Today: An Introduction to Weather, Climate, and the Environment.* 6th ed. Pacific Grove, Calif.: Brooks/Cole, 2000.

Gabler, Robert E., Robert J. Sager, and Daniel L. Wise. *Essentials of Physical Geography.* 5th ed. New York: Saunders College Publishing, 1997.

Hidore, John J., and John E. Oliver. *Climatology: An Atmospheric Science.* Englewood Cliffs, N.J.: Prentice Hall, 1993.

McKnight, Tom L., and Darrel Hess. *Physical Geography: A Landscape Appreciation.* 6th ed. Englewood Cliffs, N.J.: Prentice Hall, 2000.

Suplee, Curt. "El Niño/La Niña." *National Geographic* (March, 1999): 72-95.

_____. "Unlocking the Climate Puzzle." *National Geographic* (May, 1998): 38-71.

INFORMATION ON THE WORLD WIDE WEB

The World Climate Web site contains temperature and precipitation statistics in both metric and imperial (standard American) units for thousands of worldwide climate-reporting stations, all of which are identified by latitude, longitude, and elevation. Various maps, with enlargement and reduction capabilities, are provided to show the locations of the stations. (www.worldclimate.com/climate/)

CLOUD FORMATION

Clouds Pages 101, 102

Clouds are visible manifestations of water in the air. Cloud patterns can provide even a casual observer with much information about air movements and the processes occurring in the atmosphere. The shapes and heights of the clouds and the directions from which they have come are valuable clues in understanding weather.

IMPORTANCE OF COOLING. Clouds are formed when water vapor in the air is transformed into either water droplets or ice crystals. Sometimes large amounts of moisture are added to the air, producing clouds, but clouds generally are formed when a large amount of air is cooled. The amount of water vapor that air can hold varies with temperature: Cold air can hold less water vapor than warmer air. If air is cooled to the point at which it can hold no more water vapor, the water vapor will condense into water droplets. The temperature at which condensation begins is called the dew point. At below freezing temperatures, the water vapor will turn or deposit into ice crystals.

Cloud droplets do not necessarily form even if the air is fully saturated, that is, holding as much water vapor as possible at a given temperature. Once formed, cloud droplets can evaporate again very easily. Two factors hasten the production and growth of cloud droplets. One is the presence of particles in the atmosphere that attract water. These are called hygroscopic particles or condensation nuclei. They include salt, dust, and pollen. Once water vapor condenses on these particles, more condensation can occur. Then the droplets can grow larger and bump into other droplets, growing even larger through this process, called coalescence.

Condensation and cloud droplet growth also is hastened when the air is very cold, at about −40 degrees Farenheit (which is also −40 degrees Celsius). At this temperature ice crystals form, but some water droplets can exist as liquid water. These water droplets are said to be supercooled. The water vapor is more likely to deposit on the ice crystals than on the supercooled water. Thus the ice crystals grow larger and the supercooled water droplets evaporate, resulting in more water vapor to deposit on ice crystals. Whether the cloud droplets start as hygroscopic particles or ice crystals, they eventu-

CLOUD FORMATION

The hydrologic cycle is the continuous circulation of the earth's waters through evaporation, condensation, and precipitation. The cycle also moves water through runoff, infiltration, and transpiration.

ally can grow in size to become a raindrop; around one million cloud droplets make one raindrop.

HOW AND WHY RISING AIR COOLS. In order for air to be cooled, it must rise or be lifted. When a volume of air, or an air parcel, is forced to rise through the surrounding air, the parcel expands in size as the pressure of the air around it declines with altitude. Close to the surface, the atmospheric pressure is relatively high because the density of the atmosphere is high. As altitude increases, the atmosphere declines in density, and the still air exerts less pressure. Thus, as an air parcel rises through the atmosphere, the pressure of the surrounding air declines, and the parcel takes up more space as it expands. Since work is done by the parcel as it expands, the parcel cools and its temperature declines.

An alternative explanation of the cooling is that the number of molecules in the air parcel remains the same, but when the volume is larger, the molecules produce less frictional heat because they do not bang into each other as much. The temperature of the air parcel declines, but no

Thunderstorms Page 103

heat left the parcel—the change in temperature resulted from internal processes. The process of an air parcel rising, expanding, and cooling is called adiabatic cooling. Adiabatic means that no heat leaves the parcel. If the parcel rises far enough, it will cool sufficiently to reach its dewpoint temperature. With continued cooling, condensation will result—a cloud will be formed. At this height, which is called the lifting condensation level, an invisible parcel of air will turn into a cloud.

UPLIFT MECHANISMS. An initial force is necessary to cause the air parcel to rise and then cool adiabatically. The three major processes are convection, orographic, and frontal or cyclonic.

With certain conditions, convection or vertical movement can cause clouds to form. On a sunny day, usually in the summer, the ground is heated unevenly. Some areas of the ground become warmer and heat the air above, making it warmer and less dense. A stream of air, called a thermal, may rise. As it rises, it cools adiabatically through expansion and may reach its dewpoint temperature. With continued cooling and rising, condensation will occur, forming a cloud. Since the cloud is formed by predominantly vertical motions, the cloud will be cumulus. With continued warming of the surface, the thermals may rise even higher, perhaps producing thunderstorm, or cumulonimbus, clouds. Thus, a sunny summer day can start off without a cloud in the sky, but can be stormy with many thunderstorms by afternoon.

Clouds also can form when air is forced to rise when it meets a mountain or other large vertical barrier. This type of lifting—orographic—is especially prevalent where air moves over the ocean and then is forced to rise up a mountain, as occurs on the west coast of North and South America. As the air rises, it cools adiabatically

TYPES OF CLOUDS		
Name	*Altitude (km)*	*Altitude (miles)*
Altocumulus	2-7	6,500-23,000
Altostratus	2-7	6,500-23,000
Cirrocumulus	5-13.75	16,500-45,000
Cirrostratus	5-13.75	16,500-45,000
Cirrus	5-13.75	16,500-45,000
Cumulonimbus	to 2	to 6,500
Cumulus	to 2	to 6,500
Nimbostratus	2-7	6,500-23,000
Stratocumulus	to 2	to 6,500
Stratus	to 2	to 6,500

Source: National Oceanic and Atmospheric Administration.

Cumulonimbus clouds are huge, dense formations that rise as high as the stratosphere. Cumulonimbus clouds produce lightning and thunderstorms. (Weather Stock)

and eventually becomes so cool that it cannot hold the water vapor. Condensation occurs and clouds form. The air continues to move up the mountain, producing clouds and precipitation on the side of the mountain from which the wind came, the windward side. However, the air eventually must fall down the other side of the mountain, the leeward side. That air is warmed and moisture evaporates, resulting in no clouds.

A third lifting mechanism is frontal, or cyclonic, action. This occurs when a large mass of cold air and a large mass of warm air—often hundreds of miles in area—meet. The warm air mass and the cold air mass will not mix freely, resulting in a border or front between the two air masses. The warm, less dense, air will always rise above the cold, denser, air mass. As the warm air rises, it cools, and when it reaches its dew point, clouds will form. If the warm

air displaces the cold air, or a warm front occurs, the warm air will rise gradually, resulting in layered or stratiform clouds. The cloud types will change on an upward diagonal path, with the lowest being stratus, and nimbostratus if rain occurs, followed by altostratus, then cirrostratus, and cirrus.

On the other hand, if the cold air displaces the warm air, the warm air will be forced to rise much more quickly. The clouds formed will be puffy or cumuliform—cumulus at the lowest levels, altocumulus and cirrocumulus at the highest altitudes. Sometimes cumulonimbus clouds will also form.

Sometimes when a cold front meets a warm front, the whole warm air mass is forced off the ground. This forms a cyclone—an area of low pressure—as the warm air rises. As this air rises, it cools. If it reaches its dew point, condensation and

The cloud formations seen where tornadoes are developing are known as mammatocumulus. (National Oceanic and Atmospheric Administration)

clouds will result. In oceanic tropical areas, a cyclone can form within warm, moist air. This air also will cool and, if it reaches its dew point, will condense and form clouds. Sometimes, these tropical cyclones are the precursors of hurricanes. The clouds associated with cyclones are usually cumulus, including cumulonimbus, as they are formed by rapidly rising air.

Margaret F. Boorstein

FOR FURTHER STUDY

Lehr, Paul E., Will Burnett, and Herbert Spencer Zim. *Weather: Air Masses, Clouds, Rainfall, Storms, Weather Maps, Climate.* Rev. ed. New York: Golden Press, 1987.

Ludlum, David, Ronald L. Holle, and Richard A. Keen. National Audubon Society Pocket Guides. *Clouds and Storms.* New York: Knopf, 1995.

Rubin, Louis D., Sr., Jim Duncan, Louis D. Rubin, Jr., and Hiram J. Herbert. *The Weather Wizard's Cloud Book: How You Can Forecast the Weather Accurately and Easily by Reading the Clouds.* Chapel Hill, N. C.: Algonquin Books, 1989.

Schaefer, Vincent J., and John A. Day. Peterson Field Guide Series. *A Field Guide to the Atmosphere.* Boston: Houghton Mifflin, 1981.

Williams, Jack. *The USA Today Weather Book.* 2d and rev. ed. New York: Vintage Books, 1997.

INFORMATION ON THE WORLD WIDE WEB

The National Weather Service is the primary U.S. source of weather data, forecasts, and warnings for television weathercasters and private meteorology companies. The service's Web site provides weather reports and forecasts for the nation and the world and educational resources on meteorology, hydrology, and climatology. (www.nws.noaa.gov/)

The Weather World 2010 Web site, maintained by the Department of Atmospheric Sciences at the University of Illinois at Urbana-Champaign, features specialized information on cloud development and classification and different types of precipitation. (ww2010.atmos.uiuc.edu/(Gh)/guides/mtr/cld/home.rxml)

STORMS

A storm is an atmospheric disturbance that produces wind, is accompanied by some form of precipitation, and sometimes involves thunder and lightning. Storms that meet certain criteria are given specific names, such as hurricanes, blizzards, and tornadoes.

Stormy weather is associated with low atmospheric pressure, while clear, calm, dry weather is associated with high atmospheric pressure. Because of the way atmospheric pressure and wind direction are related, low-pressure areas are characterized by winds moving cyclonically (in a counterclockwise direction in the Northern Hemisphere; clockwise in the Southern Hemisphere) around the center of the low pressure. Storms of all kinds are associated with cyclones, but two classes of cyclones—tropical and extratropical—produce most storms.

TROPICAL CYCLONES. These storms develop during the summer and autumn in every tropical ocean except the South Atlantic and eastern South Pacific Oceans. Tropical cyclones that occur in the North Atlantic and eastern North Pacific Oceans are known as hurricanes; in the western North Pacific Ocean, as typhoons; and in the Indian and South Pacific Oceans, as cyclones.

All tropical cyclones develop in three stages. Arising from the formation of the initial atmospheric disturbance that is characterized by a cluster of thunderstorms, the first stage—tropical depression—occurs when the maximum sustained surface wind speeds (the average speed over one minute) range from 23-39 miles (37-61 km.) per hour. The second stage—tropical storm—occurs when sustained winds range from 40-73 miles (62-119 km.) per hour. At this stage, the storm is given a name. From eighty to one hundred tropical storms develop each year across the world, with about half continuing to the final stage—hurricane—at which sustained wind speeds are 74 miles (120 km.) per hour or greater. Moving over land or into colder oceans initiates the end of the hurricane after a week or so by eliminating the hurricane's fuel—warm water.

A mature hurricane is a symmetrical storm, with the "eye" at the center; the eye develops as winds increase and become

Storms Pages 104, 105

The eye of this hurricane is clearly visible at the storm's center. (PhotoDisc)

circular around the central core of low pressure. Within the eye, it is relatively warm, and there are light winds, no precipitation, and few clouds. This is caused by air descending in the center of the storm. Surrounding the eye is the "eye wall," a ring of intense thunderstorms that can extend high into the atmosphere. Within the eye wall, the strongest winds and heaviest rainfall are found; this is also where warm, moist air, the hurricane's "fuel," flows into the storm. Spiraling bands of clouds, called "rain bands," surround the eye wall. Precipitation and wind speeds decrease from the eye wall toward the edge of the rain bands, while atmospheric pressure is lowest in the eye and increases outward.

Hurricanes can be the most damaging storms because of their intensity and size. Damage is caused by high winds and the flying debris they carry, flooding from the tremendous amounts of rain a hurricane can produce, and storm surge. A storm surge, which accounts for most of the coastal property loss and 90 percent of hurricane deaths, is a dome of water that is pushed forward as the storm moves. This wall of water is lifted up onto the coast as the eye wall comes in contact with land. For example, a 25-foot (8-meter) storm surge created by Hurricane Camille in 1969 destroyed the Richelieu Apartments next to the ocean in Pass Christian, Mississippi. Ignoring advice to evacuate, twenty-five people had gathered there for a hurri-

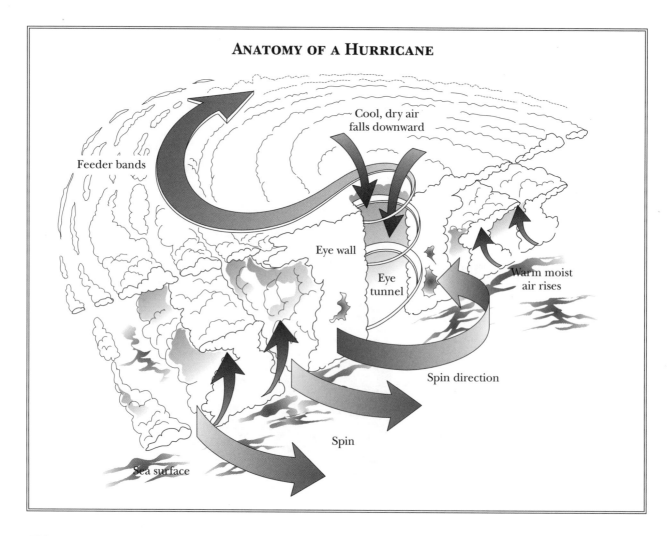

ANATOMY OF A HURRICANE

Cool, dry air falls downward

Feeder bands

Eye wall

Eye tunnel

Warm moist air rises

Spin direction

Spin

Sea surface

The immense strength of hurricane winds makes hurricanes the most damaging type of storm system. (PhotoDisc)

cane party; all but one was killed.

To help predict the damage that an approaching hurricane can cause, the Saffir-Simpson Scale was developed. A hurricane is rated from 1 (weak) to 5 (devastating), according to its central pressure, sustained wind speed, and storm surge height. Camille (1969) was a Category 5 and Andrew (1992) was a Category 4.

EXTRATROPICAL CYCLONES. Also known as midlatitude cyclones, these storms are traveling low-pressure systems that are seen on newspaper and television daily weather maps. They are created when a mass of moist, warm air from the south contacts a mass of drier, cool air from the north, causing a front to develop. At the front, the warmer air rides up over the colder air. This causes water

vapor to condense and produces clouds and rain during most of the year, and snow in the winter.

THUNDERSTORMS. Thunderstorms also develop in stages. During the cumulus stage, strong updrafts of warm air build

Hurricane Andrew Page 235

STORM CLASSIFICATIONS

Tropical Classification

Gale-force winds	>15 meters/second
Tropical depression	20-34 knots and a closed circulation
Tropical storm (named)	35-64 knots
Hurricane	65+ knots (74+ mph)

Saffir-Simpson Scale for Hurricanes

Category 1	63-83 knots (74-95 mph)
Category 2	83-95 knots (96-110 mph)
Category 3	96-113 knots (111-130 mph)
Category 4	114-135 knots (131-155 mph)
Category 5	>135 knots (>155 mph)

Notes: 1 knot = 1 nautical mile/hour = 1.152 miles/hour = 1.85 kilometers/hour.
Source: National Aeronautics and Space Administration, Office of Space Science, Planetary Data System. URL: http://atmos.nmsu.edu/jsdap/

NAMING HURRICANES

Hurricanes once were identified by their latitudes and longitudes, but this method of naming became confusing when two or more hurricanes developed at the same time in the same ocean. During World War II hurricanes were identified by radio code letters, such as Able and Baker. In 1953 the National Weather Service began using English female names in an alphabetized list. Male names and French and Spanish names were added in 1978. By 2000 six lists of names were used on a rotating basis. When a hurricane causes much death or destruction, as Hurricane Andrew did in August of 1992—its name is retired for at least ten years.

*Hail
Page 37*

*Tornado
damage
Page 104*

*Lightning
Page 103*

the storm clouds. The storm moves into the mature stage when updrafts continue to feed the storm, but cool downdrafts are also occurring in a portion of the cloud where precipitation is falling. When the warm updrafts disappear, the storm's fuel is gone and the dissipating stage begins. Eventually, the cloud rains itself out and evaporates.

Thunderstorms can also form away from a frontal system, usually during summer. This formation is related to a relatively small area of warm, moist air rising and creating a thunderstorm that is usually localized and short lived.

Wind, lightning, hail, and flooding from heavy rain are the main destructive forces of a thunderstorm. Lightning occurs in all mature thunderstorms as the positive and negative electrical charges in a cloud attempt to equal out, creating a giant spark. Most lightning stays within the clouds, but some finds its way to the surface. The lightning heats the air around it to incredible temperatures (54,000 degrees Farenheit/30,000 degrees Celsius), which causes the air to expand explosively, creating the shock wave called thunder.

Since lightning travels at the speed of light and thunder at the speed of sound, one can estimate how many miles away the lightning is by counting the seconds between the lightning and thunder and dividing by five. People have been killed by lightning while boating, swimming, biking, golfing, standing under a tree, talking on the telephone, and riding on a lawnmower.

Hail is formed in towering cumulonimbus clouds with strong updrafts. It begins as small ice pellets that grow by collecting water droplets that freeze on contact as the pellets fall through the cloud. The strong updrafts push the pellets back into the cloud, where they continue collecting water droplets until they are too heavy to stay aloft and fall as hailstones. The more an ice pellet is pushed back into the cloud, the larger the hailstone becomes. The largest authenticated hailstone in the United States fell on Coffeyville, Kansas, in September, 1970. It weighed 1.67 pounds (757 grams) and was 5.5 inches (14 centimeters) in diameter.

TORNADOES. For reasons not well understood, less than 1 percent of all thunderstorms spawn tornadoes. Called funnel clouds until they touch earth, tornadoes contain the highest wind speeds known.

Although tornadoes can occur anywhere in the world, the United States has the most, with an average of eight hundred per year. Tornadoes have occurred in every state, but the greatest number hit a portion of the Great Plains from central Texas to Nebraska, known as "Tornado Alley." There cold, Canadian air and warm, Gulf Coast air often collide over the flat land, creating the wall cloud from which most tornadoes are spawned. May is the peak month for tornado activity, but they have been spotted in every month.

Because tornado winds cannot be measured directly, the tornado is ranked according to its damage, using the Fujita Intensity Scale. The scale ranges from an F0, with wind speeds less than 72 miles (116 km.) per hour, causing light damage, to an F5, with winds greater than 260 miles (419 km.) per hour, causing incredible damage. Most tornadoes are small, but the larger ones cause much damage and death.

Kay R. S. Williams

FOR FURTHER STUDY

Arnold, Caroline. *El Niño: Stormy Weather for People and Wildlife.* New York: Clarion Books, 1998.

Freedman, D. H. "Bolts from the Blue." *Discover* 11, issue 12 (December, 1990): 50-56.

Laskin, David. *Braving the Elements: The Stormy History of American Weather.* New York: Doubleday, 1996.

Lauber, Patricia. *Hurricanes: Earth's Mightiest Storms.* New York: Scholastic Press, 1996.

"Living with Natural Hazards." *National Geographic* (July, 1998): 2-39.

Lyons, Walter A. *The Handy Weather Answer Book.* Detroit, Mich.: Visible Ink Press, 1997.

Pearce, E. A. *The Times Books World Weather Guide.* New York: Times Books/Random House, 1999.

"Unraveling the Mysteries of Twisters." *Time* 147, issue 21 (May 20, 1996): 58-64.

Watts, Alan. *The Weather Handbook.* Dobbs Ferry, New York: Sheridan House, 1994.

INFORMATION ON THE WORLD WIDE WEB

The Weather page at *USA Today*'s Web site features up-to-date national and world weather information, maps, and forecasts, as well as articles on weather-related topics, an interactive question-and-answer feature, and current information on hurricanes and tornadoes. (www.usatoday.com/weather/)

The Weather Channel's Web site, Weather.Com, features up-to-date national and world weather information, maps, and forecasts. The "Learn More" link directs viewers to the Weather Classroom, an educational series exploring weather science, with teacher and student resources, backyard projects, and severe weather safety information. (www.weather.com)

How the Weatherworks is a company dedicated to providing educational weather services to teachers and students from pre-school through adulthood. The company's Web site provides information about its services and products as well as experiments and activities and answers to frequently asked questions about the weather. (www.weatherworks.com)

BIOGEOGRAPHY
AND
NATURAL
RESOURCES

Earth's Biological Systems

Biomes

The major recognizable life zones of the continents, biomes are characterized by their plant communities. Temperature, precipitation, soil, and length of day affect the survival and distribution of biome species. Species diversity within a biome may increase its stability and capability to deliver natural services, including enhancing the quality of the atmosphere, forming and protecting the soil, controlling pests, and providing clean water, fuel, food, and drugs. Land biomes are the temperate, tropical, and boreal forests; tundra; desert; grasslands; and chaparral.

TEMPERATE FOREST. The temperate forest biome occupies the so-called temperate zones in the midlatitudes (from about 30 to 60 degrees north and south of the equator). Temperate forests are found mainly in Europe, eastern North America, and eastern China, and in narrow zones on the coasts of Australia, New Zealand, Tasmania, and the Pacific coasts of North and South America. Their climates are characterized by high rainfall and temperatures that vary from cold to mild.

Temperate forests contain primarily deciduous trees—including maple, oak, hickory, and beechwood—and, secondarily, evergreen trees—including pine, spruce, fir, and hemlock. Evergreen forests in some parts of the Southern Hemisphere contain eucalyptus trees.

The root systems of forest trees help keep the soil rich. The soil quality and color is due to the action of earthworms. Where these forests are frequently cut, soil runoff pollutes streams, which reduces fisheries because of the loss of spawning habitat. Racoons, oppossums, bats, and squirrels are found in the trees. Deer and black bear roam forest floors. During winter, small animals such as groundhogs and squirrels burrow in the ground.

TROPICAL FOREST. Tropical forests are in frost-free areas between the Tropic of Cancer and the Tropic of Capricorn. Temperatures range from warm to hot year-round, because the Sun's rays shine nearly straight down around midday. These forests are found in northern Australia, the East Indies, southeastern Asia, equatorial Africa, and parts of Central America and northern South America.

Tropical forests have high biological diversity and contain about 15 percent of the world's plant species. Animal life lives at different layers of tropical forests. Nuts and fruits on the trees provide food for birds, monkeys, squirrels, and bats. Monkeys and sloths feed on tree leaves. Roots, seeds, leaves, and fruit on the forest floor

*Biomes
Page 106*

*Temperate
rain forest
Page 107*

*Tropical
forest
Page 99*

135

feed deer, hogs, tapirs, antelopes, and rodents. The tropical forests produce rubber trees, mahogany, and rosewood. Large animals in these forests include the Asian tiger, the African bongo, the South American tapir, the Central and South American jaguar, the Asian and African leopard, and the Asian axis deer. Deforestation for agriculture and pastures has caused reduction in plant and animal diversity.

Boreal forests Pages 37, 105, 108

BOREAL FOREST. The boreal forest is a circumpolar Northern Hemisphere biome spread across Russia, Scandinavia, Canada, and Alaska. The region is very cold. Evergreen trees such as white spruce and black spruce dominate this zone, which also contains larch, balsam, pine, and fir, and some deciduous hardwoods such as birch and aspen. The acidic needles from the evergreens make the leaf litter that is changed into soil humus. The acidic soil limits the plants that develop.

Animals in boreal forests include deer, caribou, bear, and wolves. Birds in this zone include goshawks, red-tailed hawks, sapsuckers, grouse, and nuthatches. Relatively few animals emigrate from this habitat during winter. Conifer seeds are the basic winter food. The disappearing aspen habitat of the beaver has decreased their numbers and has reduced the size of wetlands.

Wetlands Pages 106, 107

TUNDRA. About 5 percent of the earth's surface is covered with Arctic tundra, and 3 percent with alpine tundra. The Arctic tundra is the area of Europe, Asia, and North America north of the boreal coniferous forest zone, where the soils remain frozen most of the year. Arctic tundra has a permanent frozen subsoil, called permafrost. Deep snow and low temperatures slow the soil-forming process. The area is bounded by a 50 degrees Fahrenheit circumpolar isotherm, known as the summer isotherm. The cold temperature north of this line prevents normal tree growth.

Deserts Pages 109, 110

The tundra landscape is covered by mosses, lichens, and low shrubs, which are eaten by caribou, reindeer, and musk oxen. Wolves eat these herbivores. Bear, fox, and lemming also live here. The larger mammals, including marine mammals and the overwintering birds, have large fat layers beneath the skin and long dense fur or dense feathers that provide protection. The small mammals burrow beneath the ground to avoid the harsh winter climate. The most common Arctic bird is the old squaw duck. Ptarmigans and eider ducks are also very common. Geese, falcons, and loons are some of the nesting birds of the area.

The alpine tundra, which exists at high altitude in all latitudes, is acted upon by winds, cold temperatures, and snow. The plant growth is mostly cushion and mat-forming plants.

DESERT. The desert biome covers about one-seventh of the earth's surface. Deserts typically receive no more than 10 inches (25 centimeters) of rainfall a year, but evaporation generally exceeds rainfall. Deserts are found around the Tropic of Cancer and the Tropic of Capricorn. As the warm air rises over the equator, it cools and loses its water content. This dry air descends in the two subtropical zones on each side of the equator; as it warms, it picks up moisture, resulting in drying the land.

Rainfall is a key agent in shaping the desert. The lack of sufficient plant cover removes the natural protection that prevents soil erosion during storms. High winds also cut away the ground.

Some desert plants obtain water from deep below the surface, for example, the mesquite tree, which has roots that are 40 feet (13 meters) deep. Other plants, such as the barrel cactus, store large amounts of water in their leaves, roots, or stems. Other plants slow the loss of water by having tiny

Barrel cactus. (Digital Stock)

leaves or shedding their leaves. Desert plants have very short growth periods, because they cannot grow during the long drought periods.

Desert animals protect themselves from the Sun's heat by eating at night, staying in the shade during the day, and digging burrows in the ground. Among the world's large desert animals are the camel, coyote, mule deer, Australian dingo, and Asian saiga. The digestive process of some desert animals produces water. A method used by some animals to conserve water is the reabsorption of water from their feces and urine.

GRASSLAND. Grasslands cover about a quarter of the earth's surface, and can be found between forests and deserts. Treeless grasslands grow in parts of central North America, Central America, and eastern South America that have between 10 and 40 inches (250-1,000 millimeters) of erratic rainfall. The climate has a high rate of evaporation and periodic major droughts. The biome is also subject to fire.

Some grassland plants survive droughts by growing deep roots, while others survive by being dormant. Grass seeds feed the lizards and rodents that become the food for hawks and eagles. Large animals include bison, coyotes, mule deer, and wolves. The grasslands produce more food than any other biome. Poor grazing and agricultural practices and mining destroy the natural stability and fertility of these lands. The reduced carrying capacity of these lands causes an increase in water pollution and erosion of the soil. Diverse natural grasslands appear to be more capable of surviving drought than are simplified manipulated grass systems. This may be due to slower soil mineralization and nitrogen turnover of plant residues in the simplified system.

Savannas are open grasslands containing deciduous trees and shrubs. They are near the equator and are associated with deserts. Grasses grow in clumps and do not form a continuous layer. The northern savanna bushlands are inhabited by oryx and gazelles. The southern savanna supports springbuck and eland. Elephants, antelope, giraffe, zebras, and black rhinoceros are found on the savannas. Lions, leopards, cheetah, and hunting dogs are the primary predators here. Kangaroos are found in the savannas of Australia. Savannas cover South America north and south of the Amazon rain forest, where jaguar and deer can be found.

CHAPARRAL. The chaparral or Mediterranean biome is found in the Mediterranean Basin, California, southern Australia, middle Chile, and Cape Province of South America. This region has a climate of wet winters and summer drought. The plants have tough leathery leaves and may contain thorns. Regional fires clear the area of dense and dead vegetation. Fire, heat, and drought shape the region. The vegetation dwarfing is due to the severe drought and extreme climate changes. The seeds from some plants, such as the California manzanita and South African fire lily, are protected by the soil during a fire and later germinate and rapidly grow to form new plants.

OCEAN. The ocean biome covers more than 70 percent of the earth's surface and includes 90 percent of its volume. The ocean has four zones. The intertidal zone is shallow and lies at the land's edge. The continental shelf, which begins where the intertidal zone ends, is a plain that slopes gently seaward. The neritic zone (continental slope) begins at a depth of about 600 feet (180 meters), where the gradual slant of the continental shelf becomes a sharp tilt toward the ocean floor, plunging about 12,000 feet (3,660 meters) to the ocean bottom, which is known as the abyss. The abyssal zone is so deep that it does not have light.

Plankton are animals that float in the ocean. They include algae and copepods, which are microscopic crustaceans. Jellyfish and animal larva are also considered plankton. The nekton are animals that move freely through the water by means of their muscles. These include fish, whales, and squid. The benthos are animals that are attached to or crawl along the ocean's floor. Clams are examples of benthos. Bacteria decompose the dead organic materials on the ocean floor.

The circulation of materials from the ocean's floor to the surface is caused by winds and water temperature. Runoff from the land contains polluting chemicals such as pesticides, nitrogen fertilizers, and animal wastes. Rivers carry loose soil to the ocean, where it builds up the bottom areas. Overfishing has caused fisheries to collapse in every world sector. In some parts of the northwestern Altantic Ocean, there has been a shift from bony fish to cartilaginous fish dominating the fisheries.

HUMAN IMPACT ON BIOMES. Human interaction with biomes has increased biotic invasions, reduced the numbers of species, changed the quality of land and water resources, and caused the proliferation of toxic compounds. Managed care of biomes may not be capable of undoing these problems.

Ronald J. Raven

FOR FURTHER STUDY

Collinson, Alan. *Grasslands.* New York: Dillon Press, 1992.

Hiscock, Bruce. *Tundra: The Arctic Land.* New York: Atheneum, 1986.

Huxley, Julian, ed. *The Atlas of World Wildlife.* New York: Portland House, 1987.

Kaplan, Elizabeth. *Temperate Forest.* New York: Marshall Cavendish, 1996.

Moore, Randy, and Darrel Vodopich. *The Living Desert.* Hillside, N.J.: Enslow, 1991.

Warburton, Lois. *Rainforests.* San Diego, Calif.: Lucent Books, 1991.

Williams, Lawrence. *Oceans.* New York: Marshall Cavendish, 1990.

FORESTS

Extending over 25 percent of the earth's land surface, forests are the world's most complex, productive, and diverse terrestrial communities. The most conspicuous plants of these ecosystems are trees. Trees are tall, woody plants that have one main stem (trunk) and a well-developed crown consisting of branches and leaves. Coniferous trees have narrow needle- or scale-like evergreen leaves and bear their seeds in cones. Flowering angiosperm trees produce seeds in a fruit, nut, or pod. They are of both evergreen broadleaf and deciduous broadleaf form. Evergreen broadleafs predominate where warm, moist conditions persist throughout the year. Deciduous trees drop their leaves in fall where adequate soil moisture is seasonally unavailable.

Numerous plants and animals coexist with trees in one or more of the canopy, shrub, herbaceous, or ground layers. The great degree of layering, or stratification, in forests produces a great variety of environmental conditions and habitat types, which promote a high diversity of animal species. The presence of photosynthesizing plants across many layers also assures high production of food for wildlife.

Forests are the natural vegetation of humid climates whose temperatures remain above 50 degrees Fahrenheit (10 degrees Celsius) for at least one month of each year. Forest communities are classified based on their adaptation to particular climates and soils. Temperate-zone rain forests, composed of large coniferous trees, occur in cool climates that remain above freezing and receive abundant rainfall. Tropical deciduous forests are found just outside the equatorial region, where a pronounced dry season prompts leaf fall. Global forest zones may begin shifting toward the poles in this century in response to climate warming.

TROPICAL RAIN FORESTS. Covering about 7 percent of the world's land area, tropical rain forests house more than half the world's species. These communities are found at low elevations in the Amazon Basin of South America, the Congo Basin of Africa, and in Southeast Asia. Moisture and energy are abundant, but competition for light produces a well-developed layering of forest plants. Broadleaf ever-

Temperate rain forest Page 107

Evergreen forest Page 108

Tropical rain forest Page 99

TROPICAL PLANTS: GOING, GOING . . . GONE?

Tropical plants are of great importance to economies worldwide. Teak, mahogany, ebony, and rosewood are used to make furniture. Brazil nuts, mangos, bananas, and breadfruit are important foods for both tropical and nontropical peoples. The rosy periwinkle of Madagascar is used to treat a rare form of leukemia. Thousands of other plants have been identified as containing possible cancer-fighting chemicals. Perhaps most importantly, tropical plants contribute to the earth's life support systems by providing clean air, preventing climate change, storing water, and preserving biodiversity.

Despite these valuable contributions, 56,000 square miles (145,000 sq. km.) of tropical forest—an area the size of Iowa—is cut down annually, and even more is fragmented or degraded. Commercial logging, plantation and small farm agriculture, cattle ranching, and mining are the main agents of destruction to tropical forests, a large amount of which will be lost within decades at the current rate of development.

greens dominate in the three canopy layers. Well-spaced, umbrella-shaped emergents soar high above the forest floor. Beneath, the main canopy contains a tremendous variety of closely spaced trees.

Although about 80 percent of the Sun's energy is absorbed in the upper layer of the rain forest, a subcanopy of slender trees and saplings survives in the shade below. Woody vines, or lianas, climb trees, competing for light in the canopy. The strangler fig emerges from a seed in the canopy and sends down roots to the ground. The roots thicken and engulf the host tree, which eventually is cut off from light and dies. After it decays, a hollow is left where the tree once stood. Epiphytic plants attach to canopy branches and obtain nutrients from the air or dead plant material fallen on them. Epiphytic bromeliads store up to two gallons of rainwater in their cup-like arrangement of leaves, providing rearing grounds for tadpoles of the poison arrow frog.

Having the most available foliage, the canopy is the most populated area of the forest. Arboreal mammals, reptiles, birds, amphibians, and insects are highly adapted to living in the canopy. Parrots use their beaks and claws to move from branch to branch. Fruit bats fly easily through the canopy. The long, grasping hands and feet of many primates help locomotion through trees. Many New World monkeys have evolved prehensile tails that serve as a fifth grasping limb and help them maintain balance.

Little light penetrates to the forest floor. Herbaceous plants possess large leaves to maximize energy reception, but are sparse. Many leaves have pointed tips to drain water and prevent fungal attack. Jaguars, bush dogs, ocelots, tapirs, ants, termites, coral snakes, and pit vipers are found on the floor of the Brazilian rain forest. Tree bases have protruding edges for support of the tall

Tree farming has become an important method of replenishing the world's diminishing forest resources. (PhotoDisc)

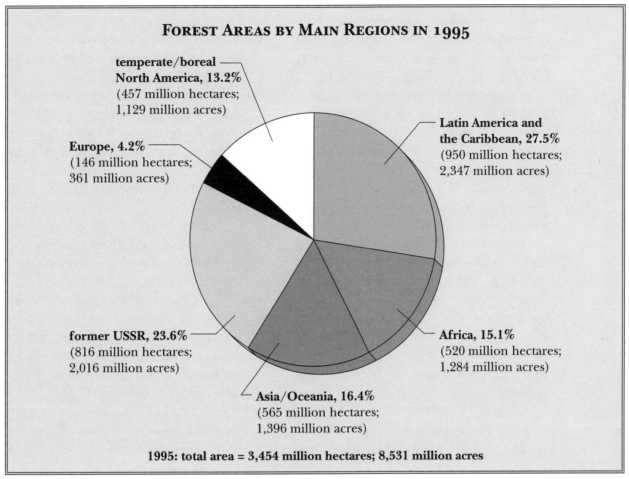

FOREST AREAS BY MAIN REGIONS IN 1995

temperate/boreal
North America, 13.2%
(457 million hectares;
1,129 million acres)

Europe, 4.2%
(146 million hectares;
361 million acres)

Latin America and
the Caribbean, 27.5%
(950 million hectares;
2,347 million acres)

former USSR, 23.6%
(816 million hectares;
2,016 million acres)

Africa, 15.1%
(520 million hectares;
1,284 million acres)

Asia/Oceania, 16.4%
(565 million hectares;
1,396 million acres)

1995: total area = 3,454 million hectares; 8,531 million acres

Source: United Nations Food and Agriculture Organization (FAOSTAT Database, 2000).

trees, and roots lie close to the soil surface to absorb nutrients from decaying vegetation. This is important since the old, weathered soils are nutrient poor.

MIDLATITUDE BROADLEAF DECIDUOUS AND BOREAL CONIFEROUS FORESTS. Midlatitude forests are found between about 30 and 60 degrees north and south of the equator, and face many climate perils. Broadleaf deciduous forests receive ample precipitation throughout the year, but must survive winter freezes when soil moisture becomes unavailable. The predominant trees, such as oak, maple, and cherry, shed their leaves in autumn to prevent dehydration. Beautiful displays of colors occur when green chlorophyll production stops before the leaves fall. The

remaining pigments (anthocyanins, carotinoids), which were always present in the leaf, then become visible.

In contrast, subarctic boreal forests consist of needle-leaf conifers, including spruces, firs, and pines. The conical-shaped trees easily shed snow, and the small surface area of the needles reduces moisture loss. Few deciduous trees grow here, since the short growing season provides insufficient energy for both growth and regeneration of leaves. Evergreens begin growing early in spring, giving them an advantage over deciduous trees.

Compared to tropical forests, midlatitude forests show less species diversity and reduced layering of vegetation. Boreal forests have only one main canopy, and a

*Boreal
forest
Page 37*

*Deciduous
trees
Page 105*

single moss, lichen, and grass layer. Rain forests have hundreds of tree species in a small area, while boreal forests are dominated by large numbers of few species.

Similar types of animal life are found in both of these forests. In North America, grizzly and black bears, bobcats, lynxes, cougars, wolves, and foxes exist where humans have not driven them out. Deer, raccoons, squirrels, rabbits, and skunks are common. Migratory birds, retreating in winter to warmer areas with more food, are attracted back by the numerous insects and fruit which reappear seasonally. Numerous adaptations, such as hibernation, food switching, and thick, wind-resistant fur aid survival during long, cold winters.

Kathleen V. Schreiber

FOR FURTHER STUDY

Acharya, Anjali. "Plundering the Boreal Forests." *World Watch* 8, no. 3 (May/June, 1995): 21-29.

Goodman, Billy. *The Rainforest.* New York: Tern Enterprise, 1991.

Greenaway, Theresa, and Geoff Dann. *Jungle: Eyewitness Books.* New York: Alfred A. Knopf, 1994.

Hirschi, Ron. *Save Our Forests.* New York: Delacorte Press, 1993.

Mitchell, John G. "Our National Forests." *National Geographic* (March, 1997): 58-87.

Nardi, James B. *Once upon a Tree: Life from Treetop to Root Tips.* Ames: Iowa State University Press, 1993.

Page, Jake, and the Editors of Time-Life Books. *Planet Earth: Forest.* Alexandria, Va.: Time-Life Books, 1983.

Pandell, Karen, Art Wolfe, and Denise Takahashi. *Journey through the Northern Rainforest.* New York: Penguin Putnam Books for Young Readers, 1999.

Schoonmaker, Peter K. *The Living Forest.* Hillside, N.J.: Enslow, 1990.

Wilson, Edward O. "Rain Forest Canopy: The High Frontier." *National Geographic* (December, 1991): 78-107.

INFORMATION ON THE WORLD WIDE WEB

The U.S. Forest Service Web site discusses forest resource management programs for U.S. national forests. The site features current Forest Service news and issues and a library of departmental publications. (www.fs.fed.us)

The Environmental Literacy Council is a nonprofit organization of scientists, economists, educators, and other experts informing environmental studies. The council's Web site provides links to a wide range of forest topics. (www.enviroliteracy.org/forest.html)

National Geographic magazine's on-line article "Congo Trek: A Journey Through the Heart of Central Africa" documents biologists' 1,200-mile hike through tropical African forest with photos and field notes. (www.nationalgeographic.com/congotrek)

Educational Web Adventures' "Amazon Interactive" site explores the geography of the Amazonian rain forest through games and activities. (www.eduweb.com/amazon.html)

Resources for the Future, a nonprofit and nonpartisan organization based in Washington, D.C., conducts independent research on environmental and natural resource issues. A section of the organization's on-line Natural Resources Library focuses on the impacts of climate change on forests. (www.rff.org/nat_resources/forests.htm)

GRASSLANDS

Grasslands cover about a quarter of the world's land surface. Since they grow on the world's richest soils, they are so intensely farmed and grazed that only small patches of natural grassland remain.

CLIMATE AND GEOGRAPHIC LOCATION. Annual precipitation between 10 and 32 inches (25-80 centimeters), often with a dry period late in the growing season, supports grassland. Grassland temperature patterns vary. Fire and grazing favor grasses and often combine with climate to maintain grasslands.

Extensive grasslands generally are found in continental interiors. In North America, grasslands occur from the eastern foothills of the Rocky Mountains to the Mississippi River, from south central Canada to northeastern Mexico, in eastern Washington and Oregon, and in California's Central Valley. Grasslands on other continents include the steppes of Europe and Asia, areas fringing the major deserts of Africa and Australia, and the Pampas of South America.

TYPES OF GRASSLANDS. Extensive grasslands are often divided into tall-grass, mixed-grass, and short-grass regions. In pre-human settlement North American grasslands, the tall-grass prairie occurred in the moist eastern zone. Big bluestem, Indian grass, and switch grass grew 6-10 feet (2-3 meters) tall in this region. The short-grass prairie or plains occupied the drier western extreme. Here, blue grama and buffalo grass seldom grew taller than 8 inches (20 centimeters). Mixed-grass prairie grew in between, with a mixture of tall, short, and middle-height grasses. Boundaries between regions were broad zones of gradual change.

GRASSES AND GRASSLANDS. Grasses are well adapted to occupy regions with intermediate annual precipitation, fires, and grazing animals. Grasses have their main center of growth at or below the ground. Their slender, widespread roots compete intensely for nutrients and moisture, especially near the surface. The above-ground parts of the plants grow densely, and the entire above-ground plant dies every year, covering the ground with a dense mulch. This combination presents difficulties for plants invading grasslands, as the grass roots usurp moisture and nutrients and the leaves and mulch intercept sunlight.

Under very dry conditions, when grasses cannot grow densely, shrubs and succulents (such as cacti) dominate and deserts occur. With heavy rainfall and infrequent dry periods, trees compete well with grasses, and forests dominate the landscape. Grasslands are often bordered by forests at their moist edges and deserts at their dry boundaries. Under intermediate rainfall conditions, however, grasses are favored over all competitors.

Fire and grazing by animals tip the balance further in favor of the grasses. The late-season dry period typical of grasslands and the mulch built up after a year or more of growth are ideal conditions for the spread of fires. Whether started by lightning or by humans, fires spread quickly through the dried mulch. The tops of plants burn to the ground, but often little damage occurs underground. Because the primary growth center of most non-grass plants is above ground and that of grasses is below ground, fire is more harmful to woody plants and nonwoody, nongrass plants (forbs).

NORTH AMERICAN GRASSLAND GRAZERS

Bison herds numbered in the millions in presettlement North American grasslands and were the center of the economies of the Plains Indians. Pronghorns may have been even more numerous than bison. Both wandered widely over the prairies and plains and undoubtedly had a great impact on plant life.

Prairie dogs are smaller and less mobile but within their "towns" they may have had an even greater impact. Many plants, especially forbs, grew primarily on soil disturbed by the burrowing of prairie dogs. Prairie dog burrowing also aerated the soil and enhanced the penetration of water. The feeding and burrowing activities of smaller mammals, such as mice and voles, also had a significant impact.

However, insects were the most influential aboveground grazers in native grasslands, and roundworms had the greatest impact underground. These invertebrates accounted for more consumption of plant material than all the mammals combined. The grazers, sustained by the grassland, were instrumental in molding its character.

Grazing buffalo Page 109

Because grazing removes the tops of plants, it does more damage to forbs and woody plants than to grasses. Many grasses actually increase growth after light grazing. Most extensive grasslands are occupied by large grazing animals, such as the bison and pronghorn of North American grasslands. These and other grazers played important roles in the maintenance of the native grasslands and in the lives of the people who lived there.

GRASSLAND SOILS. The presence of grasslands is determined by climate, fire, and grazing, but the grasses impact their environment as well. In addition to their competitive role in excluding trees, shrubs, and forbs, grasses contribute to soil formation. All the above-ground parts of grass plants die each year, become mulch, and slowly decompose into the soil. Rainfall is generally insufficient to wash nutrients out of the reach of the grass roots, so the soil accumulates both nutrients and decaying plant material. The world's richest soils develop under these conditions.

HUMAN IMPACT ON GRASSLANDS. Because of their soils, grasslands became agricultural centers. Domestic grasses became the predominant crops—corn in the tall-grass country and wheat in the mixed-grass region. The short-grass plains were too dry to support grain crops, but became an important region for grazing domestic animals.

In the process of learning what activities the grasslands could and could not support, Americans changed the grasslands of the continent forever. Farming reduced native tall-grass prairie to one of the world's rarest habitats. Although grazing had less impact on the short-grass plains, vast areas have been overgrazed severely. Grasslands in other parts of the world have been similarly abused. Given the importance of grasslands to humanity, serious conservation measures must be taken to restore their productivity.

Carl W. Hoagstrom

FOR FURTHER STUDY

Brown, Lauren. *The Audubon Society Nature Guides: Grasslands.* New York: Alfred A. Knopf, 1985.

INFORMATION ON THE WORLD WIDE WEB

The U.S. Forest Service Web site features information on the twenty publicly owned National Grasslands administered by the U.S. Forest Service. (www.fs.fed.us/grasslands)

Collinson, Alan. *Ecology Watch: Grasslands.* New York: Dillon Press, 1992.

Joern, Anthony, and Kathleen H. Keeler, eds. *The Changing Prairie.* New York: Oxford University Press, 1995.

Sampson, Fred B., and Fritz L. Knopf, eds. *Prairie Conservation.* Washington, D.C.: Island Press, 1996.

Steele, Philip. *Geography Detective: Grasslands.* Minneapolis, Minn.: Carolrhoda Books, 1996.

DESERTS

The word "desert" evokes images of searing heat and barren, windswept sand dunes. Although some deserts fit this stereotype, deserts are more than hot, dry places. Each desert is unique, but all share one important characteristic—lack of moisture. Deserts are exceptionally dry environments. The degree of their dryness is influenced by total annual rainfall, the frequency and intensity of rains, temperature, rate of evaporation, soil characteristics, and other factors.

No specific amount of rainfall serves as a criterion for deserts; however, a region is usually classified as a desert if it receives less than about 10 inches (25 centimeters) of rain per year and the rate of evaporation exceeds total annual precipitation. Each of the world's deserts is a unique environment with its own set of climatic conditions, geological characteristics, and plant and animal communities.

Most deserts experience wide shifts in daily temperature. Lack of cloud cover and low humidity allow as much as 90 percent of the Sun's heat energy striking the earth to reach the desert surface, causing daytime temperatures to climb rapidly and produce air temperatures approaching 131 degrees Fahrenheit (55 degrees Celsius) in the hottest deserts. At night, the accumulated day's heat is quickly lost to the atmosphere, and the temperature may drop to near freezing.

TYPES OF DESERTS. Deserts are found in cold as well as hot regions. Low-latitude, or subtropical, deserts occur near the latitudes of 30 degrees north and south—the Tropic of Cancer in the Northern Hemisphere and the Tropic of Capricorn in the Southern Hemisphere. The formation of low-latitude deserts is related to air circulation patterns and the physical properties of air. Warm, moist air rises at the equator, cools, and loses much of its moisture as rainfall. The cooler, drier air sinks and flows north or south toward 30 degrees north and south latitude. As the air sinks, it is compressed by the weight of the air above and warms. The resulting warm, dry air removes moisture from the land, giving rise to arid conditions and deserts. Examples of low-latitude deserts include the Sahara Desert of North Africa, the Kalahari Desert of Southern Africa, the Atacama Desert of South America, and the Victoria Desert of Australia.

Most deserts, but not all, are low-latitude regions. Rain-shadow deserts form downwind of mountain ranges. As warm,

*Sand dunes
Page 95*

*Desert
Page 109*

moist air is forced up and over a mountain range, it cools and loses its moisture as rain or snow. The cool, dry air descending down the opposite side of the range compresses and warms. As a result, little or no precipitation falls in the rain-shadow zone created leeward of the mountains. The lack of rainfall and low humidity within the rain-shadow zone create desert conditions. The deserts of the American southwest, leeward of the Sierra Nevada range, are rain-shadow deserts.

Grand Canyon Page 110

Some deserts form in the interiors of continents, principally because of their great distance from the ocean—the main source of moisture needed for precipitation. The Gobi Desert of Mongolia and northeastern China is an example of this kind of desert. Another kind of desert develops along warm tropical and subtropical coasts adjacent to cold ocean currents. The air above the ocean currents is cooled and contains little moisture. As this cool, dry air moves inland, it warms, causing high evaporation and producing little precipitation. Deserts of this kind include the Atacama Desert of northern Chile and southern Peru, and the Namib and Kala-

Namib Desert Page 110

hari Deserts along Africa's southwest coast.

Perhaps the most unusual deserts are rarely thought of as deserts. These are the polar deserts that occur in high-latitude regions, including all of Antarctica, most of Greenland, and the northernmost parts of Alaska, Canada, and Siberia. Polar deserts are bitterly cold and dry because of frigid air masses descending at the North and South Poles. Temperatures remain below freezing year-round.

DESERT LIFEFORMS. Despite their stark appearance, deserts are second only to tropical rain forests in the variety of animals and plants living there. The Sonoran Desert of the southwestern United States and northern Mexico is home to nearly twenty-five hundred species of plants and numerous animal species, thanks in part to biannual rainy seasons. In contrast, the driest portions of Africa's Sahara Desert and South America's Atacama Desert are practically devoid of life.

Desert plants and animals are adapted to arid conditions and extremes in temperature. Many plants are short-lived annuals whose life cycles are keyed to rainfall. They survive drought conditions as seeds that quickly germinate after exposure to water. For a few short days, the desert is ablaze with color, but soon the next generation of seeds is set and the blooms wither and die. Succulents, such as cacti and agave, store water in modified roots and stems or in fleshy leaves. Woody shrubs have small leaves that reduce water loss through transpiration and develop extensive root systems to take up whatever available water is in the soil. During

Typical arid desert landscape in North America. (Digital Stock)

drought conditions, some desert shrubs shed their leaves to conserve water. Spines and thorns are a common means of defense against water-seeking animals.

Some animals avoid the heat of the day by being active at night when the desert is cooler and the humidity is higher. Others spend the day in the shade or reside in burrows. Many desert animals are efficient at conserving and recycling water. Some obtain all of the water they need from the foods they eat. During dry periods, some animals enter a period of dormancy known as estivation. Because food and shelter are scarce, most mammals are small. Common animals include insects, arachnids, reptiles, and birds. Amphibians are rare as a result of the lack of permanent bodies of water.

Human beings also live in deserts. Lack of water causes many desert peoples to adopt a nomadic lifestyle. Desert soils are remarkably fertile, and irrigated crops do well if water is available. Unfortunately, human activities can negatively affect deserts and semiarid lands. The demand for water to irrigate crops or support heavily populated desert communities can severely strain limited water resources. Farming and overgrazing on semiarid lands bordering desert regions has resulted in the encroachment of deserts on productive land—a process known as desertification.

Steven D. Carey

FOR FURTHER STUDY

Larson, Peggy. *A Sierra Club Naturalist's Guide to the Deserts of the Southwest: The Deserts of the Southwest.* Vol. 1. San Francisco: Sierra Club Books, 1982.

MacMahon, James. *Deserts.* New York: Alfred A. Knopf, 1985.

Mares, Michael A., ed. *Encyclopedia of Deserts.* Norman: University of Oklahoma Press, 1999.

Pavitt, Irene, ed. *The Sierra Club Guides to the National Parks: Desert Southwest.* New York: Random House, 1995.

Phillips, Steven J., ed. *A Natural History of the Sonoran Desert.* Tucson: Arizona-Sonoran Desert Museum Press, 1999.

Wallace, Marianne D. *America's Deserts: Guide to Plants and Animals.* Golden, Colo.: Fulcrum, 1996

Zwinger, Ann Haymond. *The Mysterious Lands: A Naturalist Explores the Four Great Deserts of the Southwest.* Tucson: University of Arizona Press, 1996.

INFORMATION ON THE WORLD WIDE WEB

DesertUSA, an on-line travel and adventure guide, provides a wealth of interesting information about the deserts of the United States. Sections cover desert life, minerals and geology, and peoples and cultures. QuickTime movies show panoramic views of selected desert locations. (www.desertusa.com)

TUNDRA AND
HIGH ALTITUDE BIOMES

Tundra landscapes appear where long, cold winters, a permanently frozen subsoil, and strong winds combine to prevent the development of trees. The resulting landscapes tend to be vast plains with low-growing forbs and stunted shrubs. Vast areas of this biome encircle the northernmost portions of North America and Eurasia, constituting the Arctic tundra. Climatic conditions atop high mountains at all latitudes are similar; these smaller, more isolated areas are called the alpine tundra.

PERMAFROST. The low temperatures of the tundra cause the formation of a permanently frozen layer of soil known as permafrost. Characteristic of Arctic tundra, permafrost, which varies in depth according to latitude, thaws at the surface during the brief summers. As the permafrost below is impenetrable by both water and plant roots, it is a major factor in determining the basic nature of tundra.

The alternate freezing and thawing of soil above the permafrost creates a symmetrical patterning of the land surface characteristic of Arctic tundra. Perhaps best known are stone polygons that result when frost pushes larger rocks toward the periphery with smaller ones occupying the center of each unit. This alteration of the tundra landscape, called cryoplanation, is the major force in molding Arctic tundra landscapes.

In contrast, alpine tundra generally has little or no permafrost. Even though alpine precipitation is almost always higher than for Arctic tundra, steep grades result in a rapid runoff of water. Alpine soils are, therefore, much drier, except in the flat al-pine meadows and bogs, where conditions are more like those of Arctic areas.

VEGETATION. Both Arctic and alpine tundra regions are composed of plants that have adapted to the same generally stressful conditions. Biodiversity—the total number of species present—is low compared to most other ecosystems. Plant growth is slow because of the short growing seasons and the influence of permafrost. Most tundra plants are low-growing perennials that reproduce vegetatively rather than by seed. Often, they grow in the crevices of rocks that both shelter them in the winter and reflect heat onto them in summer.

Common plants of the low-lying Arctic tundra sites include various sedges, especially cottongrass, and sphagnum moss. On better-drained sites, biodiversity is higher, and various mosses, lichens, sedge, rush species, and herbs grow between dwarfed heath shrubs and willow. The arrangement of plants within a small area reflects the numerous microclimates resulting from the peculiar surface features.

Alpine plants possess many of the features of Arctic plants. However, because strong winds are such a prominent feature of the alpine environment, most of the plants grow flat on the ground, forming mats or cushions.

Below alpine tundra and south of Arctic tundra, there is the boreal or coniferous forest biome. Between the forest and tundra lies a transitional zone or ecotone. The ecotone is characterized by trees existing at their northern (or upper) limit. Especially in alpine regions, stunted, gnarled

Boreal forest Page 37

trees occupy an area called Krummholz. In North America, the Krummholz is much more prominent in the Appalachians of New England than in the western mountains.

ANIMAL LIFE. Biodiversity of animals, like that of plants, is relatively low in the tundra. In Arctic regions, many animal species are circumboreal; that is, they have ranges that extend around the major continents of the north. Examples are arctic hares, reindeer, muskox, and many migratory birds such as plovers, sandpipers, and waterfowl. Few insect species occupy the Arctic tundra, but some, such as flies and mosquitoes, can be locally abundant in midsummer. Except for insects, few invertebrates can endure the harsh Arctic environment. Amphibians and reptiles are almost nonexistent.

Animals of the alpine tundra are generally more like those of adjacent lowlands than those of Arctic regions. Furthermore, they differ from mountain area to mountain area. Many bird species and some mammal species, such as sheep and elk, regularly migrate from upper mountain meadows where they spend summers to lower slopes or lowlands during winters. A few hibernate in winter, finding protection under the snow. Insects of many kinds, including grasshoppers, butterflies, beetles, and springtails, are often present.

CONSERVATION. Like all world biomes, tundra regions are subject to degradation and destruction, especially as a result of human activities. Because of low human population density and their unsuitability for agriculture, tundras generally are less impacted by humans than are grasslands and forests. However, tundra ecosystems, when disturbed, recover slowly, if at all. As most tundra plants lack the ability to invade and colonize bare ground, the process of ecological succession that follows disturbances may take centuries. Even tire

tracks left by vehicles may endure for decades. The melting of permafrost also has long-lasting effects.

The discovery of oil and gas in tundra regions, such as those of Alaska and Siberia, has greatly increased the potential for disturbances. Heavy equipment used to prospect for fossil fuels and to build roads and pipelines has caused great destruction of tundra ecosystems. As the grasses and mosses are removed, the permafrost beneath melts, resulting in soil erosion. The disposal of sewage, solid wastes, and toxic chemicals poses special problems, as such pollutants tend to persist in the tundra environment longer than in warmer areas.

Animals of the Arctic tundra, such as caribou, have been hunted by the native Inuit using traditional methods for centuries without an impact on populations. The introduction of such modern inventions as snowmobiles and rifles has caused a sharp decline in caribou numbers in some areas.

Although efforts at restoring other ecosystems, especially grasslands, have been quite successful, tundra restoration poses difficult problems. Seeding of disturbed Arctic tundra sites with native grasses is only marginally successful, even with the use of fertilizers. In alpine tundra, restoration efforts have been somewhat more successful, but involve transplanting as well as seeding and fertilizing. A recognition of natural successional patterns and long-term monitoring is a necessity in such efforts.

Thomas E. Hemmerly

FOR FURTHER STUDY

Bliss, L. C., O. H. Neal, and J. J. Moore, eds. *Tundra Ecosystems.* New York: Cambridge University Press, 1981.
Bush, M. B. *Ecology of a Changing Planet.* 2d ed. Upper Saddle River, N.J.: Prentice-Hall, 2000.

Cox, G. W. *Conservation Biology, Concepts, and Applications.* Dubuque, Iowa: William C. Brown, 1997.

Hemmerly, T. E. *Appalachian Wildflowers.* Athens: University of Georgia Press, 2000.

Smith, R. L., and T. M. Smith. *Elements of Ecology.* 4th ed. San Francisco: Benjamin/Cummings, 1998.

Steele, F. L. *At Timberline: A Nature Guide to the Mountains of the Northeast.* Boston: Appalachian Mountain Club, 1982.

NATIONAL PARK SYSTEMS

Yellowstone Pages 38, 159

The world's first national parks were established as a response to the exploitation of natural resources, disappearance of wildlife, and destruction of natural landscapes that took place during the late nineteenth century. Government efforts to preserve natural areas as parks began with the establishment of Yellowstone National Park in the United States in 1872 and were soon adopted in other countries, including Australia, Canada, and New Zealand.

While the preservation of nature continues to be an important benefit provided by national parks, worldwide increases in population and the pressures of urban living have raised public interest in setting aside places that provide opportunities for solitude and interaction with nature.

Because national parks have been established by nations with diverse cultural values, land resources, and management philosophies, there is no single definition of what constitutes a national park. In some countries, areas used principally for recreational purposes are designated as national parks; other countries emphasize preservation of outstanding scenic, geologic, or biological resources. The terminology used for national parks also varies among countries. For example, protected areas that are similar to national parks may be called reserves, preserves, or sanctuaries.

Diverse landscapes are protected within national parks, including swamps, river deltas, dune areas, mountains, prairies, tropical rain forests, temperate forests, arid lands, and marine environments. Individual parks within nations form networks that vary with respect to size, accessibility, function, and the type of natural landscapes preserved. Some national park areas are isolated and sparsely populated, such as Greenland National Park; others, such as Peak District National Park in Great Britain, contain numerous small towns and are easily accessible to urban populations.

The functions of national parks include the preservation of scenic landscapes, geological features, wilderness, and plants and animals within their natural habitats. National parks also serve as outdoor laboratories for education and scientific research and as reservoirs for genetic information. Many are components of the United Nations International Biosphere Reserve Program.

National parks also play important roles in preserving cultures, by protecting archaeological, cultural, and historical sites. The United Nations recognizes several national parks that possess important cultural attributes as World Heritage Sites. Tourism to national parks has become important to the economies of many developing nations, especially in Eastern and Southern Africa, India, Nepal, Ecuador, and Indonesia. Parks are sources of local employment and can stimulate improvements to transportation and other types of infrastructure while encouraging productive use of lands that are of marginal agricultural use.

The International Union for Conservation of Nature has developed a system for classifying the world's protected areas, with Category II areas designated as national parks. Using this definition, there are 3,384 national parks in the world, with a mean average size of 457 square miles (1,183 sq. km.) each. Together, they cover an area of about 1.5 million square miles (4 million sq. km.), accounting for about 2.7 percent of the total land area on Earth.

NORTH AMERICA. In 1916 management of U.S. national parks and monuments was shifted from the U.S. Army to the newly established National Park Service (NPS). The system has since grown in size to protect fifty-five national parks, as well as other natural areas including national monuments, seashores, and preserves.

North America's second largest system of national parks is Parks Canada, created in 1930. Among the best-known Canadian parks is Banff, established in southern Alberta in 1885. Preserved within this area are glacially carved valleys, evergreen forests, and turquoise lakes. Parks Canada has the goal of protecting representative examples of each of Canada's vegetation and physiographic regions.

Mexico began providing protection for natural areas in the late ninteenth century. Among its system of forty-four national parks is Dzibilchaltún, an important Mayan archaeological site on the Yucatán Peninsula. With fewer resources available for park management, the emphasis in Mexico remains the preservation of scenic beauty for public use.

SOUTH AMERICA. Two of South America's best-known national parks are located within Argentina's park system. Nahuel Huapi National Park preserves two rare deer species of the Andes, while Iguazú National Park, located on the border with Brazil, is home to tapir, ocelot, and jaguar.

Located on a plateau of the western slope of the Andes Mountains in Chile, Lauca National Park is one of the world's highest parks, with an average elevation of more than 14,000 feet (4,267 meters)—an altitude nearly as high as the tallest mountains in the continental United States. Huascarán, another mountain park located in western Peru, boasts twenty peaks that exceed 19,000 feet (5,791 meters) in elevation. The volcanic islands of Galapagos Islands National Park, managed by Ecuador, have been of interest to biologists

Glacier National Park Page 108

STEPHEN T. MATHER AND THE U.S. NATIONAL PARK SERVICE

In 1914 businessman and conservationist Stephen T. Mather wrote to Secretary of the Interior Franklin K. Lane about the poor condition of California's Yosemite and Sequoia National Parks. Lane wrote back, "if you don't like the way the national parks are being run, come on down to Washington and run them yourself." Mather accepted the challenge and became an assistant to Lane and later the first director of the U.S. National Park Service, from 1917 to 1929.

since British naturalist Charles Darwin studied variation and adaptation in animal species there in 1835.

AUSTRALIA AND NEW ZEALAND. Established in 1886, Royal was Australia's first national park. Perhaps better known to tourists, Uluru National Park in Australia's Northern Territory protects two rock domes, Ayer's Rock and Mount Olga, that rise above the plains 15 miles (40 km.) apart.

Along with Australia and other former colonies of Great Britain, New Zealand was a leader in establishing early national parks. The first of these was Tongagiro, created in 1887 to protect sacred lands of the Maori people on the North Island. New Zealand's South Island features several national parks including Fiordland, created in 1904 to preserve high mountains, forests, rivers, waterfalls, and other spectacular features of glacial origin.

AFRICA. Game poaching continues to be a severe problem in Africa, where animals are slaughtered for ivory, meat, and hides. Many African national parks were established to protect large game. South Africa's national park system began in 1926, when the Sabie Game Preserve of the eastern Transvaal region became Kruger National Park. Among South Africa's greatest attractions to foreign visitors, Kruger is famous for its population of lions and elephants.

East Africa is also known for outstanding game sanctuaries, such as Serengeti National Park, created prior to Tanzania's independence from Great Britain. Another national park in Tanzania, Kilimanjaro, protects Africa's highest and best-known mountain. Other African countries with well-developed park systems include Kenya, Congo-Kinshasa (formerly Zaire), and Zambia. Although there is now a network of national parks in Africa that protects a wide range of habitats in various regions, there remains a need to protect additional areas in the arid northern part of the continent that includes the Sahara Desert.

EUROPE. In comparison with the United States, the national park concept spread more slowly within Europe. In 1910 Germany set aside Luneburger Heide National Park near the Elbe River, and in 1913, Sweden established Sarek, Stora Sjöfallet, Peljekasje, and Abisko National Parks. Swiss National Park was founded in Switzerland in 1914, in the Lower Engadine region. Great Britain has several national parks, including Lake District, a favorite recreation destination for English poet William Wordsworth. Spain's Doñana National Park, located on its southwestern coast, preserves the largest dune area on the European continent.

ASIA. The system of land tenure and rural economy in many Asian countries has made it difficult for national governments to set aside large areas free from human exploitation. Many national parks established by colonial powers prior to World War II were maintained or expanded by countries following independence. For example, Kaziranga National Park is a refuge for the largest heard of rhinocerous in India. Established in 1962, Thailand's Khao Yai National Park protects a sample of the country's wildlife, while Indonesia's Komodo Island National Park preserves the habitat for the large lizards known as Komodo dragons.

In Japan, high population density has made it difficult to limit human activities within large areas. Some Japanese national parks are principally recreation areas rather than wildlife sanctuaries and may contain cultural features such as Shinto shrines. One of the best known national parks in Japan is Fuji-Hakone-Izu, which contains world-famous Mount Fuji, a volcano with a nearly symmetrical shape.

THE FUTURE. National parks serve as relatively undisturbed enclaves that protect examples of the world's most outstanding natural and cultural resources. The movement to establish these areas is a relatively recent attempt to achieve an improved balance between human activities and the earth. In recent years, rising incomes and lower costs for international travel have improved the accessibility of national parks to a larger number of persons, meaning that park visitation is likely to continue to rise.

Thomas A. Wikle

FOR FURTHER STUDY

Allin, Craig W. *International Handbook of National Parks and Nature Reserves.* Westport, Conn.: Greenwood Press, 1990.

Dickinson, Mary, ed. *National Parks of North America.* Washington, D.C.: National Geographic Society, 1995.

Grazzini, Giuseppe. *National Parks of the World.* New York: Crescent Books, 1991.

McQueen, Jane B. *The Complete Guide to America's National Parks.* Washington, D.C.: National Park Foundation, 1994.

INFORMATION ON THE WORLD WIDE WEB

The World Conservation Monitoring Centre, an organization dedicated to providing information about conservation and sustainable use of the world's living resources, maintains a Web site with a searchable database of national parks and protected areas worldwide. (www.wcmc.org.uk/data/database/un_combo.html)

NATURAL RESOURCES

SOILS

Soils are the loose masses of broken and chemically weathered rock mixed with organic matter that cover much of the world's land surface, except in polar regions and most deserts. The two major solid components of soil—minerals and organic matter—occupy about half the volume of a soil. Pore spaces filled with air and water account for the other half. A soil's organic material comes from the remains of dead plants and animals, its minerals from weathered fragments of bedrock. Soil is also an active, dynamic, ever-changing environment. Tiny pores in soil fill with air, water, bacteria, algae, and fungi working to alter the soil's chemistry and speed up the decay of organic material, making the soil a better living environment for larger plants and animals.

SOIL FORMATION. The natural process of forming new soil is slow. Exactly how long it takes depends on how fast the bedrock below is weathered. This weathering process is a direct result of a region's climate and topography, because these factors influence the rate at which exposed bedrock erodes and vegetation is distributed. Global variations in these factors account for the worldwide differences in soil types.

Climate is the principal factor in determining the type and rate of soil formation. Temperature and precipitation are the two main climatic factors that influence soil formation, and they vary with elevation and latitude. Water is the main agent of weathering, and the amount of water available depends on how much falls and how much runs off. The amount of precipitation and its distribution during the year influence the kind of soil formed and the rate at which it is formed. Increased precipitation usually results in increased rates of soil formation and deep soils. Temperature and precipitation also determine the kind and amount of vegetation in a region, which determines the amount of available organics.

Topography is a characteristic of the landscape involving slope angle and slope length. Topographic relief governs the amount of water that runs off or enters a soil. On flat or gently sloping land, soil tends to stay in place and may become thick, but as the slope increases so does the potential for erosion. On steep slopes, soil cover may be very thin, possibly only a few inches, because precipitation washes it downhill; on level plains, soil profiles may be several feet thick.

TYPES OF SOIL. Typically, bedrock first weathers to form regolith, a protosoil devoid of organic material. Rain, wind, snow, roots growing into cracks, freezing and thawing, uneven heating, abrasion, and shrinking and swelling break large rock particles into smaller ones. Weathered rock particles may range in size from clay to silt, sand, and gravel, with the texture and particle size depending largely on the type of bedrock. For example, shale yields finer-textured soils than sandstone. Soils formed from eroded limestone are rich in base minerals; others tend to be acidic. Generally, rates of soil formation are largely determined by the rates at which silicate minerals in the bedrock weather: the more silicates, the longer the formation time.

In regions where organic materials, such as plant and animal remains, may be deposited on top of regolith, rudimentary soils can begin to form. When waste material is excreted, or a plant or animal dies, the material usually ends up on the earth's surface. Organisms that cause decomposition, such as bacteria and fungi, begin breaking down the remains into a beneficial substance known as humus. Humus restores minerals and nutrients to the soil. It also improves the soil's structure, helping

it to retain water. Over time, a skeletal soil of coarse, sandy material with trace amounts of organics gradually forms. Even in a region with good weathering rates and adequate organic material, it can take as long as fifty years to form 12 inches (30 centimeters) of soil. When new soil is formed from weathering bedrock, it can take from one hundred to one thousand years for less than an inch of soil to accumulate.

Water moves continually through most soils, transporting minerals and organics downward by a process called leaching. As these materials travel downward, they are filtered and deposited to form distinct soil horizons. Each soil horizon has its own color, texture, and mineral and humus content. The O-horizon is a thin layer of rotting organics covering the soil. The A-horizon, commonly called topsoil, is rich in humus and minerals. The B-horizon is a subsoil rich in minerals but poor in hu-

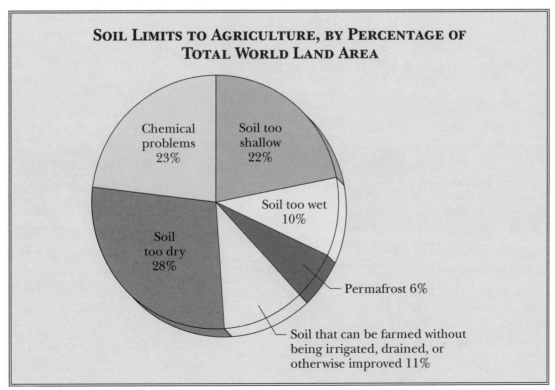

SOIL LIMITS TO AGRICULTURE, BY PERCENTAGE OF TOTAL WORLD LAND AREA

Chemical problems 23%

Soil too shallow 22%

Soil too wet 10%

Soil too dry 28%

Permafrost 6%

Soil that can be farmed without being irrigated, drained, or otherwise improved 11%

Source: United Nations Food and Agriculture Organization (FAOSTAT Database, 2000).

mus. The C-horizon consists of weathered bedrock; the D-horizon is the bedrock itself.

Because Earth's surface is made of many different rock types exposed at differing amounts and weathering at different rates at different locations, and because the availability of organic matter varies greatly about the planet due to climatic and seasonal conditions, soil is very diverse and fertile soil is unevenly distributed. Structure and composition are key

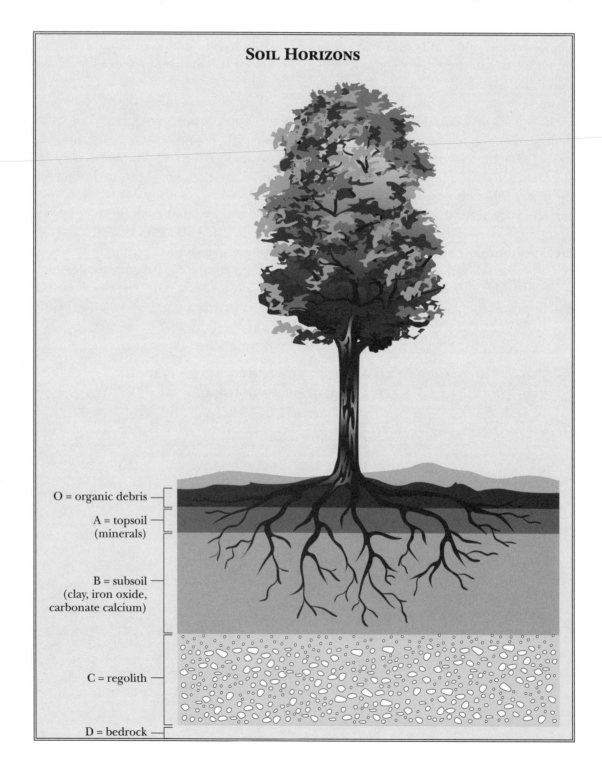

SOIL HORIZONS

O = organic debris

A = topsoil (minerals)

B = subsoil (clay, iron oxide, carbonate calcium)

C = regolith

D = bedrock

Toxic waste dump. (PhotoDisc)

factors in determining soil fertility. In a fertile soil, plant roots are able to penetrate easily to obtain water and dissolved nutrients. A loam is a naturally fertile soil, consisting of masses of particles from clays (less than 0.002 mm across), through silts (ten times larger) to sands (one hundred times larger), interspersed with pores, cracks, and crevices.

THE ROLES OF SOIL. In any ecosystem, soils play six key roles. First, soil serves as a medium for plant growth by mechanically supporting plant roots and supplying the eighteen nutrients essential for plants to survive. Different types of soil contain differing amounts of these eighteen nutrients; their combination often determines the types of vegetation present in a region, and as a result, influences the number and types of animals the vegetation can sup-

port, including humans. Humans rely on soil for crops necessary for food and fiber.

Second, the property of a particular soil is the controlling factor in how the hydrologic system in a region retains and transports water, how contaminants are stored or flushed, and at what rate water is naturally purified. Water enters the soil in the form of precipitation, irrigation, or snowmelt that falls or runs off soil. When it reaches the soil, it will either be surface water, which evaporates or runs into streams, or subsurface water, which soaks into the soil where it is either taken up by plant roots or percolates downward to enter the groundwater system. Passing through soil, organic and inorganic pollutants are filtered out, producing pure groundwater.

Soil also functions as an air-storage facility. Air is pushed into and drawn out of

*Adobe
buildings
Page 160*

the soil by changes in barometric pressure, high winds, percolating water, and diffusion. Pore spaces within soil provide access to oxygen to organisms living underground as well as to plant roots. Soil pore spaces also contain carbon dioxide, which many bacteria use as a source of carbon.

Soil is nature's recycling system, through which organic waste products and decaying plants and animals are assimilated and their elements made available for reuse. The production and assimilation of humus within soil converts mineral nutrients into forms that can be used by plants and animals, who return carbon to the atmosphere as carbon dioxide. While dead organic matter amounts to only about 1 percent of the soil by weight, it is a vital component as a source of minerals.

Soil provides a habitat for many living things, from insects to burrowing animals, from single microscopic organisms to massive colonies of subterranean fungi. Soils contain much of the earth's genetic diversity, and a handful of soil may contain billions of organisms, belonging to thousands of species. Although living organisms only account for about 0.1 percent of soil by weight, 2.5 acres (one hectare) of good-quality soil can contain at least 300 million small invertebrates—mites, millipedes, insects, and worms. Just 1 ounce (30 grams) of fertile soil can contain one million bacteria of a single type, one hundred million yeast cells, and fifty thousand fungus mycelium. Without these, soil

could not convert nitrogen, phosphorus, and sulphur to forms available to plants.

Finally, soil is an important factor in human culture and civilization. Soil is a building material used to make bricks, adobe, plaster, and pottery, and often provides the foundation for roads and buildings. Most important, soil resources are the basis for agriculture, providing people with their dietary needs.

Because the human use of soils has been haphazard and unchecked for millennia, soil resources in many parts of the world have been harmed severely. Human activities, such as overcultivation, inexpert irrigation, overgrazing of livestock, elimination of tree cover, and cultivating steep slopes, have caused natural erosion rates to increase many times over. As a result of mismanaged farm and forest lands, escalated erosional processes wash off or blow away an estimated seventy-five billion tons of soil annually, eroding away one of civilization's crucial resources.

Randall L. Milstein

FOR FURTHER STUDY

Brady, Nyle C. *The Nature and Properties of Soils.* Upper Saddle River, N.J.: Prentice Hall, 1999.

FitzPatrick, E. A. *An Introduction to Soil Science.* Essex, England: Longman Scientific and Technical, 1986.

Harpstead, Milo I., et al. *Soils Science Simplified.* Ames: Iowa State University Press, 1997.

Creation of Yellowstone National Park in 1872 inaugurated the practice of governments providing permanent protection to unique and scenically spectacular regions. (Corbis)

For thousands of years the people of the American Southwest have used the red clay soil to make adobe bricks—a building material with excellent insulating properties that is ideal for the region. (PhotoDisc)

Buildings and walls made almost entirely from mud bricks in the South American Andes. (Clyde L. Rasmussen)

PERCENTAGE OF ANNUAL DEFORESTATION BY COUNTRY, 1990-1995

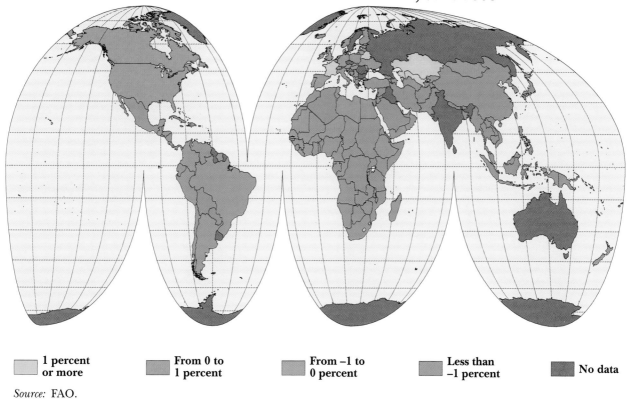

1 percent or more	From 0 to 1 percent	From −1 to 0 percent	Less than −1 percent	No data

Source: FAO.

One of the most famous ghost towns from California's pioneer mining days is Bodie, in the eastern Sierra Nevadas. (Corbis)

One of the challenges that human societies face is learning how to harness the power and resources of the natural environment without destroying it. (PhotoDisc)

162

In many mountainous parts of the world, people try to maximize the output of their land by terracing slopes to utilize every possible fragment of space. (PhotoDisc)

CAPITALS AND MAJOR CITIES OF THE WORLD

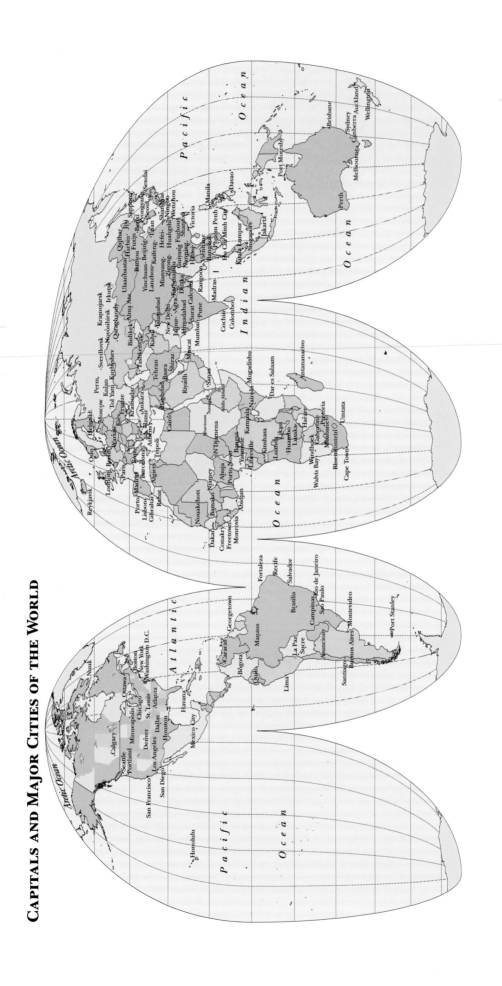

Orchard heaters are an example of a human attempt to control climate. Designed to keep valuable fruits from being spoiled during cold snaps, they make it possible to grow plants in regions where climatic conditions are generally good, but where a single period of freezing temperatures can destroy an entire season's crop. (PhotoDisc)

In populated areas, where many square miles of land are paved, rainwater flows into storm drains rather than being absorbed into the soil. When the storm drains reach capacity, flooding occurs. (PhotoDisc)

Modern chemical plant. In their rush to industrialize, many developing nations have begun generating substantial amounts of air pollution. However, the major industrial nations are the primary contributors to atmospheric pollution. North America, Europe, and East Asia produce 60 percent of the world's air pollution. (PhotoDisc)

View of Southern California's coast, from Los Angeles to San Diego (lower right) photographed from the space shuttle Discovery *in April, 1991. The haze between Catalina Island and the mainland is a smog cloud.* (Corbis)

Modern freeway interchange. Passage of the Federal-Aid Highway Act in 1956 was a turning point in the history of highway transportation in the United States. It marked the beginning of the largest peacetime public works program in the history of the world, creating a 41,000-mile National System of Interstate and Defense Highways. (PhotoDisc)

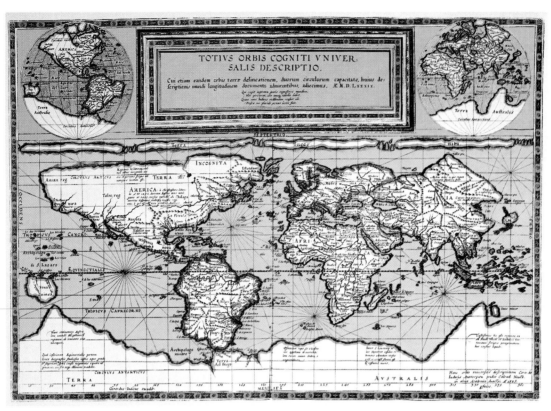

In 1589—nearly a century after Christopher Columbus opened the New World to exploration—the accuracy of details on European maps diminished with their distance from Western Europe. (Corbis)

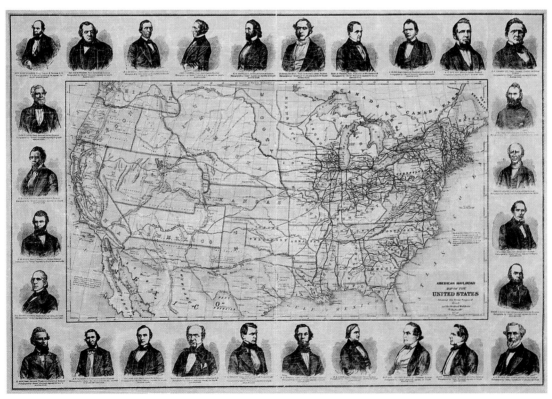

Nineteenth century map of the United States showing three different proposed routes for the first transcontinental railroad. (Corbis)

Washington, D.C., one of the last major world capitals to build a subway system, finally opened its own subway in 1976. Drawing on lessons learned in other systems, it built one of the most attractive and efficient systems in the world. (PhotoDisc)

Slash-and-burn agriculture takes its name from the practice of clearing land for planting by cutting down the trees and brush and burning the fallen materials to fertilize the soil with the ashes of the burned materials. (PhotoDisc)

Jet planes only slowly entered the commercial airline business after the mid-1950s, but by the 1970's jets accounted for the majority of passenger miles in the air. (PhotoDisc)

The romance of cowboys riding the range and driving cattle to market has become largely a thing of the past, as modern methods of livestock management have evolved. (PhotoDisc)

INCREASES IN LAND USED FOR AGRICULTURE, BY COUNTRY, 1980-1994

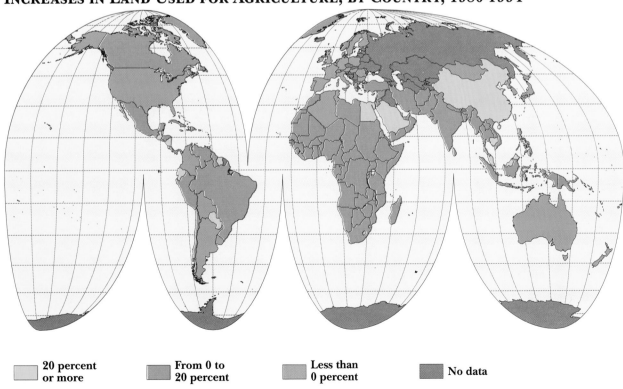

20 percent or more   From 0 to 20 percent   Less than 0 percent   No data

A distant relative of the camel, the South American llama was one of the few draft animals used in the pre-Columbian Americas and was also an important source of protein before Europeans brought cattle, sheep, and other animals. (Clyde L. Rasmussen)

Offshore oil rig. Continental margins are the principal areas in which offshore drilling is conducted; they constitute approximately 21 percent of the surface area of the oceans and may contain a majority of the world's reserves of oil and gas. (PhotoDisc)

The Soviet Union so revered nuclear fission that its government erected a monument to the splitting of the atom.
(PhotoDisc)

Firewood is the oldest and most widespread fuel and remains an important source of heat in most parts of the world.
(Digital Stock)

Solar panels used to collect sunlight to generate electric power.
(PhotoDisc)

WATER

Life on earth requires water—without it, life on earth would cease. As human populations grow, the freshwater resources of the world become scarcer and more polluted, while the need for clean water increases. Although nearly three-quarters of the earth's surface is covered with water, only about 0.3 percent of that water is freshwater suitable for consumption and irrigation. This is because more than 97 percent of the earth's water is ocean salt water, and most of the remaining freshwater is frozen in the Antarctic ice cap. Only the small amounts that remain in lakes, rivers, and groundwater is available for human use.

All of earth's water cycles between the ocean, land, atmosphere, plants, and animals over and over. On average, a molecule of surface water cycles from the ocean, to the atmosphere, to the land and back again in less than two weeks. Water consumed by plants or animals takes longer to return to the oceans, but eventually the cycle is completed.

WATER'S USES. Water supports the lives of all living creatures. People use it not only for drinking, cooking, and bathing, but also travel on it, make electricity with it, fish in it, irrigate crops with it, and use it for recreation. Globally, more than 3,240 million acre-feet of water is used, most of it for agriculture. Of the freshwater used, agriculture accounts for 69 percent, industry uses 23 percent, and domestic and municipal activities use 9 percent. Among agricultural uses, it takes 11 gallons of water to grow 1 serving of broccoli, 56 gallons to produce a serving of cheese, and 2,510 gallons for a pound of beef. In industry, 151 gallons of water are needed to make one Sunday newspaper and 65,257 gallons are used in the manufacture of the average car.

An average Westerner living in an urban setting uses approximately 159 gallons per day for personal and domestic uses, such as washing, cooking, and watering the lawn. As the world's population grows, the demand for freshwater will also increase. A study by the World Bank concluded that approximately 80 percent of human illness results from insufficient water supplies and poor water quality caused by lack of sanitation, so careful management of water resources is essential for improving the health of people in the twenty-first century.

GROUNDWATER SUPPLY AND QUALITY. The amount of groundwater in the earth is seventy times greater than all of the freshwater lakes combined. Groundwater is held within the rocks below the ground surface and is the primary source of water in many parts of the world. In the United States, approximately 50 percent of the population uses some groundwater. However, problems with both groundwater supplies and its quality threaten its future use.

The U.S. Environmental Protection Agency (EPA) found that 45 percent of the large public water systems in the United States that use groundwater were contaminated with synthetic organic chemicals that posed potential health threats. Another major problem occurs when groundwater is used faster than it is replaced by precipitation infiltrating through the ground surface. Many of the arid regions of earth are already suffering from this problem. For example, one-third of the wells in Beijing, China, have gone dry due to overuse. In the United States, the Ogallala Aquifer of the Great

THE WORLD AND NORTH AMERICA'S GREATEST RIVERS AND LAKES

	World	North America
Longest river	Nile (North Africa) 4,130 miles (6,600 km.)	Missouri-Mississippi (United States) 3,740 miles (6,000 km.)
Largest river by average discharge	Amazon (South America) 6,181,000 cubic feet/second (175,000 cubic meters/second)	Missouri-Mississippi (United States) 600,440 cubic feet/second (17,000 cubic meters/second)
Largest freshwater lake by volume	Lake Baikal (Russia) 5,280 cubic miles (22,000 cubic km.)	Lake Superior 3,000 cubic miles (12,500 cubic km.)

Plains, the largest in North America, is being severely overused. This aquifer irrigates 20 percent of U.S. farmland, and one-fourth of this groundwater resource is expected to be gone by 2020.

SURFACE WATER SUPPLY AND QUALITY. Surface water is used for transportation, recreation, electrical generation, and consumption. Ships use rivers and lakes as transport routes, people fish and boat on rivers and lakes, and dams on rivers often are used to generate electricity. The largest river on earth is the Amazon in South America, which has an aver-

The water consumed in most urban centers passes through treatment plants that remove impurities and chemically treat the water to kill potentially harmful organisms. (PhotoDisc)

age flow of 212,500 cubic meters per second, more than twelve times greater than North America's Mississippi River. Earth's largest lake—Lake Baikal in Russia—has a volume of approximately 5,280 cubic miles (22,000 cubic km.), equal to the volume of all five of North America's Great Lakes combined.

Although surface water has more uses, it is more prone to pollution than groundwater. Almost every human activity affects surface water quality. For example, water is used to create paper for books, and some of the chemicals used in the paper process are discharged into surface water sources. Most foods are grown with agricultural chemicals, which can contaminate water sources. In 1994 the EPA reported that approximately 44 percent of U.S. lakes and 37 percent of U.S. rivers are unsafe for fishing and swimming.

EARTH'S FUTURE WATER SUPPLY. Inadequate water supplies and water quality problems threaten the lives of more than one billion people worldwide. The World Health Organization estimates that polluted water causes the death of fifteen million children under five years of age each year and affects the health of 20 percent of the earth's population. As the world's population grows, these problems are likely to worsen.

The United Nations estimates that if current consumption patterns continue, two-thirds of the world's people will live in water-stressed conditions by 2025. Since access to clean freshwater is essential to health and a decent standard of living, efforts must be made to clean up and conserve the planet's freshwater, or billions of people in the twenty-first century will be negatively affected.

Mark M. Van Steeter

FOR FURTHER STUDY

Berner, E. K., and R. A. Berner. *The Global Water Cycle.* Englewood Cliffs, N.J.: Prentice-Hall, 1987.

Bowen, Robert. *Groundwater.* 2d ed. New York: Elsevier, 1986.

Clarke, Robin. *Water: The International Crisis.* Cambridge, Mass.: MIT Press, 1993.

Gardner, Gary. "From Oasis to Mirage: The Aquifers That Won't Replenish." *World Watch* (May/June, 1995): 30-36, 40-41.

Gleick, Peter H. *Water in Crisis: A Guide to the World's Fresh Water Resources.* New York: Oxford University Press, 1993.

Huang, P. M., and Iskandar Karam, eds. *Soils and Groundwater Pollution and Remediation: Asia, Africa, and Oceania.* Boca Raton, Fla.: Lewis, 2000.

Kovar, Karel, et al., eds. *Hydrology, Water Resources, and Ecology in Headwaters.* Wallingford, England: IAHS, 1998.

Leeden, F. V. D., F. L. Troise, and D. K. Todd, eds. *The Water Encyclopedia.* 2d ed. Chelsea, Mich.: Lewis, 1990.

Leopold, Luna B. *Water, Rivers, and Creeks.* Sausalito, Calif.: University Science Books, 1997.

Opie, John. *Ogallala: Water for a Dry Land.* Lincoln: University of Nebraska Press, 1993.

Perry, J., and E. Vanderklein. *Water Quality: Management of a Natural Resource.* Cambridge, Mass.: Blackwell, 1996.

Petersonn, Margaret. *River Engineering.* Englewood Cliffs, N.J.: Prentice-Hall, 1986.

Pielou, E. C. *Fresh Water.* Chicago: University of Chicago Press, 1998.

Thompson, Stephen A. *Water Use, Management, and Planning in the United States.* San Diego, Calif.: Academic Press, 1999.

RENEWABLE RESOURCES

Deforestation Page 161

Most renewable resources are living resources, such as plants, animals, and their products. With careful management, human societies can harvest such resources for their own use without imperiling future supplies. However, human history has seen many instances of resource mismanagement that has led to the virtual destruction of valuable resources.

FORESTS. Forests are large tracts of land supporting growths of trees and perhaps some underbrush or shrubs. Trees constitute probably the earth s most valuable, versatile, and easily grown renewable resource. When they are harvested intelligently, their natural environments continue to replace them. However, if a harvest is beyond the environment's ability to restore the resource that had been present, new and different plants and animals will take over the area. This phenomenon has been demonstrated many times in overused forests and grasslands that reverted to scrubby brushlands. In the worst cases, the abused lands degenerated into barren deserts.

The forest resources of the earth range from the tropical rain forests with their huge trees and broad diversity of species to the dry savannas featuring scattered trees separated by broad grasslands. Cold, subarctic lands support dense growths of spruces and firs, while moderate temperature regimes produce a variety of pines and hardwoods such as oak and ash. The forests of the world cover about 29 percent of the land surface, as compared with the oceans, which cover about 71 percent of the global surface.

Harvested wood, cut in the forest and hauled away to be processed, is termed roundwood. Globally, the cut of roundwood for all uses amounts to about 122.2 billion cubic feet (3.5 billion cubic meters). Slightly more than half of the harvested wood is used for fuel, including charcoal.

Roundwood that is not used for fuel is described as industrial wood and used to produce lumber, veneer for fine furniture, and pulp for paper products. Some industrial wood is chipped to produce such products as subflooring and sheathing board for home and other building construction. Most roundwood harvested in Africa, South America, and Asia is used for fuel. In contrast, roundwood harvested in North America, Europe, and the former Soviet Union generally is produced for industrial use.

It is easy to consider forests only in the sense of the useful wood they produce. However, many forests also yield valuable resources such as rubber, edible nuts, and what the U.S. Forest Service calls special forest products. These include ferns, mosses, and lichens for the florist trade, wild edible mushrooms such as morels and matsutakes for domestic markets and for export, and mistletoe and pine cones for Christmas decorations.

There is growing interest among the industrialized nations of the world in a unique group of forest products for use in the treatment of human disease. Most of them grow in the tropical rain forests. These medicinal plants have long been known and used by shamans (traditional medicine men). Hundreds of pharmaceutical drugs, first used by shamans, have been derived from plants, many gathered in tropical rain forests. The drugs include

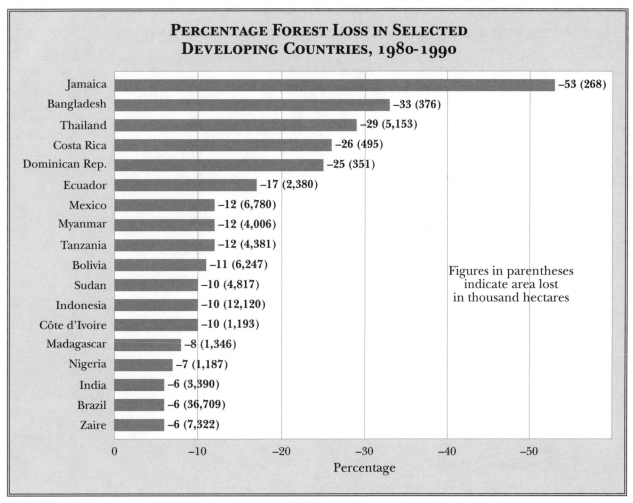

PERCENTAGE FOREST LOSS IN SELECTED DEVELOPING COUNTRIES, 1980-1990

Jamaica −53 (268)
Bangladesh −33 (376)
Thailand −29 (5,153)
Costa Rica −26 (495)
Dominican Rep. −25 (351)
Ecuador −17 (2,380)
Mexico −12 (6,780)
Myanmar −12 (4,006)
Tanzania −12 (4,381)
Bolivia −11 (6,247)
Sudan −10 (4,817)
Indonesia −10 (12,120)
Côte d'Ivoire −10 (1,193)
Madagascar −8 (1,346)
Nigeria −7 (1,187)
India −6 (3,390)
Brazil −6 (36,709)
Zaire −6 (7,322)

Figures in parentheses indicate area lost in thousand hectares

Percentage

Source: United Nations Food and Agriculture Organization (FAOSTAT Database, 2000).

quinine, from the bark of the cinchona tree, long used to combat malaria, and the alkaloid drug reserpine. Reserpine, derived from the roots of a group of tropical trees and shrubs, is used to treat high blood pressure (hypertension) and as a mild tranquilizer. It has been estimated that 25 percent of all prescriptions dispensed in the United States contain ingredients derived from tropical rain forest plants. The value of the finished pharmaceuticals is estimated at $6.25 billion per year.

Scientists screening tropical rain forest plants for additional useful medical compounds have drawn on the knowledge and experience of the shamans. In this way, the scientists seek to reduce the search time and costs involved in screening potentially useful plants. Researchers hope that somewhere in the dense tropical foliage are plant products that could treat, or perhaps cure, diseases such as cancer or AIDS.

Many as-yet-undiscovered medicinal plants may be lost forever as a consequence of deforestation of large tracts of equatorial land. The trees are cut down or burned in place and the forest converted to grassland for raising cattle. The tropical soils cannot support grasses without the input of large amounts of fertilizer. The destruction of the forests also causes flooding, leaving standing pools of water

POPULATION INCREASE, DEFORESTATION, AND RESULTS

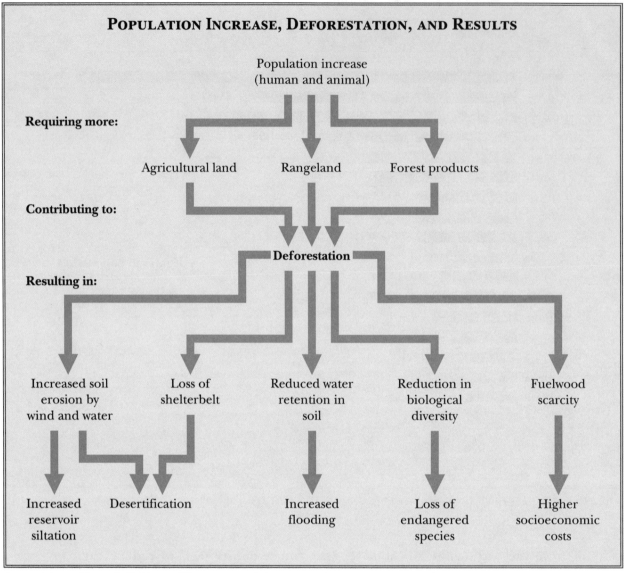

Source: Adapted from A. K. Biswas, "Environmental Concerns in Pakistan, with Special Reference to Water and Forests," in *Environmental Conservation*, 1987.

and breeding areas for mosquitoes, which can spread malaria and yellow fever.

MARINE RESOURCES. When renewable marine resources such as fish and shellfish are harvested or used, they continue to reproduce in their environment, as happens in forests and with other living natural resources. However, like overharvested forests, if the marine resource is overfished—that is, harvested beyond its ability to reproduce—new, perhaps undesirable, kinds of marine organisms will occupy the

area. This has happened to a number of marine fishes, particularly the Atlantic cod.

When the first Europeans reached the shores of what is now New England in the early seventeenth century, they encountered vast schools of cod in the local ocean waters. The cod were so plentiful they could be caught in baskets lowered into the water from a boat.

At the height of the New England cod fishery, in the 1970's, efficient, motor-driven trawlers were able to catch about 32,000 tons. The catch began to decline that year, mostly as a result of the impact of fifteen different nations fishing on the cod stocks. As a result of overfishing, rough species such as dogfish and skates constitute 70 percent of the fish in the local waters. Experts on fisheries management decided that fishing for cod had to be stopped.

The decline of the cod was attributed to two causes: a worldwide demand for more fish as food and great changes in the technology of fishing. The technique of fishing progressed from a lone fisher with a baited hook and line, to small steam-powered boats towing large nets, to huge diesel-powered trawlers towing monster nets that could cover a football field. Some of the largest trawlers were floating factories. The cod could be skinned, the edible parts cut and quick-frozen for market ashore, and the skin, scales, and bones cooked and ground for animal feed and oil. A lone fisher was lucky to be able to catch 1,000 pounds (455 kilograms) in one day. In contrast, the largest trawlers were capable of catching and processing 200 tons per day.

The world ocean population of swordfish has also declined dramatically. With a worldwide distribution, these large members of the billfish family have been eagerly sought after as a food fish. Because swordfish have a habit of basking at the surface, fishermen learned to sneak up on the swordfish and harpoon them. Advances in technology led to the doom of the swordfish. Fishermen began to catch

Indiscriminate use of huge fishing nets threatens the survival of other animal species, including dolphins. (PhotoDisc)

swordfish with fishing lines 25 to 40 miles (40 to 65 kilometers) long. Baited hooks hung at intervals on the main line successfully caught many swordfish, as well as tuna and large sharks. Whereas the harpoon fisher took only the largest (thus most valuable) swordfish, the longline gear was indiscriminate, catching and killing many swordfish too small for the market, as well as sea turtles and dolphins

As a result of the catching and killing of both sexually mature and immature swordfish, the reproductive capacity of the species was greatly reduced. Harpoons killed mostly the large, mature adults that had spawned several times. Longlines took all sizes of swordfish, including the small ones that had not yet reached sexual maturity and spawned. The decline of the swordfish population was quickly obvious in the reduced landings. In one seven-year period, swordfish landings off the east coast of the United States dropped by almost 60 percent. At the same time, the number of longline hooks set in the same area increased by 70 percent.

As a gesture of support for restoring swordfish stocks in the world's oceans, many restaurants in the United States voluntarily removed swordfish from their menus. It is hoped this action will force the United States government, and perhaps other nations, to develop an effective recovery plan for the swordfish.

Albert C. Jensen

FOR FURTHER STUDY

Kunzig, Robert. "Twilight of the Cod." *Discover* 16, no. 4 (April, 1995): 44-60.
Kurlansky, Mark. *Cod: A Biography of the Fish That Changed the World.* New York: Walker, 1997.

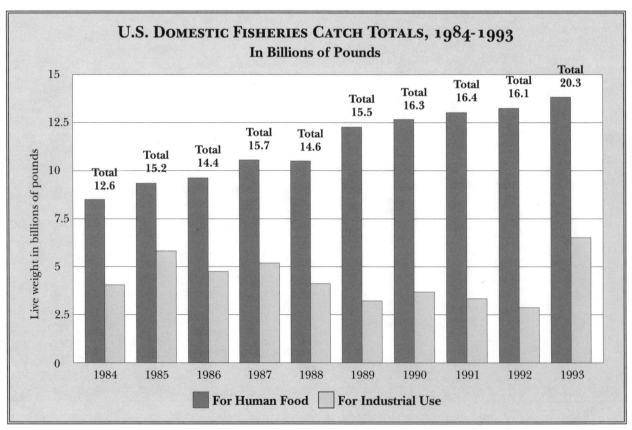

Source: U.S. Department of Commerce, *Statistical Abstract of the United States, 1996*, 1996.

McKibben, Bill. "What Good Is a Forest?" *Audubon* 98, no. 3 (May-June, 1996): 53-63.

Mark, Wesley. *The Oceans: Our Last Resource.* San Francisco: Sierra Club, 1981.

Nollman, Jim. *The Charged Border: Where Whales and Humans Meet.* New York: Henry Holt, 1998.

Safina, Carl. *Song for the Blue Ocean: Encounters with the World's Coasts and Beneath the Seas.* New York: Henry Holt, 1997.

NONRENEWABLE RESOURCES

Nonrenewable resources are useful raw materials that exist in fixed quantities in nature and cannot be replaced. They differ from renewable resources, such as trees and fish, which can be replaced if managed correctly. Most nonrenewable resources are minerals—inorganic and organic substances that exhibit consistent chemical composition and properties. Minerals are found naturally in the earth's crust or dissolved in seawater. Of roughly two thousand different minerals, about one hundred are sources of raw materials that are needed for human activities. Where useful minerals are found in sufficiently high concentrations—that is, as ores—they can be mined as profitable commercial products.

Economic nonrenewable resources can be divided into four general categories: metallic (hardrock) minerals, which are the source of metals such as iron, gold, and copper; fuel minerals, which include petroleum (oil), natural gas, coal, and uranium; industrial (soft rock) minerals, which provide materials like sulfur, talc, and potassium; and construction materials, such as sand and gravel.

Nonrenewable resources are required as direct or indirect parts of all the products that humans use. For example, metals are necessary in industrial sectors such as construction, transportation equipment, electrical equipment and electronics, and consumer durable goods—long-lasting products such as refrigerators and stoves. Fuel minerals provide energy for transportation, heating, and electrical power. Industrial minerals provide ingredients needed in products ranging from baby powder to fertilizer to the space shuttle. Construction materials are used in roads and buildings.

LOCATION. When minerals have naturally combined together (aggregated) they are called rocks. The three general rock categories are igneous, sedimentary, and metamorphic. Igneous rocks are created by the cooling of molten material (magma). Sedimentary rocks are caused when weathering, erosion, transportation, and compaction or cementation act on existing rocks.

Metamorphic rocks are created when the other two types of rock are changed by heat and pressure. The availability of nonrenewable resources from these rocks varies greatly, because it depends not only on the natural distribution of the rocks but also on people's ability to discover and

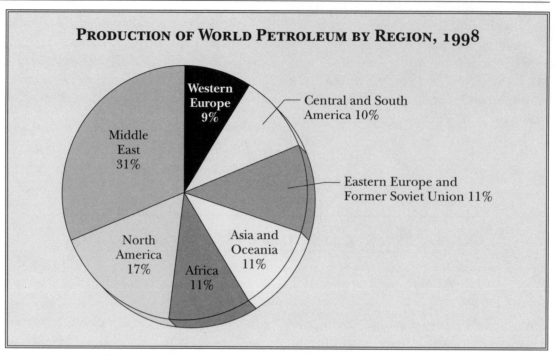

PRODUCTION OF WORLD PETROLEUM BY REGION, 1998

Western Europe 9%

Central and South America 10%

Middle East 31%

Eastern Europe and Former Soviet Union 11%

North America 17%

Asia and Oceania 11%

Africa 11%

Source: *International Energy Annual*, Energy Information Administration, U.S. Department of Energy, 1998.

process them. It is difficult to find rock formations that are covered by the ocean, material left by glaciers, or a rain forest. As a result, nonrenewable resources are distributed unevenly throughout the world.

Some nonrenewable resources, such as construction materials, are found easily around the world and are available almost everywhere. Other nonrenewable resources can only be exploited profitably when the useful minerals have an unusually high concentration compared with their average concentration in the earth's crust. These high concentrations are caused by rare geological events and are difficult to find. For example, an exceptionally rare nonrenewable resource like platinum is produced in only a few limited areas.

No one country or region is self-sufficient in providing all the nonrenewable resources it needs, but some regions have many more nonrenewable resources than others. Minerals can be found in all types of rocks, but some types of rocks are more

likely to have economic concentrations than others. Metallic minerals often are associated with shields (blocks) of old igneous (Precambrian) rocks. Important shield areas near the earth's surface are found in Canada, Siberia, Scandinavia, and Eastern Europe. Another important shield was split by the movement of the continents, and pieces of it can be found in Brazil, Africa, and Australia.

Similar rock types are in the mountain formations in Western Europe, Central Asia, the Pacific coast of the Americas, and Southeast Asia. Minerals for construction and industry are found in all three types of rocks and are widely and randomly distributed among the regions of the world.

The fuel minerals—petroleum and natural gas—are unique in that they occur in liquid and gaseous states in the rocks. These resources must be captured and collected within a rock site. Such a site needs source rock to provide the resource, a rock type that allows the resource to collect, and another surrounding rock type that traps the resource. Sedimentary rock basins are particularly good sites for fuel collection. Important fuel-producing regions are the Middle East, the Americas, and Asia.

IMPACT ON HUMAN SETTLEMENT. Nonrenewable resources have always provided raw materials for human economic development, from the flint used in early stone tools to the silicon used in the sophisticated chips in personal computers. Whole eras of human history and development have been linked with the nonrenewable resources that were key to the period and its events. For example, early human culture eras were called the Stone, Bronze, and Iron Ages.

Political conflicts and wars have occurred over who owns and controls nonrenewable resources and their trade. One recent example is the Persian Gulf War of

Gold in its native state. (U.S. Geological Survey)

1991. Many nations, including the United States, fought against Iraq over control of petroleum production and reserves in the Middle East.

Since the actual production sites often are not attractive places for human settlement and the output is transportable,

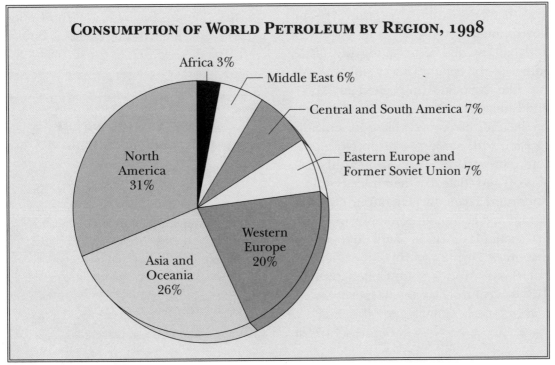

CONSUMPTION OF WORLD PETROLEUM BY REGION, 1998

- Africa 3%
- Middle East 6%
- Central and South America 7%
- Eastern Europe and Former Soviet Union 7%
- North America 31%
- Western Europe 20%
- Asia and Oceania 26%

Source: International Energy Annual, Energy Information Administration, U.S. Department of Energy, 1998.

these sites are seldom important population centers. There are some exceptions, such as Johannesburg, South Africa, which grew up almost solely because of the gold found there. However, because it is necessary to protect and work the production sites, towns always spring up near the sites. Examples of such towns can be found near the quarries used to provide the material for the great monuments of ancient Egypt and in the Rocky Mountains of North America near gold and silver mines. These towns existed because of the nonrenewable resources nearby and the needs of the people exploiting them; once the resource was gone, the towns often were abandoned, creating "ghost towns," or had to find new purposes, such as tourism.

Ghost town
Page 161

Ghost town
Page 161

More important to human settlement is the control of the trade routes for nonrenewable resources. Such controlling sites often became regions of great wealth and political power as the residents taxed the products that passed through their community and provided the necessary services and protection for the traveling traders. Just one example of this type of development is the great cities of wealth and culture that arose along the trade routes of the Sahara Desert and West Africa like Timbuktu (in present-day Mali) and Kumasi (in present-day Ghana) based on the trade of resources like gold and salt.

Even with modern transportation systems, ownership of nonrenewable resources and control of their trade is still an important factor in generating national wealth and economic development. Modern examples include Saudi Arabia's oil resources, Egypt's control of the Suez Canal, South Africa's gold, Chile's copper, Turkey's control over the Bosporus Strait, and Indonesia's metals and oil.

Gary A. Campbell

FOR FURTHER STUDY

Bailey, Ronald, ed. *The True State of the Planet.* New York: Simon & Schuster, 1995.

Brower, John. *Sustainability and Environmental Economics.* Redwood City, Calif.: Benjamin/Cummings, 1998.

Chiras, Daniel. *Environmental Science: Action for a Sustainable Future.* 4th ed. Redwood City, Calif.: Benjamin/Cummings, 1994.

Coyne, Mark, and Craig Allin, eds. *Natural Resources.* Pasadena, Calif.: Salem Press, 1998.

Craig, J. R., D. J. Vaughan, and B. J. Skinner. *Resources of the Earth.* Englewood Cliffs, N.J.: Prentice-Hall, 1988.

Crowson, Phillip. *Inside Mining: The Economics of the Supply and Demand of Minerals and Metals.* London: Mining Journal Books, 1998.

Kesler, Stephen E. *Mineral Resources, Economics, and the Environment.* New York: Maxwell Macmillan International, 1994.

Skinner, Brian J., et al. *Resources of the Earth.* Englewood Cliffs, N.J.: Prentice-Hall, 1988.

Strauss, Simon. *Trouble in the Third Kingdom.* London: Mining Journal Books, 1986.

INFORMATION ON THE WORLD WIDE WEB

A good general source of information on nonrenewable resources is the U.S. Geological Survey's (USGS) Web site for Commodity Statistics and Information, which provides articles on the worldwide supply, demand, and use of minerals and other materials. (minerals.usgs.gov/minerals/pubs/commodity/)

HUMAN GEOGRAPHY

Human Society and the Earth

The Human Environment

No person lives in a vacuum. Every human being and community is surrounded by a world of external influences with which it interacts and by which it is affected. In turn, humans influence and change their environments: sometimes intentionally, sometimes not, and sometimes with effects that are harmful to these environments, and, in turn, to humans themselves. As the only tool-creating animal, humans have always shaped the world in which they live, but developments over the past few centuries have greatly enhanced this capacity.

During the last decades of the twentieth century, people became alarmed over the effects of modern technology and accelerating human population growth in the world. Travel and transportation among the world's regions have been made surer, safer, and faster, and global communication is virtually instantaneous. The human environment is no longer a matter of local physical, biological, or social conditions, or even of merely national or regional concerns—the postmodern world has become a true global community.

Students of human geography divide the human environment into three broad areas: the physical, biological, and social environments. The study of ecology describes and analyzes the interactions of biological forms (mainly plants and animals) and seeks to uncover the optimal means of species cooperation, or symbiosis. Everything that humans do affects life and the physical world around them, and this world provides potentials for and constraints on how humans can live.

As people gained and communicated ever-greater knowledge about the world, their abilities to alter and shape it increased. Even ten thousand years ago, people cut down trees, scratched the earth's surface with simple plows, and replaced diverse plant forms with single crops. From this basic agricultural technology grew more complex human communities, and people were freed from the need to hunt and gather. The alteration of the local ecosystems could have deleterious effects, however, as gardens turned eventually to deserts in places like North Africa and what later became Iraq. Those who kept herds of animals grazed them in areas rich in grasses, and animal fertilizer helped keep them rich. Still, the herders moved on when their animals overgrazed, leaving erosion and even desertification in their wake. Modern people have a far greater ability to alter their environments than did

Mining town
Page 162

Neolithic people, and ecologists are concerned about the negative effects of modern alterations.

THE PHYSICAL ENVIRONMENT. The earth's biosphere is made up of the atmosphere—the mass of air surrounding the earth; the hydrosphere—bodies of water; and the lithosphere—the outer portion of the earth's crust. Each of these, alone and working together, affect human life and human communities.

Storm damage Pages 104, 105

Climate and weather at their most extreme can make human habitation impossible, or at least extremely uncomfortable. Desert and polar climates do not have the liquid water, vegetation, and animal life necessary to sustain human existence.

THE ENVIRONMENTAL DEBATE

Scientists, politicians, and businessmen have long debated the nature, extent, and future of environmental degradation. Most alarming was the 1972 report by the Club of Rome entitled *Limits to Growth.* Computer projections of population growth and the overuse of natural resources painted a bleak picture of overcrowding and ecological disaster. Two decades later, the group published *Beyond the Limits,* which modified the earlier projections but retained its pessimism in the absence of major changes in human behavior. U.S. vice president Al Gore, in his book *Earth in the Balance* (1992), also painted a grim picture, which he believed could be averted only by large-scale changes in the industrialized world's consumption habits.

Optimists, led by economists and social scientists and bolstered by scientific advances and forecasting, refuted or minimized the dire warnings of pessimistic ecologists. Their works emphasize the resilience of the natural and biological environments rather than their susceptibility to human interference. Julian L. Simon and Herman Kahn's book *The Resourceful Earth* (1984) emphasizes human progress in alleviating pollution, increasing food supplies, and reducing birthrates worldwide.

Elsewhere, people must adjust to even mild variations in temperature and precipitation, and do so with clothing and shelter. Excess rain can be drained off, and arid areas irrigated. Heating and, more recently, air conditioning can create healthy microclimates, whatever the external conditions. Most people live in temperate zones where weather extremes are rare or dealt with by technological adaptation. Food can be grown locally, and transportation is effective throughout the year. Local droughts, tornadoes, hurricanes, heavy winds, lightning, and hail can have devastating effects even in the most comfortable of climates.

The hydrosphere affects the atmosphere in countless ways, and provides the water so necessary for human and other life. Bodies of water provide plants and animals for food, transportation routes, and aesthetic pleasure to people, and often serve to flush away waste products. People locate near water sources for all of these reasons, but sometimes suffer from sudden shifts in the water level, as in tidal waves (tsunamis) or flooding. Encroachment of salt water into freshwater bodies (salination) is a problem that can have natural or human causes.

The lithosphere provides the solid, generally dry surface on which people usually live. It has been shaped by the atmosphere (especially wind and rain that erode rocks into soil) and the hydrosphere (for example, alluvial deposits and beach erosion). It serves as the base for much plant life and for most agriculture. People have tapped its mineral deposits and reshaped it in many places; it also reshapes itself through, for example, earthquakes and volcanic eruption. Its great variations—including vegetation—draw or repel people, who exploit or enjoy them for reasons as varied as recreation, military defense, or farming.

190

THE BIOLOGICAL ENVIRONMENT. Humans share the earth with something between five and thirty million different species of plants, animals, and microorganisms—about about two million of which have been identified and named. As part of the natural food chain, people rely upon other life forms for nourishment. Through perhaps the first 99 percent of human history, people harvested the bounty of nature in its native setting, by hunting and gathering.

Domestication of plants and animals, beginning about ten thousand years ago, provided humans a more stable and reliable food supply, revolutionizing human communities. Being omnivores, people can use a wide variety of plants and animals for food, and they have come to control or manage most important food sources through herding, agriculture, or mechanized harvesting. Which plants and animals are chosen as food, and thus which are cultivated, bred, or exploited, are matters of human culture, not, at least in the modern world, of necessity.

Huge increases in human population worldwide have, however, put tremendous strains on provision of adequate nourishment. Areas poorly endowed with foodstuffs or that suffer disastrous droughts or blights may benefit from the importation of food in the short run, but cannot sustain high populations fostered by medical advances and cultural considerations.

Human beings themselves are also hosts to myriad organisms, such as fungi, viruses, bacteria, eyelash mites, worms, and lice. While people usually can coexist with these, at times they are destructive and even fatal to the human organism. Public health and medical efforts have eradicated some of humankind's biological enemies, but others remain and baffle modern science.

The presence of these enemies to health once played a major role in locating human habitations to avoid so-called "bad air" (*mal-aria*) and the breeding grounds of tsetse flies or other pests. The use of pesticides and draining of marshy grounds have alleviated a good deal of human suffering. Human efforts can also control or eliminate biological threats to the plants and animals used for food, clothing, and other purposes.

SOCIAL ENVIRONMENTS. Human reproduction and the nurturing of young require cooperation among people. Over time, people gathered in groups that were diverse in age if not in other qualities, and the development of towns and cities eventually created an environment in which otherwise unrelated people interacted on intimate and constructive levels. Specialization, or division of labor, created a higher level of material wealth and culture and ensured interpersonal reliance.

The pooling of labor—both voluntary and forced—allowed for the creation of artificial living environments that defied the elements and met human needs for sustenance. Some seemingly basic human drives of exclusivity and territoriality may be responsible for interpersonal friction, violence and, at the extreme, war. Physical differences, such as size, skin, or hair color, and cultural differences, including language, religion, and customs, have often divided humans or communities. Even within close quarters such as cities, people often separate themselves along lines of perceived differences. Human social identity comes from shared characteristics, but which things are seen as shared, and which as differentiating, is arbitrary.

People can affect their social environment for good and ill through trade and war, cooperation and bigotry, altruism and greed. While people still are somewhat at the mercy of the biological and physical environments, technological developments have balanced the human relationship with these. Negative effects of human interaction, however, often offset the positive gains. People can seed clouds for rain, but also pollute the atmosphere around large cities, create acid rain, and perhaps contribute to global warming and depletion of the ozone layer around the earth.

Human actions can direct water to where it is needed, but people also drain freshwater bodies and increase salination, pollute streams, lakes, and oceans, and encourage flooding by modifying river beds. People have terraced mountainsides and irrigated them to create gardens in mountains and deserts, but also lose about 75 billion metric tons of soil to erosion and 15 million acres (6 million hectares) of grazing land to desertification each year. These negative effects not only jeopardize other species of terrestrial life, but also humans' ability to live comfortably, or perhaps at all.

Terraces
Page 163

GLOBALIZATION. Humankind's ability to affect its natural environments has increased enormously in the wake of the Industrial Revolution. The harnessing of steam, chemical, electrical, and atomic energy has enabled people to transform life on a global scale. Economically, the Western world has come to dominate global markets, and computer and satellite technology have made even remote parts of the globe reliant on Western information and products. Efficient transportation of goods and people over huge distances has eliminated physical barriers to travel and commerce. The power and influence of multinational corporations, and of national corporations in international mar-

kets, has become great. With the Internet, a mastery of basic English is almost essential, and global news networks based in the United States further unite the peoples of the earth.

Human environmental problems also have a global scope: Ozone depletion, changes in ocean temperatures, global warming, and the spread of disease by travelers have become planetary concerns. International agencies seek to deal with such matters, and also social and political concerns once left to nations or colonial powers, such as population growth, the provision of justice, or environmental destruction within a country. Pessimists warn of horrendous trends in population and ecological damage, and further deterioration of human life and its environments. Optimists dismiss negative reports as exaggerated and alarmist, or expect further technological advances to mitigate the negative effects of human action.

Joseph P. Byrne

FOR FURTHER STUDY

Cartledge, Bryan. *Population and the Environment.* New York: Oxford University Press, 1995.

Goodall, Brian. *Dictionary of Human Geography.* New York: Facts on File, 1987.

Gotelli, Nicholas J. *A Primer of Ecology.* Sunderland, Mass.: Sinauer Associates, 1995.

Miller, G. Tyler. *Living in the Environment.* 11th ed. Pacific Grove, Calif.: Brooks/Cole, 1992.

Southwick, Charles H. *Global Ecology in Human Perspective.* New York: Oxford University Press, 1996.

POPULATION GROWTH AND DISTRIBUTION

On October 13, 1999 the United Nations (U.N.) officially announced that the total population of the earth had reached 6 billion people. Both that number and that date were estimates, as no one could be sure exactly how many were alive on the earth that day. Indeed, the U.S. Census Bureau estimated that the world's population reached the figure of 6 billion four months earlier than the date of the U.N. estimate.

The population of the world has been growing steadily for thousands of years and has grown more in some places than in others. The population of the United States on October 13, 1999 was approximately 270 million. That meant that about 1 in 22 people on the planet lived in the United States. The populations of Canada, California, and Mexico City were each about 30 million people on that day, meaning that 1 in 200 people on the earth lived in Canada, 1 in 200 lived in California, and 1 in 200 lived in Mexico City. However, Canada is a much larger area than California, and California is much larger than Mexico City.

India's population on October 13, 1999, was just about 1 billion—1 in 6 people on the planet. China's population was about 1.25 billion—about 1 in 5 people on the planet. Although China was the most populous nation in the world in 1999, it was expected that in the twenty-first century India's population would surpass that of China.

HOW POPULATIONS ARE COUNTED. The U.S. Constitution requires that a census, or enumeration, of the population of the United States be conducted every ten years. The U.S. Census Bureau mails out millions of census forms and pays thousands of people (enumerators) to count people that did not fill out their census forms. This task cost about 4.5 billion dollars in the year 2000. Despite this great effort, millions of people are probably not counted in every U.S. census. Moreover, many countries have much less money to spend on censuses and more people to count. Therefore, information about the population of many poor or less-developed countries is even less accurate than that for the population of the United States. It is for these reasons that it is impossible to say that the population of the world reached exactly 6 billion people on exactly October 13, 1999.

Counting how many people were alive a hundred, a thousand, or hundreds of thousands of years ago is even more difficult. Estimates are made fom archaeological findings, which include human skeletons, ruins of ancient buildings, and evidence of ancient agricultural practices. Historical records of births, deaths, taxes paid, and other information are also used. Although it is not possible to estimate the global population one thousand years ago with great accuracey, it is a fascinating topic, and many people have participated in estimating the total population of the planet through the ages.

HISTORY OF HUMAN POPULATION GROWTH. Ancient ancestors of humans, known as hominids, were alive in Africa and Europe around one million years ago. It is believed that modern humans (*homo sapiens sapiens*) coexisted with the Neanderthals (*homo sapiens neandertalensis*) about 100,000 years ago. By 8000 B.C.E. (10,000 years ago) fully modern humans numbered around 8 million. If the presence of archaic *homo sapiens* is accepted as the beginning of the human population one million years ago, then the first 990,000 years of human existence are characterized by a very low population growth rate (15 persons per million per year).

Around 10,000 years ago, humans began a practice that dramatically changed their growth rate: planting food crops. This shift in human history, called the Agricultural Revolution, paved the way for the development of cities, government, and civilizations. Before the Agricultural Revolution, there were no governments to count people. The earliest censuses were conducted less than 10,000 years ago in the ancient civilizations of Egypt, Babylon, China, Palestine, and Rome. For this reason, historical estimates of the earth's total population are difficult to make. However, there is no argument that human numbers have increased dramatically in the past 10,000 years. The dramatic changes in the growth rates of the human population are typically attributed to three significant epochs of human cultural evolution: the Agricultural, Industrial, and Green Revolutions.

Before the Agricultural Revolution, the size of the human population was probably less than 10 million people, who survived primarily by hunting and gathering. After plant and animal species were domesticated, the human population increased its growth rate. By about 5000 B.C.E., gains in food production caused by the Agricultural Revolution meant that the planet could support about 50 million people. For the next several thousand years, the human population continued to grow at a rate of about 0.03 percent per year. By the first year of the common era, the planet's population numbered about 300 million.

At the end of the Middle Ages, the human population numbered about 400 million. As people lived in densely populated cities, the effects of disease increased. Starting in 1348 and continuing to 1650, the human population was subjected to massive declines caused by the bubonic plague—the Black Death. At its peak in about 1400, the Black Death may have killed 25 percent of Europe's population in just over fifty years. By the end of the last great plague in 1650, the human population numbered 600 million.

The Industrial Revolution began between 1650 and 1750. Since then, the growth of the human population has increased greatly. In just under three hundred years, the earth's population went from 0.5 billion to 6 billion people, and the annual rate of increase went from 0.1 percent to 1.8 percent. This population growth was not because people were having more babies, but because more babies lived to become adults and the average adult lived a longer life.

The Green Revolution occurred in the 1960's. The development of various vaccines and antibiotics in the twentieth century and the spread of their use to most of the world after World War II caused big drops in the death rate, increasing population growth rates. Feeding this growing population has presented a challenge. This third revolution is called the Green Revolution because of the technology used to increase the amount of food produced by farms. However, the Green Revolution was really a combination of improvements in health care, medicine, and sanitation, in addition to an increase in food production.

GEOGRAPHY OF HUMAN POPULATION GROWTH. The present-day human race traces its lineage to Africa. Humans migrated from Africa to the Middle East, Europe, Asia, and eventually to Australia, North and South America, and the Pacific Islands. It is believed that during the last Ice Age, the world's sea levels were lower because much of the world's water was trapped in ice sheets. This lower sea level created land bridges that facilitated many of the major human migrations across the world.

Patterns of human settlement are not random. People generally avoid living in deserts because they lack water. Few humans are found above the Arctic Circle because of that region's severely cold climate. Environmental factors, such as the availability of water and food and the livability of climate, influence where humans choose to live. How much these factors influence the evolution and development of human societies is a subject of debate.

The domestication of plants and animals that resulted from the Agricultural Revolution did not take place everywhere on the earth. In many parts of the world, humans remained as hunter-gatherers while agriculture developed in other parts of the world. Eventually, the agriculturalists outbred the hunter-gatherers, and few hunter-gatherers remain in the twenty-first century. Early agricultural sites have been found in many places, including Central and South America, Southeast Asia and China, and along the Tigris and Euphrates Rivers in what is now Iraq. The practice of agriculture spread from these areas throughout most of the world.

By the time Christopher Columbus reached the Americas in the late fifteenth century, there were millions of Native Americans living in towns and villages and practicing agriculture. Most of them died from diseases that were brought by European colonists. Colonization, disease, and war are major mechanisms that have changed the composition and distribution of the world's population in the last three hundred years.

The last few centuries also produced another change in the geography of the human population. During this period, the concentration of industry in urban areas and the efficiency gains of modern agricultural machinery caused large numbers of people to move from rural areas to cities to find jobs. From 1900 to 2000 the percentage of people living in cities went from 14 percent to just about 50 percent. Demographers estimate that by the year 2025, more than 60 percent of the earth's population will live in cities. Scientists estimate that the human population will continue to increase until the year 2050, at which time it will level out at between eight and fifteen billion.

EARTH'S CARRYING CAPACITY. Many people are concerned that the earth can-

In countries with large populations and limited resources, bicycles are a sensible alternative to automobiles for transportation. (PhotoDisc)

not grow enough food or provide enough other resources to support fifteen billion people. There is great debate about the concept of the earth's carrying capacity—the maximum human population that the earth can support indefinitely. Answers to questions about the earth's carrying capacity must account for variations in human behavior. For example, the earth could support more bicycle-riding vegetarians than car-driving carnivores. Questions about carrying capacity and the environmental impacts of the human race on the planet are fundamental to the United Nations' goals of sustainable development. Dealing with these questions will be one of the major challenges of the twenty-first century.

Paul C. Sutton

FOR FURTHER STUDY

Cohen, Joel. *How Many People Can the Earth Support?* London: W. W. Norton, 1996.

Diamond, Jared. *Guns, Germs, and Steel: The Fate of Human Societies.* London: W. W. Norton, 1998.

Population (Special Millenium Series Issue). *National Geographic* (October, 1998).

Weeks, John. *Population: An Introduction to Concepts and Issues.* Belmont, Calif.: Wadworth, 1994.

GLOBAL URBANIZATION

World cities Page 164

Urbanization is the process of building and living in cities. Although the human impulse to live in groups sharing a "home base" probably dates back to cave-dweller times or before, the creation of towns and cities with a few hundred to many thousands to millions of inhabitants, required several other developments.

Foremost of these was the invention of agriculture. Tilling crops requires a permanent living place near the cultivated land. The first agricultural villages were small. Jarmo, a village site from c. 7000 B.C.E., located in the Zagros Mountains of present-day Iran, appears to have had only twenty to twenty-five houses. Still, farmers' crops and livestock provided a food surplus that could be stored in the village or traded for other goods. Surplus food also meant surplus time, enabling some people to specialize in producing other useful items, or to engage in less tangible things like religious rituals or recordkeeping.

Given these conditions, it took people with foresight and political talents to lead the process of city formation. Once in cities, however, the inhabitants found many benefits. Walls and guards provided more security than the open country. Cities had regular markets where local craftsmen and traveling merchants displayed a variety of goods. City governments often provided amenities like primitive street lighting and sanitary facilities. The faster pace of life, and the exchange of ideas from diverse people interacting, made city life more interesting and speeded up the processes of social change and invention. Writing, law, and money all evolved in the earliest cities.

ANCIENT AND MEDIEVAL CITIES. Cities seem to have appeared almost simulta-

196

neously, around 3500 B.C.E., in three separate regions. In the Fertile Crescent, a wide curve of land stretching from the Persian gulf to the northwest Mediterranean Sea, the cities of Ur, Akkad, and Babylon rose, flourished, and succeeded one another. In Egypt, a connected chain of cities grew, soon unified by a ruler using Memphis, just south of the Nile River's delta, as his strategic and ceremonial base. On the Indian subcontinent, Mohenjo-Daro and Harappa oversaw about a hundred smaller towns in the Indus River valley. Similar developments took place about a thousand years later in northern China.

These first city sites were in the valleys of great river systems, where rich alluvial soil boosted large-scale food production. The rivers served as a "water highway" for ships carrying commodities and luxury items to and from the cities. They also furnished water for drinking, irrigation, and waste disposal. Even the rivers' rampages promoted civilization, as making flood control and irrigation systems required practical engineering, an organized workforce, and ongoing political authority to direct them.

Eurasia was still full of peoples who were not urbanized, however, and who lived by herding, pirating, or raiding. Early cities declined or disappeared, in some cases destroyed by invasions from such forces around 1200 B.C.E. Afterward, the cities of Greece became newly important. Their surrounding land was poor, but their access to the sea was an advantage. Greek cities prospered from fishing and trade. They also developed a new idea, the city-state, run by and for its citizens.

Rome, the Greek cities' successor to power, reached a new level of urbanization. Its rise owed more to historical accident and its citizens' political and military talents than to location, but some geographical features are salient. In some

ways, the fertile coastal plain of Latium was an ideal site for a great city, central to both the Italian peninsula and the Mediterranean Sea. There, the Tiber River becomes navigible and crossable.

In other ways, Rome's site was far from ideal. Its lower areas were swampy and mosquito-ridden. The seven hills, with their sacred sites later filled with public buildings and luxury houses, imposed a crazy-quilt pattern on the city's growth. Romans built cities with a simple rectangular plan all over Europe and the Middle East, but their home city grew in a less rational way.

At its peak, Rome had a million residents, a size no other city reached before nineteenth century London. It provided facilities found in modern cities: a piped water supply, a sewage disposal system, a police force, public buildings, entertainment districts, shops, inns, restaurants, and taverns. The streets were crowded and noisy; to control traffic, wheeled wagons could make deliveries only at night. Fire and building collapse were constant risks in the cheaply built apartment structures that housed the city's poorer residents. Still, few wanted to live anywhere but in Rome, their world's preeminent city.

In the Dark Ages after the western Roman Empire collapsed, feudalism, based on land holdings, eclipsed urban life. Cities never disappeared, but their populations and services declined drastically. Urban life still flourished for another millenium in the eastern capital of Constantinople. When Islam spread across the Middle East, it caused the growth of new cities, centered around a mosque and a marketplace.

In the twelfth and thirteenth centuries, life revived in Western Europe. As in the Islamic cities, the driving forces were religious—the building of cathedrals—and commercial—merchants and artisans expanding the reach of their activities. Medieval cities were usually walled, with narrow twisting streets and a lack of basic sanitary measures, but they drew ambitious people and innovative forces together. Italy's cities revived the concept of the city-state with its outward reach. Venice sent its merchant fleet all over the known world. Farther north, Paris and Bologna hosted the first universities. As the feudal system slowly gave way to nation-states ruled by one king, the cities generally supported the latter.

MODERN CITIES. Modern cities differ from earlier ones because of changes wrought by technology, but most of today's cities arose before the Industrial Revolution. Until the early nineteenth century, travel within a city was by foot or on horse, which limited street widths and city sizes. The first effect of railroads was to shorten travel time between cities. This helped country residents moving to the cities, and speeded raw materials going into and manufactured goods coming out of the factories that increasingly dotted urban areas. Rail transit soon caused the growth of a suburban ring. Prosperous city workers could live in more spacious homes outside the city, riding rail lines to work every day. This pattern was common in London and New York City.

Factories, the lifeblood of the Industrial Revolution, were built in pockets of existing cities. Smaller cities like Glasgow, Scotland, and Pittsburgh, Pennsylvania, grew as ironworking industries, using nearby or easily transported coal and ore resources, built large foundries there. Neither industrialists nor city authorities worried about where the people working there would live. Workers took whatever housing they could find in tenements or subdivided old mansions.

Beginning in the 1880's, metal-framed construction made taller buildings possible. These skyscrapers towered over stately three- to eight-story structures of an earlier period. Because this technology enabled expensive central-city ground space to house many profitable office suites, up through the 1930's, city cores became quite compacted. Many people believed such skyward growth was the wave of the future and warned that city streets were becoming sunless, dangerous canyons.

Automobiles kept these predictions from fully coming true. As car ownership became widespread, more roads were built or widened to carry the traffic. Urban areas began to decentralize. The car, like rail transit before it, allowed people to flee the urban core for suburban living. Because roads could be built almost anywhere, built-up areas around cities came to resemble large patches filling a circle, rather than the spokes-of-a-wheel pattern introduced by rail lines. Cities born during the automotive age tend to have an indistinct city center, surrounded by large areas of diffuse urban development. The prime example is Los Angeles: It has a small downtown area, but a consolidated metropolitan area of about 34,000 square miles (88,000 sq. km.).

Almost everywhere, urban sprawl has created satellite cities with major manufacturing, office, and shopping nodes. These cause an increasing portion of daily travel within metropolitan areas to be between one edge city and another, rather than to and from downtown. Since these journeys have an almost limitless variety of start points and destinations within the urban region, mass transit is only a partial solution to highway crowding and air pollution problems.

The above trends typify the so-called developed world, especially the United States. Many cities in poor nations have grown even more rapidly but with a different mix of patterns and problems. How-

URBANIZATION AND DEVELOPING NATIONS

The urban population, or number of people living in cities, in North America accounts for about 75 percent of its total population. In Europe, about 90 percent of the population lives in cities. In developing countries, the urban population is often less than 30 percent. The term "urbanization" refers to the rate of population growth of cities. Urbanization mainly results from people moving to cities from elsewhere. In developing countries, the urbanization rate is very high compared to those of North America or Europe. The high rate of urbanization of these countries makes it difficult for their governments to provide housing, water, sewers, jobs, schools, and other services for their fast-growing urban populations.

ever, the basic pattern can be detected around the globe, as urban dwellers seek to better their own circumstances.

MEGACITIES AND THE FUTURE. In the year 2000 the world had twenty-one megacities, defined as urban areas with a population of 10 million or more. The largest was Tokyo, with an estimated 27 million people in 1995, predicted to grow to around 29 million in 2015. Second largest was Mexico City, with more than 16 million in 1995 and annual growth at 1.81 percent. New York at 16 million and Los Angeles at 12 million are North America's other megacities. In the first half of the 1990's, Los Angeles grew 1.6 percent annually, much of it from international migration to the region.

Megacities profoundly affect the air, weather, and terrain of their surrounding territory. Smog is a feature of urban life almost everywhere, but is worse where the exhaust from millions of cars mixes with industrial pollution. Some megacities have slowed the problem by regulating combustion technology; none have solved it. Huge expanses of soil preempted by buildings and pavements can turn heavy rains into floods almost instantly, and the ambient heat in large cities stays several degrees higher than in comparable rural areas. Recent engineering studies suggest that megacities create instability in the ground beneath, compressing and undermining it.

Orchard heaters Page 165

Smog Page 166

URBAN HEAT ISLANDS

Large cities have distinctly different climates from the rural areas that surround them. The most important climatic characteristic of a city is the urban heat island, a concentration of relatively warmer temperatures, especially at nighttime. Large cities are frequently at least 11 degrees Fahrenheit (6 degrees Celsius) warmer than the surrounding countryside.

The urban heat island results from several factors. Primary among these are human activities, such as heating homes and operating factories and vehicles, that produce and release large quantities of energy to the atmosphere. Most of these activities involve the burning of fossil fuels such as oil, gas, and coal. A second factor is the abundance of heat-absorbing urban materials, such as brick, concrete, and asphalt. A third factor is the surface dryness of a city. Urban surface materials normally absorb little water and therefore quickly dry out after a storm. In contrast, the evaporation of moisture from wet soil and vegetation in rural areas uses a large quantity of solar energy—often more than is converted directly to heat—resulting in cooler air temperatures and higher relative humidities.

INFORMATION ON THE WORLD WIDE WEB

Peopleandplanet.net, a Web site derived from the work of the quarterly journal *People & the Planet*, focuses on issues of population, poverty, health, consumption and the environment. An on-line article, "How the Cities Grow," features graphs and startling facts on recent and future urban growth. (oneworld.org/patp/vol6/livernas.html)

How will cities evolve? Barring an unforeseen technological or social breakthrough—which could happen in the twenty-first century—the current growth and problems will probably continue. The process of megapolis—metropolitan areas blending together along the corridors between them—is well underway in many areas. Predictions that the computer will so change the nature of work as to cause massive population shifts away from cities have proven premature. Despite its drawbacks, increasing numbers of people are drawn to urban life, seeking the economic opportunities and wider social world that cities offer.

Emily Alward

FOR FURTHER STUDY

Hall, Peter. *Cities in Civilization*. New York: Pantheon Books, 1998.

Lo, Fu-Chen, and Yue-Man Yeung. *Globalization and the World of Large Cities*. New York: United Nations University Press, 1998.

Mumford, Lewis. *The City in History*. New York: Harvest/HBJ Books, 1961.

Scott, Allen J. *City: Los Angeles and Urban Theory at the End of the Twentieth Century*. Berkeley: University of California Press, 1998.

The Rise of Cities. Alexandria, Va.: Time-Life Books, 1991.

GLOBAL TIME AND TIME ZONES

Before the nineteenth century, people kept time by local reckoning of the position of the Sun; consequently, thousands of local times existed. In medieval Europe, "hours" varied in length, depending upon the seasons: Each hour was determined by the Roman Catholic Church. In the sixteenth century, Holy Roman emperor Charles V was the first secular ruler to decree hours to be of equal length. As the industrial and scientific revolutions swept Europe, North America, and other areas, some form of time standardization became necessary as communities and regions increasingly interacted. In 1780 Geneva, Switzerland, was the first locality known to employ a standard time, set by the town-hall clockkeeper, throughout the town and its immediate vicinity.

The growth and expansion of railroads,

providing the first relatively fast movement of people and goods from city to city, underscored the need for a standard system in Great Britain. As early as 1828, Sir John Herschel, Astronomer Royal, called for a national standard time system based on instruments at the Royal Observatory at Greenwich. That practice began in 1852, when the British telegraph system had developed sufficiently for the Greenwich time signals to be sent instantly to any point in the country.

As railroads expanded through North America, they exposed a problem of local time variation similar to that in Great Britain but on a far larger scale, since the distances between the East and West Coasts were much greater than in Great Britain. In order for long-distance train schedules to work, different parts of the country had to coordinate their clocks. The first to suggest a standard time framework for the United States was Charles F. Dowd, president of Temple Grove Seminary for Women in Saratoga Springs, New York. Initially, Dowd proposed putting all U.S. railroads on a single standard time, based on the time in Washington, D.C. When he realized that the time in California would be behind such a standard by almost four hours, he produced a revised system, establishing four time zones in the United States. Dowd's plan, published in 1870, included the first known map of a time zone system for the country.

Not everyone was happy with the designation of Washington, D.C., as the administrative center of time in the United States. Northeastern railroad executives urged that New York, the commercial capital of the nation, be used instead: Many cities and towns in the region already had standardized to New York time out of practical necessity. Dowd proposed a compromise: to set the entire national time zone system in the United States using the Greenwich prime meridian, already in use in many parts of the world for maritime and scientific purposes. In 1873 the American Association of Railways (AAR) flatly rejected the proposal.

In the end, Dowd proved to be a visionary. In 1878 Sandford Fleming, chief engineer of the government of Canada, proposed a worldwide system of twenty-four time zones, each fifteen degrees of longitude in width, and each bisected by a meridian, beginning with the prime meridian of Greenwich. William F. Allen, general secretary of the AAR and armed with a deep knowledge of railroad practices and politics, took up the crusade and persuaded the railroads to agree to a system. At noon on Sunday, November 18, 1883, most of the more than six hundred U.S. railroad lines dropped the fifty-three arbitrary times they had been using and adopted Greenwich-indexed meridians that defined the times in each of four times zones: eastern, central, mountain, and Pacific. Most major cities in the United States and Canada followed suit.

TIME SYSTEM FOR THE WORLD. Almost at the same time that American railroads adopted a standard time zone system, the State Department, authorized by the United States Congress, invited governments from around the world to assemble delegates in Washington, D.C., to adopt a global system. The International Meridian Conference assembled in the autumn of 1884, attended by representatives of twenty-five countries. Led by Great Britain and the United States, most favored adoption of Greenwich as the official prime meridian and Greenwich mean time as universal time.

There were other contenders: The French wanted the prime meridian to be set in Paris, and the Germans wanted it in Berlin; others proposed a mountaintop in the Azores or the tip of the Great Pyramid in Egypt. Greenwich won handily. The conference also agreed officially to start the universal day at midnight, rather than at noon or at sunrise, as practiced in many parts of the world. Each time zone in the world eventually came to have a local name, although technically, each goes by a letter in the alphabet in order eastward from Greenwich.

Once a global system was in place, there was a new issue: Many jurisdictions wanted to adjust their clocks for part of the year to account for differences in the number of hours of daylight between summer and winter months. In 1918 Congress decreed a system of daylight saving time for the United States but almost immediately abolished it, leaving state governments and communities to their local options. Daylight saving time, or a form of it, returned in the United States and many Allied nations during World War II. In the Uniform Time Act of 1966, Congress finally established a national system of daylight saving time, although with an option for states to abstain.

To the extent that it indicates how human communities want to manipulate time for social, political, or economic reasons, the issue of daylight saving time, rather than the establishment of a system of world time zones, is a better clue to the geographical issues involved in time administration. Both the history and the present format of the world time zone system show that the mathematically precise arrangement envisioned by many of the pioneers of time zones is not as important as things on the ground.

In the United States, the railroad time system adopted in 1883 drew the boundary between eastern time and central time more or less between the thirteen original states and the trans-Appalachian West: The entire Midwest, including Ohio, Indiana, and Michigan, fell in the central time

zone. As the center of population migrated westward, train speeds increased, highways developed, and New York emerged as the center of mass media in the United States, the boundary between the eastern and central time zones marched steadily westward. In 1918 it ran down the middle of Ohio; by the 1960's, it was at the outskirts of Chicago.

One of the principal reasons for the popularity of Greenwich as the site of the prime meridian (zero degrees longitude), is that it places the international date line (180 degrees longitude)—where, in effect, time has to move forward to the next day rather than the next hour—far out in the Pacific Ocean where few people are affected by what otherwise would be an awkward arrangement. However, even this line is somewhat irregular, to avoid placing a small section of eastern Russia and some of the Aleutian Islands of the United States in different days.

By 1950 most nations had adopted the universal time zone system, although a few followed later: Saudi Arabia in 1962, Liberia in 1972. Despite adhering to the system in principle, many nations take considerable liberties with the zones, especially if their territory spans several. All of Western Europe, despite covering an area equivalent to two zones, remains on a single standard. The People's Republic of China, which stretches across five different time zones, arbitrarily sets the entire country officially on Beijing time, eight hours behind Greenwich. Iran, Afghanistan, India, and Myanmar, each of which straddle time zone boundaries, operate on half-hour compromise systems as their time standards (as does Newfoundland). As late as 1978, Guyana's standard time was three hours, forty-five minutes in advance of Greenwich.

It can be argued that adoption of a worldwide system of time zones in the late nineteenth century was one of the earliest manifestations of the emergence of a global economy and society, and has been a crucial factor in the unfolding of this process throughout the twentieth century and beyond.

Ronald W. Davis

FOR FURTHER STUDY

Bartky, Ian R., and Elizabeth Harrison. "Standard and Daylight-Saving Time." *Scientific American* 240, no. 5 (May, 1979): 46-53.

Bartky, Ian R. "The Invention of Railroad Time." *Railroad History*, no. 148 (Spring, 1983): 13-22.

Davies, Alun C. "Greenwich and Standard Time." *History Today* 27 (1978): 194-199.

Doane, Doris Chase. *Time Changes in the U.S.A.* Tempe, Ariz.: National Federation of Astrologers, 1980.

Howse, Derek. *Greenwich Time and the Longitude*. Greenwich, England: National Maritime Museum, 1997.

Movahedi, Siamak. "Cultural Preconceptions of Time: Can We Use Operational Time to Meddle in God's Time?" *Comparative Studies in Society and History* 27 (1985): 385-400.

Zerubavel, Eviatar. "The Standardization of Time: A Sociohistorical Perspective." *American Journal of Sociology* (1982): 1-23.

INFORMATION ON THE WORLD WIDE WEB

The Time Service Department of the United States Naval Observatory in Washington, D.C., maintains a Web site on time zones in the United States and around the world. Visitors can access a map of current world time zones, review the U.S. law on standard time zones, and inquire about local standard time for any country or region in the world.
(tycho.usno.navy.mil/tzones.html)

CLIMATE AND HUMAN SOCIETIES

CLIMATE AND HUMAN SETTLEMENT

Weather hazards Page 165

"Everyone talks about the weather," goes an old saying, "but nobody does anything about it." If everyone talks about the weather, it is because it is important to them—to how they feel and to how their bodies and minds function. There is plenty they can do about it, from going to a different location to creating an artificial indoor environment.

CLIMATE. The term "climate" refers to average weather conditions over a long period of time and to the variations around that average from day to day or month to month. Temperature, air pressure, humidity, wind conditions, sunshine, and rainfall—all are important elements of climate and differ systematically with location. Temperatures tend to be higher near the equator and are so low in the polar regions that very few people live there. In any given region, temperatures are lower at higher altitudes. Areas close to large bodies of water have more stable temperatures. Rainfall depends on topography: The Pacific Coast of the United States receives a great deal of rain, but the nearby mountains prevent it from moving very far inland. Seasonal variations in temperature are larger in temperate zones.

Throughout human history, climate has affected where and how people live.

People in technologically primitive cultures, lacking much protective clothing or housing, needed to live in mild climates, in environments favorable to hunting and gathering. As agricultural cultivation developed, populations located where soil fertility, topography, and climate were favorable to growing crops and raising livestock. Areas in the Middle East and near the Mediterranean Sea flourished before 1000 B.C.E. Many equatorial areas were too hot and humid for human and animal health and comfort, and too infested with insect pests and diseases.

Improvements in technology allowed settlement to range more widely north and south. Sturdy houses and stables, internal heating, and warm clothing enabled people to survive and be active in long cold winters. Some peoples developed nomadic patterns, moving with herds of animals to adapt to seasonal variations.

A major challenge in the evolution of settled agriculture was to adapt production to climate and soil conditions. In North America, such crops as cotton, tobacco, rice, and sugarcane have relatively restricted areas of cultivation. Wheat, corn, and soybeans are more widely grown, but usually further north. Winter wheat is an ingenious adaptation to climate. It is sown and germinates in autumn, then matures and is harvested the following spring. Rice, which generally grows in standing water, requires special environmental conditions.

TROPICAL PROBLEMS. Some scholars argue that tropical climates encourage life to flourish but do not promote quality of life. In hot climates, people do not need much caloric intake to maintain body heat. Clothing and housing do not need to protect people from the cold. Where temperatures never fall below freezing, crops can be grown all year round. Large numbers of people can survive even where productivity is not high. However, hot humid conditions are not favorable to human exertion nor (it is claimed) to mental, spiritual, and artistic creativity. Some tropical areas, such as South India, Bangladesh, Indonesia, and Central Africa, have developed large populations living at relatively low levels of income.

SLAVERY. Efforts to develop tropical regions played an important part in the rise of the slave trade after 1500 C.E. Black Africans were kidnapped and forceably transported to work in hot, humid regions. The West Indian islands became an important location for slave labor, particularly in sugar production. On the North American continent, slave labor was important for producing rice, indigo, and tobacco in colonial times. All these were eclipsed by the enormous growth of cotton production in the early years of U.S. independence. It has been estimated that the forced migration of Africans to the Americas involved about 1,800 Africans per year from 1450 to 1600, 13,400 per year in the seventeenth century, and 55,000 per year from 1701 to 1810. Estimates vary wildly, but at least 7.7 million Africans were forced to migrate in this process.

EUROPEAN MIGRATION. Migration of European peoples also accelerated after the discovery of the New World. They settled mainly in temperate-zone regions, particularly North America. Although Great Britain gained colonial dominion over India, the Netherlands over present-day Indonesia, and Belgium over a vast part of central Africa, few Europeans went to those places to live. However, many Chinese migrated throughout the Nanyang (South Sea) region, becoming commercial leaders in present-day Malaysia, Thailand, Indonesia, and the Philippines, despite the heat and humidity. British emigrants settled in Australia and New Zealand, Spanish and Italians in Argentina, Dutch (Boers) in South Africa—all temperate regions.

IRELAND'S POTATO FAMINE AND EUROPEAN EMIGRATION

Mass migration from Europe to North America began in the 1840's after a serious blight destroyed a large part of the potato crop in Ireland and other parts of Northern Europe. The weather played a part in the famine; during the autumns of 1845 and 1846 climatic conditions were ideal for spreading the potato blight. The major cause, however, was the blight itself, and the impact was severe on low-income farmers for whom the potato was the major food.

The famine and related political disturbances led to mass emigration from Ireland and from Germany. By 1850 there were nearly a million Irish and more than half a million Germans in the United States. Combined, these two groups made up more than two-thirds of the foreign-born U.S. population of 1850. The settlement patterns of each group were very different. Most Irish were so poor they had to work for wages in cities or in construction of canals and railroads. Many Germans took up farming in areas similar in climate and soil conditions to their homelands, moving to Wisconsin, Minnesota, and the Dakotas.

CLIMATE AND ECONOMICS. Most of the economic progress of the world between 1492 and 2000 occurred in the temperate zones, primarily in Europe and North America. Climatic conditions favored agricultural productivity. Some scholars believe that these areas had climatic conditions that were stimulating to intellectual and technological development. They argue that people are invigorated by seasonal variation in temperature, sunshine, rain, and snow. Storms—particularly thunderstorms—can be especially stimulating, as many parents of young children have observed for themselves.

Mediterranean
Page 233

Climate has contributed to the great economic productivity of the United States. This productivity has attracted a flow of immigrants, which averaged about one million a year from 1905 to 1914. Immigration approached that level again in the 1990's, as large numbers of Mexicans crossed the southern border of the United States, often coming for jobs as agricultural laborers in the hot conditions of the Southwest—a climate that made such work unattractive to many others.

Unpredictable climate variability was important in the peopling of North America. During the 1870's and 1880's, unusually favorable weather encouraged a large flow of migration into the grain-producing areas just west of the one-hundredth meridian. Then came severe drought and much agrarian distress. Between 1880 and 1890, the combined population of Kansas and Nebraska increased by about a million, an increase of 72 percent. During the 1890's, however, their combined population was virtually constant, indicating that a large out-migration was offsetting the natural increase. Much of the area reverted to pasture, as climate and soil conditions could not sustain the grain production that had attracted so many earlier settlers.

Los Angeles
smog
Page 166

Climate variability can be a serious hazard. Freezing temperatures for more than a few hours during spring can seriously damage fruits and vegetables. A few days of heavy rain can produce serious flooding.

RECREATION AND RETIREMENT. Whenever people have been able to separate decisions about where to live from decisions about where to work, they have gravitated toward pleasant climatic conditions. Vacationers head for Caribbean islands, Hawaii, the Crimea, the Mediterranean Coast, even the Baltic coast. "The mountains" and "the seashore" are attractive the world over. Paradoxically, some of these areas (the Caribbean, for instance) have monotonous weather year-round and thus have not attracted large inflows of permanent residents. Winter sports have created popular resorts such as Vail and Aspen in Colorado, and numerous older counterparts in New England. Large numbers of Americans have retired to the warm climates in Florida, California, and Arizona. These areas then attract working-age adults who earn a living serving vacationers and retirees. Since these locations are uncomfortably hot in summer, their attractiveness for residence had to await the coming of air conditioning in the latter part of the twentieth century.

HUMAN IMPACT ON CLIMATE. Climate interacts with pollution. Bad-smelling factories and refineries have long relied on the wind to disperse atmospheric pollutants. The city of Los Angeles, California, is uniquely vulnerable to atmospheric pollution because of its topography and wind currents. Government regulations of automobile emissions have had to be much more stringent there than in other areas to keep pollution under control.

Human activities have sometimes altered the climate. Development of a large city substitutes buildings and pavements for grass and trees, raising summer temperatures and changing patterns of water evaporation. Atmospheric pollutants have contributed to acid rain, which damages vegetation and pollutes water resources. Many observers have also blamed human activities for a trend toward global warming. Much of this has been blamed on carbon dioxide generated by combustion, particularly of fossil fuels. A widespread rise in temperatures could be expected to raise water levels in the oceans as polar icecaps melt and change the relative attractiveness of many locations.

Paul B. Trescott

FOR FURTHER STUDY

Burroughs, William James. *Does the Weather Really Matter?: The Social Implications of Climate Change.* Cambridge, England: Cambridge University Press, 1997.

Huntington, Ellsworth. *Mainsprings of Civilization.* New York: John Wiley, 1945.

Lamb, H. H. *Climate, History, and the Modern World.* 2d ed. New York: Routledge, 1995.

McNeill, William H., and Ruth S. Adams, eds. *Human Migration: Patterns and Policies.* Bloomington: Indiana University Press, 1978.

Schneider, Stephen H., and Randi Londer. *The Coevolution of Climate and Life.* San Francisco: Sierra Club, 1984.

Rising sea level
Page 233

FLOOD CONTROL

Flood control presents one of the most daunting challenges humanity faces. The regions that human communities have generally found most desirable, for both agriculture and industry, have also been the lands at greatest risk of experiencing devastating floods. Early civilization developed along river valleys and in coastal floodplains because those lands contained the most fertile, most easily irrigated soils for agriculture, combined with the convenience of water transportation.

The Nile River in North Africa, the Ganges River on the Indian subcontinent, and the Yangtze River in China all witnessed the emergence of civilizations that relied on those rivers for their growth. People learned quickly that living in such areas meant living with the regular occurrence of life-threatening floods.

Knowledge that floods would come did not lead immediately to attempts to prevent them. For thousands of years, attempts at flood control were rare. The people living along river valleys and in floodplains often developed elaborate systems of irrigation canals to take advantage of the available water for agriculture and became adept at using rivers for transportation, but they did not try to control the river itself. For millennia, people viewed periodic flooding as inevitable, a force of nature over which they had no control. In Egypt, for example, early people learned how far out over the riverbanks the annual flooding of the Nile River would spread and accommodated their society to the river's seasonal patterns. Villagers built their homes on the edge of the desert, beyond the reach of the flood waters, while the land between the towns and the river became the area where farmers planted crops or grazed livestock.

In other regions of the world, buildings were placed on high foundations or built with two stories on the assumption that the local rivers would regularly overflow their banks. In Southeast Asian countries such as Thailand and Vietnam, it is common to see houses constructed on high wooden posts above the rivers' edge. The inhabitants have learned to allow for the water levels' seasonal changes.

FLOOD CONTROL STRUCTURES. Eventually, societies began to try to control floods rather than merely trying to survive them. Levees and dikes—earthen embankments constructed to prevent water from flowing into low-lying areas—were

Marker showing flood levels previously reached in an urban area. (PhotoDisc)

built to force river waters to remain within their channels rather than spilling out over a floodplain. Flood channels or canals that fill with water only during times of flooding, diverting water away from populated areas, are also a common component of flood control systems. Areas that are particularly susceptible to flash floods have constructed numerous flood channels to prevent flooding in the city. For example, for much of the year, Southern California's Los Angeles River is a small stream flowing down the middle of an enormous, 20- to 30-foot-deep (6-9 meters) concrete-lined channel, but winter rains can fill its bed from bank to bank. Flood channels prevent the river from washing out neighborhoods and freeways.

Engineers designed dams with reservoirs to prevent annual rains or snowmelt entering the river upstream from running into populated areas. By the end of the twentieth century, extremely complex flood control systems of dams, dikes, levees, and flood channels were common. Patterns of flooding that had existed for thousands of years ended as civil engineers attempted to dominate natural forces.

The annual inundation of the Egyptian delta by the flood waters of the Nile River ceased in 1968 following construction of the 365-foot-high (111 meters) Aswan High Dam. The reservoir behind the 3,280-foot-long (1,000-meter) dam forms a lake almost ten miles (16 km.) wide and almost three hundred miles (480 km.) long. Flood waters are now trapped behind the dam and released gradually over a year's time.

ENVIRONMENTAL CONCERNS. Such high dams are increasingly being questioned as a viable solution for flood control. The Three Gorges Dam being constructed in China at the end of the twentieth century may be the last high dam constructed for the purpose of flood

Urban flooding. (PhotoDisc)

control. As human understanding of both hydrology and ecology have improved, the disruptive effects of flood control projects such as high dams, levees, and other engineering projects are being examined more closely.

Hydrologists and other scientists who study the behavior of water in rivers and soils have long known that vegetation and soil types in watersheds can have a profound effect on downstream flooding. The removal of forest cover through logging or clearing for agriculture can lead to severe flooding in the future. Often that flooding will occur many miles downstream from the logging activity. Devastating floods in the South Asian country of Bangladesh, for example, have been blamed in part on clear-cutting of forested hillsides in the Himalaya Mountains in India and Nepal. Monsoon rains that once were absorbed

or slowed by forests now run quickly off mountainsides, causing rivers to reach unprecedented flood levels. Concerns about cause-and-effect relationships between logging and flood control in the mountains of the United States were one reason for the creation of the U.S. Forest Service in the nineteenth century.

In populated areas, even seemingly trivial events such as the construction of a shopping center parking lot can affect flood runoff. When thousands of square feet of land are paved, all the water from rain runs into storm drains rather than being absorbed slowly into the soil and then filtered through the watertable. Engineers have learned to include catch basins, either hidden underground or openly visible but disguised as landscaping features such as ponds, when planning a large paving project.

Urban flooding Page 165

*Wetlands
Pages 106,
107*

WETLANDS AND FLOODING. Less well known than the influence of watersheds on flooding is the impact of wetlands along rivers. Many river systems are bordered by long stretches of marsh and bog. In the past, flood control agencies often allowed farmers to drain these areas for use in agriculture and then built levees and dikes to hold the river within a narrow channel. Scientists now know that these wetlands actually serve as giant sponges in the flood cycle. Flood waters coming down a river would spread out into wetlands and be held there, much like water is trapped in a sponge.

Draining wetlands not only removes these natural flood control areas but worsens flooding problems by allowing floodwater to precede downstream faster. Even if life-threatening or property-damaging floods do not occur, faster-flowing water significantly changes the ecology of the river system. Waterborne silt and debris will be carried farther. Trying to control floods on the Mississippi River has had the unintended consequence of causing waterborne silt to be carried farther out into the Gulf of Mexico by the river, rather than its being deposited in the delta region. This, in turn, has led to the loss of shore land as ocean wave actions washes soil away, but no new alluvial deposits arrive to replace it.

In any river system, some species of aquatic life will disappear and others replace them as the speed of flow of the water affects water temperature and the amount of dissolved oxygen available for fish. Warm-water fish such as bass will be replaced by cold-water fish such as trout, or vice versa. Biologists estimate that more than twenty species of freshwater mussels have vanished from the Tennessee River since construction of a series of flood control and hydroelectric power generation dams have turned a fast-moving river into a series of slow-moving reservoirs.

FUTURE OF FLOOD CONTROL. By the end of the twentieth century, engineers increasingly recognized the limitations of human interventions in flood control. Following devastating floods in the early 1990's in the Mississippi River drainage, the U.S. Army Corps of Engineers recommended that many towns that had stood right at the river's edge be moved to higher ground. That is, rather than trying to prevent a future flood, the Corps advised citizens to recognize that one would inevitably occur, and that they should remove themselves from its path. In the United States and a number of other countries, land that has been zoned as floodplains can no longer be developed for residential use. While there are many things humanity can do to help prevent floods, such as maintaining well-forested watersheds and preserving wetlands, true flood control is probably impossible. Dams, levees, and dikes can slow the water down, but eventually, the water always wins.

Nancy Farm Männikkö

FOR FURTHER STUDY

Barry, John M. *Rising Tide: The Great Mississippi Flood of 1927 and How It Changed America.* New York: Simon & Schuster, 1997.

Frank, Arthur D. *Development of the Federal Program of Flood Control on the Mississippi River.* New York: AMS Press, 1999.

Kelley, Robert Lloyd. *Battling the Inland Sea: Floods, Public Policy, and the Sacramento Valley.* Berkeley: University of California Press, 1998.

Peterson, Elmer T. *Big Dam Foolishness: The Problems of Modern Flood Control and Water Storage.* New York: Basic Books, 1990.

Saul, A. J., ed. *Floods and Flood Management.* Dordrecht, Netherlands: Kluwer, 1992.

ATMOSPHERIC POLLUTION

Pollution of the earth's atmosphere comes from many sources. Some forces are natural, such as volcanoes and lightning-caused forest fires, but most sources of pollution are byproducts of industrial society. Atmospheric pollution cannot be confined by national boundaries; pollution generated in one country often spills over into another country, as is the case for acid deposition, or acid rain, generated in the midwestern states of the United States that affects lakes in Canada.

MAJOR AIR POLLUTANTS. Each of eight major forms of air pollution has an impact on the atmosphere. Often two or more forms of pollution have a combined impact that exceeds the impact of the two acting separately. These eight forms are:

1. Suspended particulate matter: This is a mixture of solid particles and aerosols suspended in the air. These particles can have a harmful impact on human respiratory functions.

2. Carbon monoxide (CO): An invisible, colorless gas that is highly poisonous to air-breathing animals.

3. Nitrogen oxides: These include several forms of nitrogen-oxygen compounds that are converted to nitric acid in the atmosphere and are a major source of acid deposition.

4. Sulfur oxides, mainly sulfur dioxide: This sulfur-oxygen compound is converted to sulfuric acid in the atmosphere and is another source of acid deposition.

5. Volatile organic compounds: These include such materials as gasoline and organic cleaning solvents, which evaporate and enter the air in a vapor state. VOCs are a major source of ozone formation in the lower atmosphere.

6. Ozone and other petrochemical oxidants: Ground-level ozone is highly toxic to animals and plants. Ozone in the upper atmosphere, however, helps to shield living creatures from ultraviolet radiation.

*Mining town
Page 162*

U.S. AIR POLLUTION TRENDS, 1950-1994
In Thousands of Tons

Year	PM-10	PM-10, Fugitive Dust	Sulfur Dioxide	Nitrogen Dioxides	Volatile Organic Compounds	Carbon Monoxide
1950	17,133	(NA)	22,358	10,093	20,936	102,609
1960	15,558	(NA)	22,227	14,140	24,459	109,745
1970	13,044	(NA)	31,161	20,625	30,646	128,079
1980	7,050	(NA)	25,905	23,281	25,893	115,625
1985	4,094	40,889	23,230	22,860	25,798	114,690
1990	3,882	39,451	22,433	23,038	23,599	100,650
1994	3,705	41,726	21,118	23,615	23,174	98,017

Source: U.S. Department of Commerce, *Statistical Abstract of the United Sates, 1996,* 1996.

Note: PM-10 emissions consist of particulate matter smaller than 10 microns in size. Lead emissions are not included in the table. Lead emissions in 1970 were 219,471 tons; in 1980, 74,956 tons; in 1990, 5,666 tons; in 1994, 4,956 tons.

7. Lead and other heavy metals: Generated by various industrial processes, lead is harmful to human health even at very low concentrations.

8. Air toxics and radon: Examples include cancer-causing agents, radioactive materials, or asbestos. Radon is a radioactive gas produced by natural processes in the earth.

All eight forms of pollution can have adverse effects on human, animal, and plant life. Some, such as lead, can have a very harmful effect over a small range. Others, such as sulfur and nitrogen oxides, can cross national boundaries as they enter the atmosphere and are carried many miles by prevailing wind currents. For example, the radioactive discharge from the explosion of the Chernobyl nuclear plant in the former Soviet Union in 1986 had harmful impacts in many countries. Atmospheric radiation generated by the explosion rapidly spread over much of the Northern Hemisphere, especially the countries of northern Europe.

IMPACTS OF ATMOSPHERIC POLLUTION. Atmospheric pollution not only has a direct impact on the health of humans, animals, and plants but also affects life in more subtle, often long-term, ways. It also affects the economic well-being of people and nations and complicates political life.

Atmospheric pollution can kill quickly, as was the case with the killer smog, brought about by a temperature inversion, that struck London in 1952 and led to more than four thousand pollution-related deaths. In the late 1990's, the atmosphere of Mexico City was so polluted from automobile exhausts and industrial pollution that sidewalk stands selling pure oxygen to people with breathing problems became thriving businesses. Many of the heavy metals and organic constituents of air pollution can cause cancer when people are exposed to large doses or for long periods of time. Exposure to radioactivity in the atmosphere can also increase the likelihood of cancer.

In some parts of Germany and Scandinavia in the 1990's, as well as places in southern Canada and the southern Appalachians in the United States, certain types of trees began dying. There are several possible reasons for this die-off of forests, but one potential culprit is acid deposition. As noted above, one byproduct of burning fossil fuels (for example, in coal-fired electric power plants) is the sulfur and nitrous oxides emitted from the smokestacks. Once in the atmosphere, these gases can be carried for many miles and produce sulfuric and nitric acids.

These acids combine with rain and snow to produce acidic precipitation. Acid deposition harms crops and forests and can make a lake so acidic that aquatic life can-

Ukraine's Chernobyl nuclear power plant, shortly after its 1986 accident. The plant was not completely shut down until late 2000. (AP/Wide World Photos)

Emissions from motor vehicles are the leading contributors to atmospheric pollution in most urban centers. (PhotoDisc)

not exist in it. Forests stressed by contact with acid deposition can become more susceptible to damage by insects and other pathogens. Ozone generated from automobile emissions also kills many plants and causes human respiratory problems in urban areas.

Air pollution also has an impact on the quality of life. Acid pollutants have damaged many monuments and building facades in urban areas in Europe and the United States. By the late 1990's, the distance that people could see in some regions, such as the Appalachians, was reduced drastically because of air pollution.

The economic impact of air pollution may not be as readily apparent as dying trees or someone with a respiratory ail-

ment, but it is just as real. Crop damage reduces agricultural yield and helps to drive up the cost of food. The costs of repairing buildings or monuments damaged by acid rain are substantial. Increased health-care claims resulting from exposure to air pollution are hard to measure but are a cost to society nevertheless.

It is impossible to predict the potential for harm from rapid global warming arising from greenhouse gases and the destruction of the ozone layer by chlorofluorocarbons (CFCs), but it could be catastrophic. Rapid global warming would cause the sea level to rise because of the melting of the polar ice caps. Low-lying coastal areas would be flooded, or, in the case of Bangladesh, much of the country.

Rising sea level
Page 233

GLOBAL WARMING

An aspect of atmospheric pollution is the potential impact that several pollutants have on the world's climate. Carbon dioxide (CO_2), methane, water vapor, and other trace gases are labeled "greenhouse gases" because they act like glass in a greenhouse, blanketing and insulating the earth and slowing radiational cooling. Atmospheric carbon dioxide has increased, largely because of the burning of fossil fuels, which also contributes to other forms of atmospheric pollution.

Trace gases of particular importance are synthetic chlorofluorocarbons (CFCs),

by-products of aerosols and some forms of refrigerants used for air conditioning. CFCs deplete ozone in the stratosphere, allowing increased ultraviolet radiation to reach the earth. The amount of CFCs in the atmosphere has been declining since the industrial nations signed the Montreal Accord of 1987, calling for a dramatic reduction in their use. However, CFCs still pose a problem because they remain in the stratosphere for many years. Presently, there are holes in the ozone layer of the stratosphere over both the Arctic and the Antarctic.

Global warming would also change crop patterns for much of the world.

SOLUTIONS FOR ATMOSPHERIC POLLUTION. Although there is still some debate, especially among political leaders, most scientists recognize that air pollution is a problem that affects both the industrialized and less-industrialized world. In their rush to industrialize, many nations begin generating substantial amounts of air pollution; China's extensive use of coal-fired power plants is just one example.

The major industrial nations are the primary contributors to atmospheric pollution. North America, Europe, and East Asia produce 60 percent of the world's air pollution and 60 percent of its food supply. Because of their role in supplying food for many other nations, anything that damages their ability to grow crops hurts the rest of the world.

Many industrialized nations are making efforts to control air pollution, for example, the Clean Air Act of 1970 in the United States or the international Montreal Accord to curtail CFC production. Progress is slow and the costs of reducing

air pollution are often high. In the year 2000 the record of the nations of the world in dealing with air pollution was a mixed one. There were some signs of progress, such as reduced automobile emissions and sulfur and nitrous oxides in industrialized nations, but acid deposition remains a problem in some areas. CFC production has been halted, but the impact of CFCs on the ozone layer will continue for many years. However, more nations are becoming aware of the health and economic impact of air pollution and are working to keep the problem from getting worse.

John M. Theilmann

Chemical plant Page 166

INFORMATION ON THE WORLD WIDE WEB

The Sierra Club's Web site features a Global Warming home page, with links to current news and publications. (www.sierraclub.org/globalwarming/resources/innactio.htm)

FOR FURTHER STUDY

Graedal, Thomas E., and Paul J. Crutzen. *Atmosphere, Climate, and Change.* New York: Scientific American Library, 1995.

McDonald, Alan. "Combating Acid Deposition and Climate Change: Priorities for Asia." *Environment* 41, no. 3 (April, 1999): 4-11, 34-41.

Munton, Don. "Dispelling the Myths of the Acid Rain Story." *Environment* 40, no. 6 (July/August, 1998): 4-7, 27-34.

Somerville, Richard C. J. *The Forgiving Air.* Berkeley: University of California Press, 1996.

Soroos, Marvin S. "The Thin Blue Line: Preserving the Atmosphere as a Global Commons." *Environment* 40, no. 2 (March, 1998): 6-13, 32-35.

DISEASE AND CLIMATE

Climate influences the spread and persistence of many diseases, such as tuberculosis and influenza, which thrive in cold climates, and malaria and encephalitis, which are limited by the warmth and humidity that sustains the mosquitoes carrying them. Because the earth is warming as a result of the generation of carbon dioxide and other "greenhouse gases" from the burning of fossil fuels, there is intensified scientific concern that warm-weather diseases will reemerge as a major health threat in the near future.

SCIENTIFIC FINDINGS. The question of whether the earth is warming as a result of human activity was settled in scientific circles in 1995, when the Second Assessment Report of the Intergovernmental Panel on Climate Change, a worldwide group of about twenty-five hundred experts, was issued. The panel concluded that the earth's temperature had increased between 0.5 to 1.1 degrees Farenheit (0.3 to 0.6 degrees Celsius) since reliable worldwide records first became available in the late nineteenth century. Furthermore, the intensity of warming had increased over time. By the 1990's, the temperature was rising at the most rapid rate in at least ten thousand years.

The Intergovernmental Panel concluded that human activity—the increased generation of carbon dioxide and other "greenhouse gases"—is responsible for the accelerating rise in global temperatures. The amount of carbon dioxide in the atmosphere has risen nearly every year because of increased use of fossil fuels by ever-larger human populations experiencing higher living standards.

In 1998, Paul Epstein of the Harvard School of Public Health described the spread of malaria and dengue fever to higher altitudes in tropical areas of the earth as a result of warmer temperatures. Rising winter temperatures have allowed disease-bearing insects to survive in areas that could not support them previously. According to Epstein, frequent flooding, which is associated with warmer temperatures, also promotes the growth of fungus and provides excellent breeding grounds for large numbers of mosquitoes. Some experts cite the flooding caused by Hurri-

215

cane Floyd and other storms in North Carolina during 1999 as an example of how global warming promotes conditions ideal for the spread of diseases imported from the Tropics.

HEAT, HUMIDITY, AND DISEASE. During the middle 1990's, an explosion of termites, mosquitoes, and cockroaches hit New Orleans, following an unprecedented five years without frost. At the same time, dengue fever spread from Mexico across the border into Texas for the first time since records have been kept. Dengue fever, like malaria, is carried by a mosquito that is limited by temperature and humidity. Colombia was experiencing plagues of mosquitoes and outbreaks of the diseases they carry, including dengue fever and encephalitis, triggered by a record heat wave followed by heavy rains. In 1997 Italy also had an outbreak of malaria.

The global temperature is undeniably rising. According to the National Oceanic and Atmospheric Administration, July, 1998, was the hottest month since reliable worldwide records have been kept, or about 150 years. The previous record had been set in July, 1995.

The rising incidence of some respiratory diseases may be related to a warmer, more humid environment. The American Lung Association reported that more than fifty-six hundred people died of asthma in the United States during 1995, a 45.3 percent increase in mortality over ten years, and a 75 percent increase since 1980. Roughly a third of those cases occurred in children under the age of eighteen. Asthma is now one of the leading diseases among the young. Since 1980, there has been a 160 percent increase in asthma in children under the age of five.

HEAT WAVES AND HEALTH. A study by the Sierra Club found that air pollution, which will be enhanced by global warming, could be responsible for many human health problems, including respiratory diseases such as asthma, bronchitis, and pneumonia.

According to Joel Schwartz, an epidemiologist at Harvard University, air pollution concentrations in the late 1990's were responsible for 70,000 early deaths per year and more than 100,000 excess hospitalizations for heart and lung disease in the United States. Global warming could cause these numbers to increase 10 to 20 percent in the United States, with significantly greater increases in countries that are more polluted to begin with, according to Schwartz.

Studies indicate that global warming will directly kill hundreds of Americans from exposure to extreme heat during summer months. The U.S. Centers for Disease Control and Prevention have found that extreme heat is responsible for an average of at least 240 deaths a year in the United States. Heat waves can double or triple the overall death rates in large cities. The death toll in the United States from a heat wave during July, 1999, surpassed two hundred people. As many as six hundred people died in Chicago alone during the 1990's due to heat waves. The elderly and very young have been most at risk.

Respiratory illness is only part of the picture. The Sierra Club study indicated that rising heat and humidity would broaden the range of tropical diseases, resulting in increasing illness and death from diseases such as malaria, cholera, and dengue fever, whose range will spread as mosquitoes and other disease vectors migrate.

The effects of El Niño in the 1990's indicate how sensitive diseases can be to changes in climate. A study conducted by Harvard University showed that warming waters in the Pacific Ocean likely contributed to the severe outbreak of cholera that led to thousands of deaths in Latin American countries. Since 1981, the number of cases of dengue fever has risen significantly in South America and has begun to spread into the United States. According to health experts cited by the Sierra Club study, the outbreak of dengue near Texas shows the risks that a warming climate might pose. Epstein and the Sierra Club study concur that if tropical weather expands, tropical diseases will expand.

In many regions of the world, malaria is already resistant to the least expensive, most widely distributed drugs. Worldwide, malaria already causes two million deaths a year, as well as 350 million new infections. The increased incidence of diseases will add to society's expenditures for hospitalization and other health care, the cost of lost productivity, and the trauma of illness and death.

Bruce E. Johansen

FOR FURTHER STUDY

Abrahamson, Dean Edwin. *The Challenge of Global Warming.* Washington, D.C.: Island Press, 1989.

Christianson, Gale E. *Greenhouse: The 200-Year Story of Global Warming.* New York: Walker, 1999.

Cline, William R. *The Economics of Global Warming.* Washington, D.C.: Institute for International Economics, 1992.

Lyman, Francesca. *The Greenhouse Trap: What We Are Doing to the Atmosphere and How We Can Slow Global Warming.* Boston: Beacon Press, 1990.

Nance, John J. *What Goes Up: The Global Assault on Our Atmosphere.* New York: William Morrow, 1991.

EXPLORATION AND TRANSPORTATION

EXPLORATION AND HISTORICAL TRADE ROUTES

The world's exploration was shaped and influenced substantially by economic needs. Lacking certain resources and outlets for trade, many societies built ships, organized caravans, and conducted military expeditions to protect their frontiers and obtain new markets.

Over the last five thousand years, the world evolved from a cluster of isolated communities into a firmly integrated global community and capitalist world system. By the beginning of the twentieth century, explorers had successfully navigated the oceans, seas, and landmasses and gathered many regional economies into the beginnings of a global economy.

EARLY TRADE SYSTEMS. Trade and exploration accompanied the rise of civilization in the Middle East. Egyptian pharaohs, looking for timber for shipbuilding, established trade relations with Mediterranean merchants. Phoenicians probed for new markets off the coast of North Africa and built a permanent settlement at Carthage. By 513 B.C.E., the Persian Empire stretched from the Indus River in India to the Libyan coast, and it controlled the pivotal trade routes in Iran and Anatolia. A regional economy was tak-

ing shape, linking Africa, Asia, and Europe into a blended economic system.

Alexander the Great's victory against the Persian Empire in 330 B.C.E. thrust Greece into a dominant position in the Middle Eastern economy. Trade between the Mediterranean and the Middle East increased, new roads and harbors were constructed, and merchants expanded into sub-Saharan Africa, Arabia, and India. The Romans later benefited from the Greek foundation. Through military and political conquest, Rome consolidated its control over such diverse areas as Arabia and Britain and built a system of roads and highways that facilitated the growth of an expanding world economy. At the apex of Roman power in 200 C.E., trade routes provided the empire with Greek marble, Egyptian cloth, seafood from Black Sea fisheries, African slaves, and Chinese silk.

The emergence of a profitable Eurasian trade route linked people, customs, and economies from the South China Sea to the Roman Empire. Although some limited activity occurred during the Hellenistic period, East-West trade flourished following the rise of the Han Dynasty in China. With the opening of the Great Silk

Road from 139 B.C.E. to 200 C.E., goods and services were exchanged between people from three different continents.

The Great Silk Road was an intricate network of middlemen stretching from China to the Mediterranean Sea. Eastern merchants sold their products at markets in Afghanistan, Iran, and even Syria, and exchanged a variety of commodities through the use of camel caravans. Chinese spices, perfumes, metals, and especially silk were in high demand. The Parthians from central Asia added their own sprinkling of merchandise, introducing both the East and the West to various exotic fruits, rare birds, and ostrich eggs.

Romans peddled glassware, statuettes, and acrobatic performing slaves. Since communication lines were virtually nonexistent during this period, trade routes were the only means by which ideas regarding art, religion, and culture could mix. The contacts and exchanges enacted along the Great Silk Road initiated a process of cultural diffusion among a diversity of cultures and increased each culture's knowledge of the vast frontiers of world geography.

THE ATLANTIC SLAVE TRADE. Beginning in the fifteenth century, European navigators explored the West African coastline seeking gold. Supplies were difficult to procure, because most of the gold mines were located in the interior along the Senegal River and in the Ashanti forests. Because mining required costly investments in time, labor, and security, the Europeans quickly shifted their focus toward the slave trade. Although slavery had existed since antiquity, the Atlantic slave trade generated one of the most significant movements of people in world history. It led to the forced migration of more than ten million Africans to South America, the Caribbean islands, and North America. It ensured the success of

several imperial conquests, and it transformed the demographic, cultural, and political landscape on four continents.

Originally driven by their quest to circumnavigate Africa and open a lucrative trade route with India, the Portuguese initiated a systematic exploration of the West African coastline. The architect of this system, Henry the Navigator, pioneered the use of military force and naval superiority to annex African islands and open up new trade routes, and he increased Portugal's southern frontier with every acquisition. In 1415 his ships captured Ceuta, a prosperous trade center located on the Mediterranean coast overlooking North African trade routes. Over the next four decades, Henry laid claim to the Madeira Islands, the Canary Islands, the Azores, and Cape Verde. After his death, other Portuguese explorers continued his pursuit of circumnavigation of Africa.

Diego Cão reached the Congo River in 1483 and sent several excursions up the river before returning to Lisbon. Two explorers completed the Portuguese mission at the end of the fifteenth century. Vasco da Gama, sailing from 1497 to 1499, and Bartholomeu Dias, from 1498 to 1499, sailed past the southern tip of Africa and eventually reached India. Since Muslims had already created a number of trade links between East Africa, Arabia, and India, Portuguese exploration furthered the integration of various regions into an emerging capitalist world system.

When the Portuguese shifted their trading from gold to slaves, the other European powers followed suit. The Netherlands, Spain, France, and England used their expanding naval technology to explore the Atlantic Ocean and ship millions of slaves across the ocean. A highly efficient and organized trade route quickly materialized. Since the Europeans were unwilling to venture beyond the walls of

Ferdinand Magellan. (Library of Congress)

Sixteenth century world map Page 168

their coastal fortresses, merchants relied on African sources for slaves, supplying local kings and chiefs with the means to conduct profitable slave-raiding parties in the interior. In both the Congo and the Gold Coast region, many Africans became quite wealthy trading slaves.

In 1750 merchants paid the king of Dahomey 250,000 pounds for nine thousand slaves, and his income exceeded the earnings of many in England's merchant and landowning class. After purchasing slaves, dealers sold them in the Americas to work in the mines or on plantations. Commodities such as coffee and sugar were exported back to Europe for home consumption. Merchants then sold alcohol, tobacco, textiles, and firearms to Africans

in exchange for more slaves. This practice was abolished by the end of the nineteenth century, but not before more than ten million Africans had been violently removed from their homeland. The Atlantic slave trade, however, joined port cities from the Gold Coast and Guinea in Africa with Rio de Janeiro, Hispaniola, Havana, Virginia, Charleston, and Liverpool, and constituted a pivotal step toward the rise of a unified global economy.

MAGELLAN AND ZHENG HE. The Portuguese explorer Ferdinand Magellan generated considerable interest in the Asian markets when he led an expedition that sailed around the world from 1519 to 1522. Looking for a quick route to Asia and the Spice Islands, he secured financial backing from the king of Spain. Magellan sailed from Spain in 1519, canvassed the eastern coastline of South America, and visited Argentina. He ultimately traversed the narrow straits along the southern tip of the continent and ventured into the uncharted waters of the Pacific Ocean.

Magellan explored the islands of Guam and the Philippines but was killed in a skirmish on Mactan in 1521. Some of his crew managed to return to Spain in 1522, and one member subsequently published a journal of the expedition that drastically enhanced the world's understanding of the major sea lanes that connected the continents.

China also opened up new avenues of trade and exploration in Southeast Asia during the fifteenth century. Under the direction of Chinese emperor Yongle, explorer Zheng He organized seven overseas trips from 1405 to 1433 and investigated economic opportunities in Korea, Vietnam, the Indian Ocean, and Egypt. His first voyage consisted of more than twenty-eight thousand men and four hundred ships and represented the largest naval force assembled prior to World War I.

Zheng's armada carried porcelains, silks, lacquerware, and artifacts to Malacca, the vital port city in Indonesia. He purchased an Arab medical text on drug therapy and had it translated into Chinese. He introduced giraffes and mahogany wood into the mainland's economy, and his efforts helped spread Chinese ideas, customs, diet, calendars, scales and measures, and music throughout the global economy. Zheng He's discoveries, coupled with all the material gathered by the European explorers, provided cartographers and geographers with a credible store of knowledge concerning world geography.

EMERGING GLOBAL TRADE NET-WORKS. From 1400 to 1900, several regional economic systems facilitated the exchange of goods and services throughout a growing world system. Building on the triangular relationships produced by the slave trade, the Atlantic region helped spread new foodstuffs around the globe. Plants and plantation crops provided societies with a plentiful supply of sweet potatoes, squash, beans, and maize. This system, often referred to as the Columbian exchange, also assisted development in other regions by supplying the global economy with an ample money supply in gold and silver. Europeans sent textiles and other manufactures to the Americas. In return, they received minerals from Mexico; sugar and molasses from the Caribbean; money, rum, and tobacco from North America; and foodstuffs from South America. Trade routes also closed the distance between the Pacific coastline in the Americas and the Pacific Rim.

Additional thriving trade routes existed in the African-West Asian region. Linking Europe and Africa with Arabia and India, this area experienced a considerable amount of trade over land and through the sea lanes in the Persian Gulf and Red Sea. Europeans received grains, timber, furs, iron, and hemp from Russia in exchange for wool textiles and silver. Central Asians secured stores of cotton textiles, silk, wheat, rice, and tobacco from India and sold silver, horses, camel, and sheep to the Indians. Ivory, blankets, paper, saltpeter, fruits, dates, incense, coffee, and wine were regularly exchanged among merchants situated along the trade route connecting India, Persia, the Ottoman Empire, and Europe.

Finally, a Russian-Asian-Chinese market provided Russia's ruling czars with arms, sugar, tobacco, and grain, and a sufficient supply of drugs, medicines, livestock, paper money, and silver moved eastward. Overall, this system linked the economies of three continents and guaranteed that a nation could acquire essential foodstuffs, resources, and money from a variety of sources.

Several profitable trade routes existed in the Indian Ocean sector. After Malacca emerged as a key trading port in the sixteenth century, this territory served as an international clearinghouse for the global economy. Indians sent tin, elephants, and wood into Burma and Siam. Rice, silk, and sugar were sold to Bengal. Pepper and other spices were shipped westward across the Arabian Sea, while Ceylon furnished India with vital quantities of jewels, cinnamon, pearls, and elephants. The booming interregional trade routes positioned along the Indian coastline ensured that many of the vast commodities produced in the world system could be obtained in India.

The final region of crucial trade routes was between Southeast Asia and China. While the extent of Asian overseas trade prior to the twentieth century is usually downplayed, an abundance of products flowed across the Bay of Bengal and the South China Sea. Japan procured silver, copper, iron, swords, and sulphur from Cantonese merchants, and Japanese-finished textiles, dyes, tea, lead, and manufactures were in high demand on the mainland. The Chinese also purchased silk and ceramics from the Philippines in exchange for silver. Burma and Siam traded pepper, sappan wood, tin, lead, and saltpeter to China for satin, velvet, thread, and labor. As goods increasingly moved from the Malabar coast in India to the northern boundaries of Korea and Japan, the Pacific Rim played a prominent role in the global economy.

Robert D. Ubriaco, Jr.

FOR FURTHER STUDY

Blaut, J. M. *The Colonizer's Model of the World.* New York: Guilford Press, 1993.

Frank, Andre Gunder. *ReOrient: Global Economy in the Asian Age.* Berkeley: University of California Press, 1998.

Frank, Andre Gunder, and Barry K. Bills, eds. *The World System: Five Hundred or Five Thousand?* Reprint. New York: Routledge, 1996.

Smith, Alan K. *Creating a World Economy: Merchant Capital, Colonialism, and World Trade, 1400-1825.* Boulder, Colo.: Westview Press, 1991.

Thomas, Hugh. *The Slave Trade.* New York: Simon & Schuster, 1997.

Wallerstein, Immanuel. *The Capitalist World-Economy.* Reprint. New York: Cambridge University Press, 1993.

Wolf, Eric R. *Europe and the People Without History.* Berkeley: University of California Press, 1982.

Canal in the Netherlands. Canal construction is an ancient engineering technology that was perfected in the mid-eighteenth century. Before modern engines, canals were simple waterbeds of uniform width and depth bordered by towpaths on which animals or men towed barges on the water. (PhotoDisc)

Large-scale strip mine. Strip mining is simpler and cheaper to undertake than underground mining but does heavy—and often irreparable—damage to the environment. (PhotoDisc)

A worker in Thailand prepares piles of salt for marketing. Sodium chloride—which we know as ordinary table salt— is an important mineral throughout the world. (Clyde L. Rasmussen)

This cemetery in Lithuania's capital city, Vilnius, recalls the years of Soviet occupation. The tombs of former Soviet officials are tiered according to the officials' status, with the most important people buried at the highest levels. (Charles F. Bahmueller)

During the Persian Gulf War of 1991, Saudi Arabia served as the major staging ground for allied military operations against Saddam Hussein's Iraq. Here troops of the U.S. First Cavalry Division deploy across a Saudi desert in late 1990, two months before the war began. (AP/Wide World Photos)

Gulf of Aden (lower right) and Red Sea (upper left) photographed from space. Yemen's capital city, Aden, is on the north coast of the gulf, immediately to the right of the smooth semicircular indentation in the coastline. (PhotoDisc)

CURRENT POLITICAL BOUNDARIES OF THE WORLD

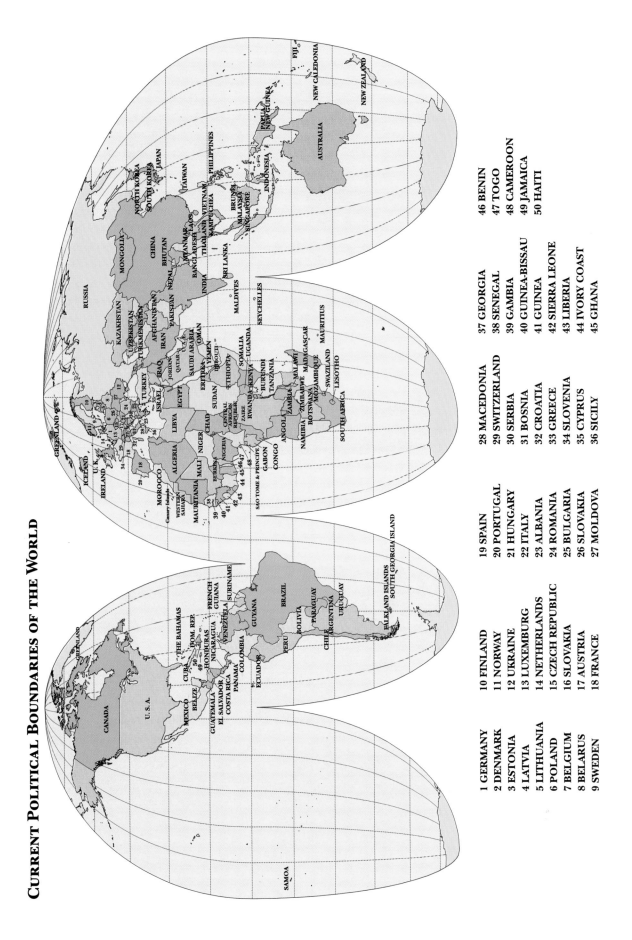

1 GERMANY
2 DENMARK
3 ESTONIA
4 LATVIA
5 LITHUANIA
6 POLAND
7 BELGIUM
8 BELARUS
9 SWEDEN

10 FINLAND
11 NORWAY
12 UKRAINE
13 LUXEMBURG
14 NETHERLANDS
15 CZECH REPUBLIC
16 SLOVAKIA
17 AUSTRIA
18 FRANCE

19 SPAIN
20 PORTUGAL
21 HUNGARY
22 ITALY
23 ALBANIA
24 ROMANIA
25 BULGARIA
26 SLOVAKIA
27 MOLDOVA

28 MACEDONIA
29 SWITZERLAND
30 SERBIA
31 BOSNIA
32 CROATIA
33 GREECE
34 SLOVENIA
35 CYPRUS
36 SICILY

37 GEORGIA
38 SENEGAL
39 GAMBIA
40 GUINEA-BISSAU
41 GUINEA
42 SIERRA LEONE
43 LIBERIA
44 IVORY COAST
45 GHANA

46 BENIN
47 TOGO
48 CAMEROON
49 JAMAICA
50 HAITI

The Arabian Sea seen from space, looking approximately to the southeast, in April, 1984. Iran is at the left, the Gulf of Oman at the lower center, and Oman at the right. (Corbis)

OCEANS AND CONTINENTS

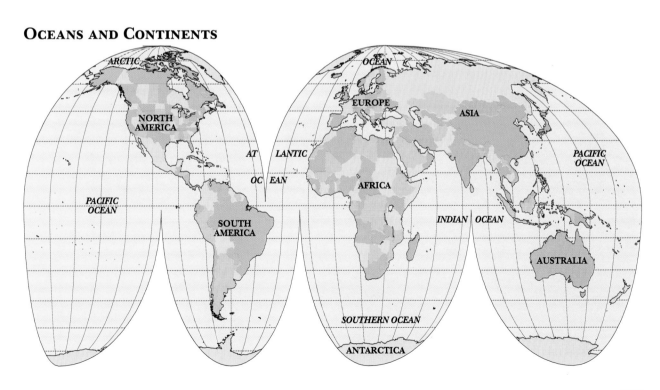

Asia's, and the world's, tallest mountain, Everest rises to a height of 29,035 feet (8,850 meters). That distinction may well have made it the world's single most-famous landform. (AP/Wide World Photos)

The so-called "Bermuda Triangle" is a loosely defined region off the Atlantic coast of Florida. It roughly encompasses an imaginary triangle that connects Melbourne, Florida; Bermuda; and Puerto Rico. A large number of ships and airplanes have disappeared within this region, giving rise to unscientific speculation about extraterrestrial intervention and supernatural forces. However, scientists counter that the number of missing ships and aircraft is not unusual, given the region's heavy air and sea traffic and often stormy weather conditions. (PhotoDisc)

228

Early eighteenth century map of the Atlantic ocean by Manoel Ferreira. (Corbis)

The Great Barrier Reef, the largest coral formation in the world, lies in the Coral Sea off the east coast of Australia. The reef system and its small islands stretch for more than 1,100 miles (1,750 km.).

Fish in the Great Barrier Reef. (Corbis)

Coral in the Great Barrier Reef. (Corbis)

Typical California coastline, where wave erosion and landslides threaten structures and roadways. (Ray Sumner)

The Hatteras Abyssal Plain, which forms part of the floor of the northwest Atlantic Basin, rises to form shallow sandbars around Cape Hatteras, North Carolina, that are navigational hazards. Cape Hatteras itself is mostly a sandbar. (PhotoDisc)

Of all the countries in the world threatened by the effects of global warming, the tiny Indian Ocean Maldive archipelago, off the southern tip of India, stands to lose the most. Many of the country's hundreds of islands rise barely six feet (two meters) above sea level, making them peculiarly susceptible to the hazards of rising water. In an effort to protect beaches from tidal erosion, Japan helped the country erect concrete-block barriers on Maale in the late 1990's. However, the cost of building similar barriers on all the islands was prohibitive. (AP/Wide World Photos)

The clear, warm Mediterranean waters off the coasts of Spain, France, and Italy attract many vacationers. (PhotoDisc)

View of Sicily, southern Italy, and Malta (lower left) in the Mediterranean Sea, photographed from the shuttle Columbia *in June, 1991.* (Corbis)

Highway over the Florida Keys, a chain of small islands extending about 150 miles (240 km.) from the southern tip of Florida into the Gulf of Mexico. (PhotoDisc)

Hurricane Andrew moving through the Gulf of Mexico in 1992. (PhotoDisc)

Pacific Ocean waves breaking on the coast of Hawaii. (PhotoDisc)

235

Pacific Ocean sunset, viewed from the West Coast of the United States. (PhotoDisc)

The opening of the Panama Canal in 1914 created a direct link between the Pacific Ocean and the Atlantic Ocean, by way of the Caribbean Sea. (PhotoDisc)

Bora-Bora, in the Society Island group of French Polynesia, is surrounded by perhaps the most spectacular coral reef in the South Pacific. The island is known to many readers of James Michener, whose popular novel Hawaii *(1959) is partly set there.* (American Stock Photography)

Satellite view of a dust plume blowing onto the Red Sea from the east coast of Sudan, near Port Sudan. (PhotoDisc)

The term Ring of Fire has been given to the zone of volcanic and earthquake activity marking the edges of tectonic plates around the Pacific Ocean. The zone includes the coastal regions of California, where in 1989 a violent earthquake caused the collapse of this freeway in Oakland. (National Oceanic and Atmospheric Administration)

ROAD TRANSPORTATION

Roads—the most common surfaces on which people and vehicles move—are a key part of human and economic geography. Transportation activities form part of a nation's economic product: They strengthen regional economy, influence land and natural resource use, facilitate communication and commerce, expand choices, support industry, aid agriculture, and increase human mobility. The need for roads closely correlates with the relative location of centers of population, commerce, industry, and other transportation.

HISTORY OF ROAD MAKING. The great highway systems of modern civilization have their origin in the remote past. The earliest travel was by foot on paths and trails. Later, pack animals and crude sleds were used. The development of the wheel opened new options. As various ancient civilizations reached a higher level, many of them realized the importance of improved roads.

The most advanced highway system of the ancient world was that of the Romans. When Roman civilization was at its peak, a great system of military roads reached to the limits of the empire. The typical Roman road was bold in conception and construction, built in a straight line when possible, with a deep multilayer foundation, perfect for wheeled vehicles.

After the decline of the Roman Empire, rural road building in Europe practically ceased, and roads fell into centuries of disrepair. Commerce traveled by water or on pack trains that could negotiate the badly maintained roads. Eventually, a commercial revival set in, and roads and wheeled vehicles increased.

Interest in the art of road building was revived in Europe in the late eighteenth century. P. Trésaguet, a noted French engineer, developed a new method of lightweight road building. The regime of French dictator Napoleon Bonaparte (1800-1814) encouraged road construction, chiefly for military purposes. At about the same time, two Scottish engineers, Thomas Telford and John McAdam, also developed road-building techniques.

Modern British road built on the bed of an ancient Roman road, whose original stone siding is still visible. (PhotoDisc)

ROADS IN THE UNITED STATES. Toward the end of the eighteenth century, public demand in the United States led to the improvement of some roads by private enterprise. These improvements generally took the form of toll roads, called "turnpikes" because a pike was rotated in each road to allow entry after the fee was paid, and generally were located in areas adjacent to larger cities. In the early nineteenth century, the federal government paid for an 800-mile-long macadam road from Cumberland, Maryland, to Vandalia, Illinois.

With the development of railroads, interest in road building began to wane. By 1900, however, demand for better roads came from farmers, who wanted to move their agricultural products to market more easily. The bicycle craze of the 1890's and the advent of motorized vehicles also added to the demand for more and better roads. Asphalt and concrete technology was well developed by then; now, the problem was financing. Roads had been primarily a local issue, but the growing demand led to greater state and federal involvement in funding.

Freeway interchange Page 167

The Federal-Aid Highway Act of 1956 was a milestone in the development of highway transportation in the United States; it marked the beginning of the largest peacetime public works program in the history of the world, creating a 41,000-mile National System of Interstate and Defense Highways, built to high standards. Later legislation expanded funding, improved planning, addressed environmental concerns, and provided for more balanced transportation. Other developed countries also developed highway programs but were more restrained in construction.

ROADS AND DEVELOPMENT. Transportation presents a severe challenge for sustainable development. The number of motor vehicles at the beginning of the twenty-first century—estimated at more than 600 million worldwide—is growing almost everywhere at higher rates than either population or the gross domestic product. Overall road traffic grows even more quickly. Americans own the most cars—one for every 1.7 residents—but even in crowded Japan there is one car for every 2.1 people. In Great Britain, there is one car for every 5.3 people.

Highways around the world have been built to help strengthen national unity. The Trans-Canada Highway, the world's longest national road, for example, extends east-west across the breadth of the country. Completed in the 1960's, it had the same goal as the Canadian Pacific Railroad a century before, to improve east-west commerce within Canada.

Sometimes, existing highways need to be upgraded; in less-developed countries,

HIGHWAY CLASSIFICATION

Modern roads can be classified by roadway design or traffic function. The basic type of roadway is the conventional, undivided two-way road. Divided highways have median strips or other physical barriers separating the lanes going in opposite directions.

Another quality of a roadway is its right-of-way control. The least expensive type of system controls most side access and some minor at-grade intersections; the more expensive type has side access fully controlled and no at-grade intersections. The amount of traffic determines the number of lanes. Two or three lanes in each direction is typical, but some roads in Los Angeles have five lanes, while some sections of the Trans-Canada Highway have only one lane. Some highways are paid for entirely from public funds; if users pay directly when they use the road, the roads are called tollways or turnpikes.

Roads are classified as expressway, arterial, collector, and local in urban areas, with a similar hierarchy in rural areas. The highest level—expressway—is intended for long-distance travel.

this can simply mean paving a road for all-weather operation. An example of a late-1990's project of this nature was the Brazil-Venezuela Highway project, which had this description: Improve the Brazil-Venezuela highway link by completion of paving along the BR-174, which runs northward from Manaus in the Amazon, through Boa Vista and up to the frontier, so opening a route to the Caribbean. Besides the investment opportunities in building the road itself, the highway would result in investment opportunities in mining, tourism, telecommunications, soy and rice production, trade with Venezuela, manufacturing in the Manaus Free Trade Zone, ecotourism in the Amazon, and energy integration.

Growing road traffic has required increasingly significant national contributions to road construction. Beginning in the 1960's, the World Bank began to finance road construction in several countries. It required that projects be organized to the highest technical and economic standards, with private contracting and international competitive bidding rather than government workers. Still, there were questions as to whether these economic assessments had a road-sector bias and properly incorporated environmental costs. Sustainability was also a question—could the facilities be maintained once they were built?

In the 1990's, the World Bank financed a program to build an asphalt road network in Mozambique. Asphalt makes very smooth roads but is very maintenance-intensive, requiring expensive imported equipment and raw materials. By the end of the decade, the roads required resurfacing but the debt was still outstanding. Alternative materials would have given a rougher road, but it could have been built with local materials and labor.

The European Investment Bank has be-

come a major player in the construction of highways linking Eastern and Western Europe to further European integration. Some of the fastest growth in the world in ownership of autos has been in Eastern Europe. There is a two-way feedback effect between highway construction and auto ownership.

ENVIRONMENT CONSEQUENCES. Highways and highway vehicles have social, economic, and environmental consequences. Compromise is often necessary to balance transportation needs against these constraints. For example, in Israel, there has been a debate over construction of the Trans-Israel highway, a $1.2 billon, six-lane highway stretching 180 miles (300 km.) from Galilee to the Negev.

Demand on resources for worldwide road infrastructure far exceeds available funds; governments increasingly are looking to external sources such as tolls. Private toll roads, common in the nineteenth century, are making a comeback. This has spread from the United States to Europe, where private and government-owned highway operators have begun to sell shares on the stock market. Private companies are not only operating and financing roads in Europe, they are also designing and building them. In Eastern Europe, where road construction languished under communism, private financing and toll collecting are seen as the means of supporting badly needed construction.

Industrial development in poor countries is adversely affected by limited transportation. Costs are high—unreliable delivery schedules make it necessary to maintain excessive inventories of raw materials and finished goods. Poor transport limits the radius of trade and makes it difficult for manufacturers to realize the economies of large-scale operations to compete internationally.

In more difficult terrain, roads become

more expensive because of a need for cuts and fills, bridges, and tunnels. To save money, such roads often have steeper grades, sharper curves, and reduced width than might be desired. Severe weather changes also damage roads, further increasing maintenance costs.

Stephen B. Dobrow

INFORMATION ON THE WORLD WIDE WEB

Information on road transportation can be found at the Web sites of professional organizations involved in highways, such as the Institute of Transportation Engineers (www.ite.org) and the American Association of State Highway and Transportation Officials (www.aashto.org).

The Institute for Transportation and Development Policy is an organization concerned with worldwide sustainable transportation; its Web site features information on programs and publications and links to relevant sites. (www.itdp.org)

FOR FURTHER STUDY

Edwards, John D., Jr, ed. *Transportation Planning Handbook.* 2d ed. Washington, D.C.: Institute of Traffic Engineers, 1999.

Hawkes, Nigel. *Structures.* New York: Macmillan, 1990.

Lay, M. G. *Ways of the World.* New Brunswick, N.J.: Rutgers University Press, 1992.

Owen, Wilfred. *Transportation and World Development.* Baltimore: Johns Hopkins University Press, 1987.

Wright, Paul. *Highway Engineering.* New York: Wiley, 1995.

RAILWAYS

Modern subway Page 169

Railroads were the first successful attempts by early industrial societies to develop integrated communication systems. At the start of the twenty-first century, global societies are linked by Internet systems dependent upon communication satellites orbiting around Earth. The speed by which information and ideas can reach remote places breaks down isolation and aids in the developing of a world community. In the nineteenth century, railroads had a similar impact. Railroads were critical for the creation of an urban-industrial society: They linked regions and remote places together, were important contributors in developing nation-states, and revolutionized the way business was conducted through the creation of corporations. Although alternative forms of transportation exist at the beginning of the twenty-first century, railroads remain important.

THE INDUSTRIAL REVOLUTION AND THE RAILROAD. Development of the steam engine gave birth to the railroad. Late in the eighteenth century, James Watt perfected his steam engine in England.

Water was superheated by a boiler and vaporized into steam, which was confined to a cylinder behind a piston. Pressure from expanding steam pushes the cylinder forward, causing it to do work if it is attached to wheels. Watt's engine was used in the manufacturing of textiles, thus beginning the Industrial Revolution whereby machine technology mass produced goods for mass consumption. Robert Fulton was the first innovator to commercially apply the steam engine to water transportation. His steamboat *Clermont* made its maiden voyage up the Hudson River in 1807.

Not until the 1820's was a steam engine used for land transportation. Rivers and lakes were natural features where no road needed to be built. Applying steam to land movement required some type of roadbed. In England, George Stephenson ran a locomotive over iron strips attached to wooden rails. Within a short time, England's forges were able to roll rails made completely of iron shaped like an inverted "U."

How much profit a manufacturer could make was determined partially by the cost of transportation. The lower the cost of moving cargo and people, the higher the profitability. Several alternatives existed before the emergence of railroads. Toll roads were too slow. A loaded wagon pulled by four horses could average 15 miles (25 km.) a day.

Canals were more efficient than early railroads, because barges pulled by mules moved faster over waterways. However, canals could not be built everywhere, especially over mountains. The application of railroad technology, using steam as a power source, made it possible to overcome obstacles in moving goods and people over considerable distances and at profitable costs. Railroads transformed the way goods were purchased by reducing the costs for consumers, thus raising the living standards in industrial societies. Railroads transformed the human landscape by strengthening the link between farm and city, changed commercial cities into industrial centers, and started early forms of suburban growth well before automobiles arrived.

FINANCING RAILROADS. Constructing railroads was costly. Tunnels had to be blasted through mountains, and rivers had to be crossed by bridges. Early in the building of U.S. railroads, the nation's iron foundries could not meet the demands for rolled rails. Rails had to be imported from England until local forges developed more efficient technologies. Once a railway completed, there was a constant need to maintain the right-of-way so that traffic flow would not be disrupted. Accidents were frequent, and it was an early practice to burn damaged cars because salvaging them was too expensive.

In some countries, railroads were built and operated by national governments. In the United States, railroads were privately owned; however, it was impossible for any single individual to finance and operate a rail system with miles of track. Businessmen raised money by selling stocks and bonds. Just as investors buy stocks in modern high-technology companies, investors purchased stocks and bonds in railroads.

Investing in railroads was good as long as they earned profits and returned money to their investors, but not all railroads made sufficient profits to reward their investors. Competition among railroads was heavy in the United States, and some railroads charged artificially low fares to attract as much business as they could. When ambitious investment schemes collapsed, railroads went bankrupt and were taken over by financiers.

Selling shares of common stock and bonds was made possible by creating cor-

Dutch canal Page 223

porations. Railroads were granted permission from state governments to organize a corporation. Every investor owned a portion of the railroad. Stockholders' interests were served by boards of directors, and all business transactions were opened for public inspection. One important factor of the corporation was that it relieved individuals from the responsibilities associated with accidents. The railroad, as a corporation, was held accountable, and any compensation for claims made against the company came out of corporate funds, not from individual pockets. This had an impact on the law profession, as law schools began specializing in legal matters relevant to railroads and interstate commerce.

Nineteenth century U.S. railroads Page 168

THE SUCCESS OF RAILROADS. Railroads usually began by radiating outward from port cities where merchants engaged in transoceanic trade. A classic example, in the United States, is the country's first regional railroad—the Baltimore and Ohio. Construction commenced from Baltimore in 1828; by 1850, the railroad had crossed the Appalachian Mountains and was on the Ohio River at Wheeling, Virginia.

Once trunk lines were established, rail networks became more intensive as branch lines were built to link smaller cities and towns. Countries with extremely large continental dimensions developed interior articulating cities where railroads from all directions converged. Chicago and Atlanta are two such cities in the United States. Chicago was surrounded by three circular railroads (belts) whose only function was to interchange cars. Railroads from the Pacific Coast converged with lines from the Atlantic Coast as well as routes moving north from the Gulf Coast.

Mechanized farms and heavy industries developed within the network. Railroads made possible the extraction of fossil fuels and metallic ores, the necessary ingredients for industrial growth. Extension of railroads deep into Eastern Europe helped to generate massive waves of immigration into both North and South America, creating multicultural societies.

Building railroads in Africa and South Asia made it possible for Europe to increase its political control over native populations. The ultimate aim of the colonial railroad was to develop a colony's economy according to the needs of the mother country. Railroads were usually single-line routes transhipping commodities from interior centers to coastal ports for exportation. Nairobi, Kenya, began as a rail hub linking British interests in Uganda with Kenya's port city of Mombasa. Similar examples existed in Malaysia and Indonesia.

Railroads generated conflicts among colonial powers as nations attempted to acquire strategic resources. In 1904-1905 Russia and Japan fought a war in the Chinese province of Manchuria over railroad rights; Imperial Germany attempted to get around British interests in the Middle East by building a railroad linking Berlin with Baghdad to give Germany access to lucrative oil fields. India was a region of loosely connected provinces until British railroads helped establish unification. The resulting sense of national unity led to the termination of British rule in 1947 and independence for India and Pakistan.

In the United States, private railroads discontinued passenger service among cities early in the 1970's and the responsibility was assumed by the federal government (Amtrak). Most Americans riding trains do so as commuters traveling from the suburbs to jobs in the city. High-speed train service is planned along the rail corridor between Washington and New York, Amtrak's most popular route. Passenger service remains popular in Japan and

Japanese bullet trains, which move commuters in and out of major cities at speeds well in excess of one hundred miles per hour. (Corbis)

Europe. France, Germany, and Japan operate high-speed luxury trains with speeds averaging above 100 miles (160 km.) per hour.

Railroads are no longer the exclusive means of land transportation as they were early in the twentieth century. Although competition from motor vehicles and air freight provide alternate choices, railroads have remained important. France and England have direct rail linkage beneath the English Channel. In the United States, great railroad mergers and the application of computer technology have reduced operating costs while increasing profits. Transoceanic container traffic has been aided by railroads hauling trailers on flatcars. Railroads began the process of bringing regions within a nation together

in the nineteenth century just as the computer and the World Wide Web began uniting nations throughout the world at the end of the twentieth century.

Sherman E. Silverman

FOR FURTHER STUDY

Banister, David, et al. *European Transport and Communication Networks.* New York: Wiley, 1995.

Daniels, Rudolph. *Trains Across the Continent: North American Railroad History.* Bloomington: Indiana University Press, 2000.

Dilts, James D. *The Great Road. The Building of the Baltimore & Ohio, the Nation's First Railroad, 1828-1855.* Stanford, Calif.: Stanford University Press, 1993.

Jensen, Oliver. *The American Heritage His-*

tory of Railroads in America. New York: American Heritage, 1975.

Martin, Albro. *Railroads Triumphant.* New York: Oxford University Press, 1992.

Pindell, Terry. *Making Tracks: An American Rail Odyssey.* New York: Grove Weidenfeld, 1990.

Vance, James E., Jr. *The North American Railroad.* Baltimore: Johns Hopkins University Press, 1995.

INFORMATION ON THE WORLD WIDE WEB

The World Rail and Transit site lists different types of worldwide railway transit systems by both country and city, with definitions and source materials. (home.cc.umanitoba.ca/~wyatt/rail-transit-list.html)

AIR TRANSPORTATION

The movement of goods and people among places is an important field of geographic study. Transportation routes form part of an intricate global network through which commodities flow. Speed and cost determine the nature and volume of the materials transported, so air transportation has both advantages and disadvantages when compared with road, rail, or water transport.

EARLY FLYING MACHINES. The transport of people and freight by air is less than a century old. Although hot-air balloons were used in the late eighteenth century for military purposes, aerial mapping, and even early photography, they were never commercially important as a means of transportation. In the late nineteenth century, the German count Ferdinand von Zeppelin began experimenting with dirigibles, which added self-propulsion to lighter-than-air craft. These aircraft were used for military purposes, such as the bombing of Paris in World War I. However, by the 1920's zeppelins had become a successful means of passenger transportation. They carried thousands of passengers on trips in Europe or across the Atlantic Ocean and also were used for exploration. Nevertheless, they had major problems and were soon superseded by flying machines heavier than air. The early term for such a machine was an aeroplane, which is still the word used for airplane in Great Britain.

Following pioneering advances with the internal combustion engine and in aerodynamic theory using gliders, the development of powered flight in a heavier-than-air machine was achieved by Wilbur and Orville Wright in December, 1903. From that time, the United States moved to the forefront of aviation, with Great Britain and Germany also making significant contributions to air transport. World War I saw the further development of aviation for military purposes, evidenced by the infamous bombing of Guernica.

EARLY COMMERCIAL SERVICE. Two decades after the Wright brothers' brief flight, the world's first commercial air service began, covering the short distance from Tampa to St. Petersburg in Florida. The introduction of airmail service by the U.S. Post Office provided a new, regular source of income for commercial airlines in the United States, and from these beginnings arose the modern Boeing Company, United Airlines, and American Airlines. Europe, however, was the home of the world's first commercial airlines. These include the Deutsche Luftreederie in Germany, which connected Berlin, Leipzig, and Weimar in 1919; Farman in France, which flew from Paris to London; and KLM in the Netherlands (Amsterdam to London), followed by Qantas—the Queensland and Northern Territory Aerial Services, Limited—in Australia. The last two are the world's oldest still operating airlines.

Aircraft played a vital role in World War II, as a means of attacking enemy territory, defending territory, and transporting people and equipment. A humanitarian use of air power was the Berlin Air Lift of 1948, when Western nations used airplanes to deliver food and medical supplies to the people of West Berlin, which the Soviet Union briefly blockaded on the ground.

CARGO AND PASSENGER SERVICE. The jet engine was developed and used for fighter aircraft during World War II by the Germans, the British, and the United States. Further research led to civil jet transport, and by the 1970's, jet planes accounted for most of the world's air transportation. Air travel in the early days was extremely expensive, but technological advances enabled longer flights with heavier loads, so commercial air travel became both faster and more economical.

Although people in the United States still use personal vehicles for most of

A national memorial to the Wright brothers on North Carolina's Outer Banks commemorates the site of their first heavier-than-air flight. (PhotoDisc)

Jetliner
Page 170

their travel, they prefer air travel for longer trips. Almost three-quarters of trips in excess of 1,000 miles (1,600 km.) that Americans undertook during the 1990's were made by airplane. Most air travel is made for business purposes. The United States had more than 1,500 airports in the year 2000, but more than half of these were short, privately owned, unpaved airstrips. There are 180 commercial or military airports with runways more than 9,800 feet (3,000 meters) long.

Between 1960 and 2000 the number of passengers carried by air grew at an annual rate of 9 percent. Air travel is also extremely safe. In 1999 there were 674 airline fatalities, which is only 0.36 percent of the number of passengers. At the same time, the volume of freight increased by 7 percent and the volume of mail by 11 percent. Some 30 percent of the world's trade, by value, is carried by air. In 1999 more than 1,337 million passengers were carried on scheduled domestic or international flights.

More than seven thousand airplanes fly for airlines around the world on any given day, logging almost 5 billion miles (8 billion km.) each year. North Americans dominate the world in use of commercial flights, accounting for almost 40 percent of all passenger miles flown. In the year 2000 almost half of all air passengers boarded flights in America. In 1999 Delta Airlines carried the greatest number of passengers in the world—105.5 million passengers on its domestic and international routes. British Airways and Lufthansa were the major international carriers, with 30.3 and 27.3 million international passengers respectively. Seven of the top ten of the world's major domestic airlines were U.S. carriers, with the other three being Japanese.

The biggest air cargo carriers in 1999 were Federal Express, which carried more than 5 million tons of cargo, and United Parcel Service (3 million tons). For international air freight, the leaders in 1999 were Federal Express (1.25 million tons), Lufthansa (1.1 million tons), and Korean Air (0.9 tons). Federal Express and UPS also led in domestic air freight transport.

The first commercial supersonic airliner, the British-French Concorde, which could fly at more than twice the speed of sound, began regular service in early 1976. However, the fleet was grounded after a Concorde crash in France in mid-2000. The first space shuttle flew in 1981, and the hundredth space shuttle launch took place in October, 2000. The shuttles have transported 600 people and 3 million pounds (1.36 million kilograms) of cargo into space.

HEALTH PROBLEMS TRANSPORTED BY AIR. The high speed of intercontinental air travel and the increasing numbers of air travelers have increased the risk of exotic diseases being carried into destination countries, thereby globalizing diseases previously restricted to certain parts of the world. Passengers traveling by air might be unaware that they are carrying infections or viruses. The worldwide spread of HIV/AIDS after the 1980's was accelerated by international air travel.

Disease vectors such as flies or mosquitoes can also make air journeys unnoticed inside airplanes. At some airports, both airplane interiors and passengers are subjected to spraying with insecticide upon arrival and before deplaning. The West Nile virus (West Nile encephalitis) was previously found only in Africa, Eastern Europe, and West Asia, but in the 1990's it appeared in the northeastern United States, transported there by birds, mosquitos, or people.

It was feared in the mid-1990's that the highly infectious and deadly Ebola virus, which originated in tropical Africa, might spread to Europe and the United States, by air passengers or through the importing of monkeys. The devastation of native bird communities on the island of Guam has been traced to the emergence there of a large population of brown tree snakes, whose ancestors are thought to have arrived as accidental stowaways on a military airplane in the late 1940's.

Ray Sumner

FOR FURTHER STUDY

Bilstein, Roger E. *Flight in America: 1900-1983, from the Wrights to the Astronauts.* Baltimore: Johns Hopkins University Press, 1984.

Christy, Joe, with Alexander Wells. *American Aviation: An Illustrated History.* Blue Ridge Summit, Pa.: TAB Books, 1987.

Davies, Ronald E. G. *A History of the World's Airlines.* New York: Oxford University Press, 1983.

Graham, Brian. *Geography and Air Transport.* New York: John Wiley & Sons, 1995.

Kane, Robert M. *Air Transportation.* 13th ed. Dubuque, Iowa: Kendall/Hunt, 1999.

Morrison, Steven A., and Clifford Winston. *The Evolution of the Airline Industry.* Washington, D.C.: Brookings Institution, 1995.

Richter, William L. *The ABC-Clio Companion to Transportation in America.* Santa Barbara, Calif.: ABC-Clio, 1995.

ECONOMIC GEOGRAPHY

AGRICULTURE

TRADITIONAL AGRICULTURE

Two agricultural practices that are widespread among the world's traditional cultures, slash-and-burn and nomadism, share several common features. Both are ancient forms of agriculture, both involve farmers not remaining in a fixed location, and both can pose serious environmental threats if practiced in a nonsustainable fashion. The most significant difference between the two forms is that slash-and-burn generally is associated with raising field crops, while nomadism as a rule involves herding livestock.

SLASH-AND-BURN AGRICULTURE. Farmers have practiced slash-and-burn agriculture, which is also referrred to as shifting cultivation or swidden agriculture, in almost every region of the world where the climate makes farming possible. Although at the end of the twentieth century slash-and-burn agriculture was most commonly found in tropical areas such as the Amazon River basin in South America, swidden agriculture also once dominated agriculture in more temperate regions, such as northern Europe. Swidden agriculture was, in fact, common in Finland and northern Russia well into the early decades of the twentieth century.

Slash-and-burn acquired its name from the practice of farmers who cleared land for planting crops by cutting down the trees or brush on the land and then burning the fallen timber on the site. The farmers literally slash and burn. The ashes of the burnt wood add minerals to the soil, which temporarily improves its fertility. Crops the first year following clearing and burning are generally the best crops the site will provide. Each year after that, the yield diminishes slightly as the fertility of the soil is depleted.

Farmers who practice swidden cultivation do not attempt to improve fertility by adding fertilizers such as animal manures but instead rely on the soil to replenish itself over time. When the yield from one site drops below acceptable levels, the farmers then clear another piece of land, burn the brush and other vegetation, and cultivate that site while leaving their previous field to lie fallow and its natural vegetation to return. This cycle will be repeated over and over, with some sites being allowed to lie fallow indefinitely while others may be revisited and farmed again in five, ten, or twenty years.

Farmers who practice shifting cultivation do not necessarily move their dwelling places as they change the fields they cultivate. In some geographic regions, farmers live in a central village and farm cooperatively, with the fields being alternately allowed to remain fallow, and the fields being farmed making a gradual circuit around the central village. In other cases, the village itself may move as new fields are cultivated. Anthropologists

Slash-and-burn fields Page 170

studying indigenous peoples in Amazonia, for example, discovered that village garden sites were on a hundred-year cycle. Villagers farmed cooperatively, with the entire village working together to clear a garden site. That garden would be used for about five years, then a new site was cleared. When the garden moved an inconvenient distance from the village, about once every twenty years, the entire village would move to be closer to the new garden. Over a period of approximately one hundred years, a village would make a circle through the forest, eventually ending up close to where it had been located long before any of the present villagers had been born.

In more temperate climates, individual farmers often owned and lived on the land on which they practiced swidden agriculture. Farmers in Finland, for example, would clear a portion of their land, burn the brush and other covering vegetation, grow grains for several years, and then allow that land to remain fallow for from five to twenty years. The individual farmer rotated cultivation around the land in a fashion similar to that practiced by whole villages in other areas, but did so as an individual rather than as part of a communal society.

Although slash-and-burn is frequently denounced as a cause of environmental degradation in tropical areas, the problem with shifting cultivation is not the practice itself but the length of the cycle. If the cycle of shifting cultivation is long enough, forests will grow back, the soil will regain its fertility, and minimal adverse effects will occur. In some regions, a piece of land may require as little as five years to regain its maximum fertility; in others, it may take one hundred years. Problems arise when growing populations put pressure on traditional farmers to return to fallow land too soon. Crops are smaller than needed,

leading to a vicious cycle in which the next strip of land is also farmed too soon, and each site yields less and less. As a result, more and more land must be cleared.

NOMADISM. Nomadic peoples have no permanent homes. They earn their livings by raising herd animals, such as sheep, cattle, or horses, and they spend their lives following their herds from pasture to pasture with the seasons. Most nomadic animals tend to be hardy breeds of goats, sheep, or cattle that can withstand hardship and live on marginal lands. Traditional nomads rely on natural pasturage to support their herds and grow no grains or hay for themselves. If a drought occurs or a traditional pasturing site is unavailable, they can lose most of their herds to starvation.

In many nomadic societies, the herd animal is almost the entire basis for sustaining the people. The animals are slaughtered for food, clothing is woven from the fibers of their hair, and cheese and yogurt may be made from milk. The animals may also be used for sustenance without being slaughtered. Nomads in Mongolia, for example, occasionally drink horses' blood, removing only a cup or two at a time from the animal. Nomads go where there is sufficient vegetation to feed their animals.

In mountainous regions, nomads often spend the summers high up on mountain meadows, returning to lower altitudes in the autumn when snow begins to fall. In desert regions, they move from oasis to oasis, going to the places where sufficient natural water exists to allow brush and grass to grow, allowing their animals to graze for a few days, weeks, or months, then moving on. In some cases, the pressure to move on comes not from the depletion of food for the animals but from the depletion of a water source, such as a spring or well. At many natural desert oases, a natural water seep or spring provides only enough water to support a nomadic group for a few days at a time.

In addition to true nomads—people who never live in one place permanently—a number of cultures have practiced seminomadic farming: The temperate months of the year, spring through fall, are spent following the herds on a long loop, sometimes hundreds of miles long, through traditional grazing areas, then the winter is spent in a permanent village.

Nomadism has been practiced for millennia, but there is strong pressure from several sources to eliminate it. Pressures generated by industrialized society are increasingly threatening the traditional cultures of nomadic societies, such as the Bedouin of the Arabian Peninsula. Traditional grazing areas are being fenced off or developed for other purposes. Environmentalists are also concerned about the ecological damage caused by nomadism.

Nomads generally measure their wealth by the number of animals they own and so will try to develop their herds to be as large as possible, well beyond the numbers required for simple sustainability. The herd animals eat increasingly large amounts of vegetation, which then has no

opportunity to regenerate, and desertification may occur. Nomadism based on herding goats and sheep, for example, has been blamed for the expansion of the Sahara Desert in Africa. For this reason, many environmental policymakers have been attempting to persuade nomads to give up their roaming lifestyle and become sedentary farmers.

Nancy Farm Männikkö

FOR FURTHER STUDY

Colfer, Carol J., with Nancy Peluso and Chin See Chung. *Beyond Slash and Burn: Building on Indigenous Management of Borneo's Tropical Rain Forest.* New York: New York Botanical Garden, 1994.

Goldstein, Melvyn C. *The Changing World of Mongolia's Nomads.* Berkeley: University of California Press, 1994.

Keohane, Allen. *Bedouin: Nomads of the Desert.* London: Trafalgar Square, 1995.

Posey, D. A., and W. Balee, eds. *Resource Management in Amazonia: Indigenous and Folk Strategies.* New York: New York Botanical Garden, 1989.

Van Offelen, Marion, and Carol Beckwith. *Nomads of Niger.* New York: Academic Press, 1987.

COMMERCIAL AGRICULTURE

Commercial farmers are those who sell substantial portions of their output of crops, livestock, and dairy products for cash. In some regions, commercial agriculture is as old as recorded history, but only in the twentieth century did the ma-

jority of farmers come to participate in it. For individual farmers, this has offered the prospect of larger income and the opportunity to buy a wider range of products. For society, commercial agriculture has been associated with specialization and increased productivity. Commercial agriculture has enabled world food production to increase more rapidly than world population, improving nutrition levels for millions of people.

STEPS IN COMMERCIAL AGRICULTURE. In order for commercial agriculture to exist, products must move from farmer to ultimate consumer, usually through six stages:

1. Processing, packaging, and preserving to protect the products and reduce their bulk to facilitate shipping.

2. Transport to specialized processing facilities and to final consumers.

3. Networks of merchant middlemen who buy products in bulk from farmers and processors and sell them to final consumers.

4. Specialized suppliers of inputs to farmers, such as seed, livestock feed, chemical inputs (fertilizers, insecticides, pesticides, soil conditioners), and equipment.

5. A market for land, so that farmers can buy or lease the land they need.

6. Specialized financial services, especially loans to enable farmers to buy land and other inputs before they receive sales revenues.

Improvements in agricultural science and technology have resulted from extensive research programs by government, business firms, and universities.

INTERNATIONAL TRADE. Products such as grain, olive oil, and wine moved by ship across the Mediterranean Sea in ancient times. Trade in spices, tea, coffee, and cocoa provided powerful stimulus for exploration and colonization around 1500 C.E. The coming of steam locomotives and steamships in the nineteenth century greatly aided in the shipment of farm products and spurred the spread of population into potentially productive farmland all over the world. Beginning with Great Britain in the 1840's, countries were

Tokyo market. (Corbis)

THE HERITAGE SEED MOVEMENT

Modern hybrid seeds have increased yields and enabled the tremendous productivity of the modern mechanized farm. However, the widespread use of a few hybrid varieties has meant that almost all plants of a given species in a wide area are almost identical genetically. This loss of biodiversity, or the range of genetic difference in a given species, means that a blight could wipe out an entire season's crop. Historical examples of blight include the nineteenth century Great Potato Famine of Ireland and the 1971 corn blight in the United States.

In response to the concern for biodiversity, there has been a movement in North America to preserve older forms of crops with different genes that would otherwise be lost to the gene pool. Nostalgia also motivates many people to keep alive the varieties of fruits and vegetables that their grandparents raised. Many older recipes do not taste the same with modern varieties of vegetables that have been optimized for commercial considerations such as transportability. Thus, raising heritage varieties also can be a way of continuing to enjoy the foods one's ancestors ate.

willing to relinquish agricultural self-sufficiency to obtain cheap imported food, paid for by exporting manufactured goods.

Most of the leaders in agricultural trade were highly developed countries, which typically had large amounts of both imports and exports. These countries are highly productive both in agriculture and in other commercial activities. Much of their trade is in high-value packaged and processed goods. Although the vast majority of China's labor force works in agriculture, their average productivity is low and the country showed an import surplus in agricultural products. The same was true for Russia. India, similar to China in size, development, and population, had relatively little agricultural trade. Australia and Argentina are examples of countries with large export surpluses, while Japan and South Korea had large import surpluses. Judged by volume, trade is dominated by grains, sugar, and soybeans. In contrast, meat, tobacco, cotton, and coffee reflect much higher values per unit of weight.

THE UNITED STATES. Blessed with advantageous soil, topography, and climate, the United States has become one of the most productive agricultural countries in the world. Technological advances have enabled the United States to feed its own residents and export substantial quantities with only 2 percent of its labor force engaged directly in farming. In the 1990's there were about two million farms cultivating about one billion acres. They produced about $200 billion worth of products. After expenses, this yielded about $50 billion of net farm income—an average of only about $25,000 per farm. However, most farm families derive substantial income from nonfarm employment.

There is a great deal of agricultural specialization by region. Corn, soybeans, and wheat are grown in many parts of the United States (outside New England). Some other crops have much more limited growing areas. Cotton, rice, and sugarcane require warmer temperatures. Significant production of cotton occurred in seventeen states, rice in six, and sugarcane in four. Twelve states were leaders in agri-

Farmland in Iowa, which like other regions has its own specializations. (PhotoDisc)

culture in 1998: Iowa in corn, soybeans, and hogs; Illinois in corn and soybeans; Texas and Nebraska in cattle; California in fruits, vegetables, and dairy products; Florida in fruits and vegetables; Wisconsin in dairy products; Georgia and Arkansas in broiler chickens; North Carolina in hogs; and North Dakota and Kansas in wheat. Typically the top two states in a category account for about 30 percent of sales. Fruits and vegetables are the main exception; the great size, diversity, and mild climate of California gives it a dominant 45 percent.

SOCIALIST EXPERIMENTS. Under the dictatorship of Joseph Stalin, the communist government of the Soviet Union established a program of compulsory collectivized agriculture in 1929. Private ownership of land, buildings, and other assets was abolished. There were some state farms, "factories in the fields," operated on a large scale with many hired workers. Most, however, were collective farms, theoretically run as cooperative ventures of all residents of a village, but in practice directed by government functionaries. The arrangements had disastrous effects on productivity and kept the rural residents in poverty. Nevertheless, similar arrangements were established in China in 1950 under the rule of Mao Zedong. A restoration of commercial agriculture after Mao's death in 1976 enabled China to achieve greater farm output and farm incomes.

Most Western countries, including the United States, subsidize agriculture and restrict imports of competing farm products. Objectives are to support farm incomes, reduce rural discontent, and slow the downward trend in the number of farmers. In 1998 the European Union spent nearly $150 billion in farm support, and Japan spent $50 billion. Restricting imports kept prices high for consumers. Such policies led to bitter disputes with the United States, which wanted to open world markets for U.S. farm exports.

PROBLEMS FOR FARMERS. Farmers in a system of commercial agriculture are vulnerable to changes in market prices as well as the universal problems of fluctuating weather. Congress tried to reduce farm subsidies through the Freedom to Farm Act of 1996, but serious price declines in 1997-1999 led to backtracking. Efforts to increase productivity by genetic alterations, radiation, and feeding synthetic hormones to livestock have drawn critical responses from some consumer groups. Environmentalists have been concerned about soil depletion and water pollution resulting from chemical inputs.

PRODUCTIVITY AND WORLD HUNGER. Despite advances in agricultural production, the problem of world hunger persists. Even in countries that store surpluses of farm commodities, there are still people who go hungry. In less-developed countries, the prices of imported food from the West are too low for local producers to compete and too high for the poor to buy them.

Paul B. Trescott

FOR FURTHER STUDY

Bonanno Alessandro, and Lawrence Busch. *From Columbus to Conagra: The Globalization of Agriculture and Food.* Lawrence: University Press of Kansas, 1994.

Ilbery, Brian, Quentin Chiotti, and Timothy Rickard. *Agricultural Restructuring and Sustainability: A Geographic Perspective.* New York: CAB International, 1997.

Kahn, E. J., Jr., *Supermarketer to the World.* New York: Warner Books, 1991.

Norton, George W., and Jeffrey Alwang. *Introduction to Economics of Agricultural Development.* New York: McGraw-Hill, 1993.

Rapp, David. *How the U.S. Got into Agriculture and Why It Can't Get Out.* Washington: Congressional Quarterly, 1988.

Strange, Marty. *Family Farming: A New Economic Vision.* Lincoln: University of Nebraska Press, 1988.

MODERN AGRICULTURAL PROBLEMS

Ever since human societies started to grow their own food, there have been problems to solve. Much of the work of nature was disrupted by the work of agriculture as many as ten thousand years ago. Nature took care of the land and made it productive in its own intricate way, through its own web of interdependent systems. Agriculture disrupts those systems with the hope of making the land even more productive, growing even more food to feed even more people. Since the first spade of soil was turned over and the first plants domesticated, farmers have been trying to figure out how to care for the land as well as nature did before.

Many modern problems in agriculture are not really modern at all. Erosion and pollution, for example, have been around as long as agriculture. However, agriculture has changed drastically within those ten thousand years, especially since the dawn of the Industrial Revolution in the seventeenth century. Erosion and pollution are now bigger problems than before and have been joined by a host of others that are equally critical—not all related to physical deterioration. Modern farmers use many more machines than did farmers of old, and modern machines require advanced sources of energy to unleash their power. The machines do more work than could be accomplished before, so fewer farmers are needed, which causes economic problems.

Modern cowboys Page 171

Cities continue to grow bigger as land—usually the best farmland around—is converted to homes and parking lots for shopping centers. The farmers that remain on the land, needing to grow ever more food, turn to the research and engineering industries to improve their seeds. These industries have responded with recombinant technologies that move genes from one species to another; for example, genes cut from peanuts may be spliced into chickens. This creates another set of cultural problems, which are even more difficult to solve because most are still "potential"—their impact is not yet known.

EROSION. Soil loss from erosion continues to be a huge problem all over the world. As agriculture struggles to feed more millions of people, more land is plowed. The newly plowed lands usually are considered more marginal, meaning they are either too steep, too thin, or too sandy; are subject to too much rain; or suffer some other deficiency. Natural vegetative cover blankets these soils and protects them from whatever erosive agents are active in their regions: water, wind, ice, or gravity. Plant cover also increases the amount of rain that seeps downward into the soil rather than running off into rivers. The more marginal land that is turned over for crops, the faster the erosive agents will act and the more erosion will occur.

Expansion of land under cultivation is not the only factor contributing to erosion. Fragile grasslands in dry areas also are being used more intensively. Grazing more livestock than these pastures can handle decreases the amount of grass in

---

DESERTIFICATION

Desertification is the extension of desert conditions into new areas. Typically, this term refers to the expansion of deserts into adjacent nondesert areas, but it can also refer to the creation of a new desert. Land that is susceptible to prolonged drought is always in danger of losing its vegetative ground cover, thereby exposing its soil to wind. The wind carries away the smaller silt particles and leaves behind the larger sand particles, stripping the land of its fertility. This naturally occurring process is assisted in many areas by overgrazing.

In the African Sahel, south of the Sahara, the impact of desertification is acute. Recurring drought has reduced the vegetation available for cattle, but the need for cattle remains high to feed populations that continue to grow. The cattle eat the grass, the soil is exposed, and the area becomes less fertile and less able to support the population. The desert slowly encroaches, and the people must either move or die.

the pasture and exposes more of the soil to wind—the primary erosive agent in dry regions.

Overgrazing can affect pastureland in tropical regions too. Thousands of acres of tropical forest have been cleared to establish cattle-grazing ranges in Latin America. Tropical soils, although thick, are not very fertile. Fertility comes from organic waste in the surface layers of the soil. Tropical soils form under constantly high temperatures and receive much more rain than soils in moderate, midlatitude climates; thus, tropical organic waste materials rot so fast they are not worked into the soil at all. After one or two growing seasons, crops grown in these soils will yield substantially less than before.

Tropical fields require fallow periods of about ten years to restore themselves after they are depleted. That is why tropical cultures using slash-and-burn methods of agriculture move to new fields every other year in a cycle that returns them to the same place about every ten years, or however long it takes those particular lands to regenerate. The heavy forest cover protects these soils from exposure to the massive amounts of rainfall and provides enough organic material for crops—as long as the forest remains in place. When the forest is cleared, however, the resulting grassland cannot provide the adequate protection, and erosion accelerates. Grasslands that are heavily grazed provide even less protection from heavy rains, and erosion accelerates even more.

The use of machines also promotes erosion, and modern agriculture relies on machinery: tractors, harvesters, trucks, balers, ditchers, and so on. In the United States, Canada, Europe, Russia, Brazil, South Africa, and other industrialized areas, machinery use is intense. Machinery use is also on the rise in countries such as India, China, Mexico, and Indonesia, where traditional nonmechanized methods are practiced widely. Farming machines, in gaining traction, loosen the topsoil and inhibit vegetative cover growth, especially when they pull behind them any of the various farm implements designed to rid the soil of weeds, that is, all vegetation except the desired crop. This leaves the soil more exposed to erosive weather, so more soil is carried away in the runoff of water to streams.

Eco-fallow farming has become more popular in the United States and Europe as a solution to reducing erosion. This method of agriculture, which leaves the crop residue in place over the fallow (nongrowing) season, does not root the soil in place, however. Dead plants do not "grab" the soil like live plants that need to extract from it the nutrients they need to live. So erosion continues, even though it is at a slower rate. Eco-fallow methods also require heavier use of chemicals, such as herbicides, to "burn down" weed growth at the start of the growing season, which contributes to accelerated erosion and increases pollution.

Slash-and-burn fields Page 170

POLLUTION. Pollution, besides being a problem in general, continues to grow as an agricultural problem. With the onset of the Green Revolution, the use of herbicides, insecticides, and pesticides has increased dramatically all over the world. These chemicals are not used completely in the growth of the crop, so the leftovers (residue) wash into, and contaminate, surface and groundwater supplies. These supplies then must be treated to become useful for other purposes, a job nature used to do on its own. Agricultural chemicals reduce nature's ability to act as a filter by inhibiting the growth of the kinds of plant life that perform that function in aquatic environments. The chemical residues that are not washed into surface supplies contaminate wells.

As soil fertility decreases, applications of chemical fertilizers are increased, intensifying the toxicity of cyclical chemical dependency. (PhotoDisc)

As chemical use increases, contamination accumulates in the soil and fertility decreases. The microorganisms and animal life in the soil, which had facilitated the breakdown of soil minerals into usable plant products, are no longer nourished because the crop residue on which they feed is depleted, or they are killed by the active ingredients in the chemical. As a result, soil fertility must be restored to maintain yield. Chemical replacement is usually the method of choice, and increased applications of chemical fertilizers intensify the toxicity of this cyclical chemical dependency.

Chemicals, although problematic, are not as difficult to contend with as the increasingly heavy silt load choking the life out of streams and rivers. Accelerated erosion from water runoff carries silt particles into streams, where they remain suspended and inhibit the growth of many beneficial forms of plant and animal life.

The silt load in U.S. streams has become so heavy that the Mississippi River delta is growing faster than it used to. The heavy silt load, combined with the increased load of chemical residues, is seriously taxing the capabilities of the ecosystems around the delta that filter out sediments, absorb nutrients, and stabilize salinity levels for ocean life, creating an expanding dead zone.

This general phenomenon is not limited to the Mississippi delta—it is widespread. Its impact on people is high, because most of the world's population lives in coastal zones and comes in direct contact with the sea. Additionally, eighty percent of the world's fish catch comes from the coastal waters over continental shelves that are most susceptible to this form of pollution.

MONOCULTURE. Modern agriculture emphasizes crop specialization. Farmers, especially in industrialized regions, often grow a single crop on most of their land, perhaps rotating it with a second crop in successive years: corn one year, for example, then soybeans, then back to corn. Such a strategy allows the farmer to reduce costs, but it also makes the crop, and, thus, the farmer and community, susceptible to widespread crop failure. When the crop is infested by any of an ever-changing number and variety of pests—worms, molds, bacteria, fungi, insects, or other diseases—the whole crop is likely to die quickly, unless an appropriate antidote is immediately applied. Chemical antidotes can do the job but increase pollution. Maintaining species diversity—growing several different crops instead of one or two—allows for crop failures without jeopardizing the entire income for a farm or region that specializes in a particular monoculture, such as tobacco, coffee, or bananas.

Chemicals are not the only modern methods of preventing crop loss. Genetically engineered seeds are one attempt at replacing post-infestation chemical treatments. For example, splicing genes into varieties of rice or potatoes from wholly unrelated species—say, hypothetically, a grasshopper—to prevent common forms of blight is occurring more often. Even if the new genes make the crop more resistant, however, they could trigger unknown side effects that have more serious long-term environmental and economic consequences than the problem they were used to solve. Genetically altered crops are essentially new life-forms being introduced into nature with no observable precedents to watch beforehand for clues as to what might happen.

URBAN SPRAWL. As more farms become mechanized, the need for farmers is being drastically reduced. There were more farmers in the United States in 1860 than there were in the year 2000. From a peak in 1935 of about 6.8 million farmers farming 1.1 billion acres, the United States at the end of the twentieth century counted fewer than 2 million farmers farming 950 million acres. As fewer people care for land, the potential for erosion and pollution to accelerate is likely to increase, causing land quality to decline.

As farmers are displaced and move into towns, the cities take up more space. The resulting urban sprawl converts a tremendous amount of cropland into parking lots, malls, industrial parks, or suburban neighborhoods. If cities were located in marginal areas, then the concern over the loss of farmland to commercial development would be nominal. However, the cities attracting the greatest numbers of people have too often replaced the best cropland. Taking the best cropland out of primary production imposes a severe economic penalty.

James Knotwell and Denise Knotwell

FOR FURTHER STUDY

Baskin, Yvonne. *The Work of Nature: How the Diversity of Life Sustains Us.* Washington, D.C.: Island Press, 1997.

INFORMATION ON THE WORLD WIDE WEB

The Web site of the Consortium for Sustainable Agriculture Research and Education contains information about programs and publications focused on sustainable agriculture and food systems research. (www.csare.org)

Physicians and Scientists for Responsible Application of Science and Technology maintains a Web site devoted to the discussion of safety problems of genetically engineered food. (www.psrast.org/ctglobal.htm)

Jackson, Wes. *New Roots for Agriculture.* Lincoln: University of Nebraska Press, 1980.

Reid, T. R. "Feeding the Planet." *National Geographic* (October, 1998): 56-75.

Shreeve, James. "Secrets of the Gene." *National Geographic* (October, 1999): 42-75.

Union of International Associations, ed. *Encyclopedia of World Problems and Human Potential.* 2d ed. New York: K. G. Saur, 1986.

Yergin, Daniel. *The Prize: The Epic Quest for Oil, Money, and Power.* New York: Simon & Schuster, 1991.

WORLD FOOD SUPPLIES

All living things need food to begin the life process and to live, grow, work, and survive. Almost all foods that humans consume come from plants and animals. Not all of Earth's people eat the same foods, however, nor do they require the same caloric intakes. The types, combinations, and amounts of food consumed by different peoples depend upon historic, socioeconomic, and environmental factors.

THE HISTORY OF FOOD CONSUMPTION. Early in human history, people ate what they could gather or scavenge. Later, people ate what they could plant and harvest and what animals they could domesticate and raise. Modern people eat what they can grow, raise, or purchase. Their diets or food composition are determined by income, local customs, religion or food biases, and advertising. There is a global food market, and many people can select what they want to eat and when they eat it according to the prices they can pay and what is available.

Historically, in places where food was plentiful, accessible, and inexpensive, humans devoted less time to basic survival needs and more time to activities that led to human progress and enjoyment of leisure. Despite a modern global food system, instant telecommunications, the United Nations, and food surpluses at places, however, the problem of providing food for everyone on Earth has not been solved.

In 1996 leaders from 186 countries gathered in Rome, Italy, and agreed to reduce by half the number of hungry people in the world by the year 2015. United Nations data for 1998 revealed that more than 790 million people in the developing parts of the world did not have enough food to eat. This is more people than the total population of North America and Europe at that time. The number of undernourished people has been decreasing since 1990. At the current pace of hunger reduction in the world, 600 million will suffer from "acute food insecurity" and go to sleep hungry in 2015. Despite efforts being made to feed the world, outbreaks of food deficiencies, mass starvation, and famine are a certainty in the twenty-first century.

WORLD FOOD SOURCE REGIONS. Agriculture and related primary food production activities, such as fishing, hunting, and gathering, continue to employ more than one-third of the world's labor force. Agriculture's relative importance in the

World food supplies Page 171

world economic system has declined with urbanization and industrialization, but it still plays a vital role in human survival and general economic growth. Agriculture in the third millennium must supply food to an increasing world population of nonfood producers. It must also produce food and nonfood crude materials for industry, accumulate capital needed for further economic growth, and allow workers from rural areas to industrial, construction, and expanding intraurban service functions.

Soil types, topography, weather, climate, socioeconomic history, location, population pressures, dietary preferences, stages in modern agricultural development, and governmental policies combine to give a distinctive personality to regional agricultural characteristics. Two of the most productive food-producing regions of the world are North America and Europe. Countries in these regions export large amounts of food to other parts of the world.

North America is one of the primary food-producing and food-exporting continents. After 1940 food output generally increased as cultivated acreage declined. Progress in improving the quantity and quality of food production is related to mechanization, chemicalization, improved breeding, and hybridization. Food output is limited more by market demands than by production obstacles.

Western Europe, although a basic food-deficit area, is a major producer and exporter of high-quality foodstuffs. After 1946 its agriculture became more profit-driven. Europe's agricultural labor force grew smaller, its agriculture became more mechanized, its farm sizes increased, and capital investment per acre increased.

FOODS FROM PLANTS. Most basic staple foods come from a small number of plants and animals. Ranked by tonnage pro-

duced, the most important food plants throughout the world are wheats, corn (maize), rice, potatoes, cassava (manioc), barley, soybeans, sorghums and millets, beans, peas and chickpeas, and peanuts (groundnuts).

Wheat and rice are the most important plant foods. More than one-third of the world's cultivated land is planted with these two crops. Wheat is the dominant food staple in North America, Western and Eastern Europe, northern China, and the Middle East and North Africa. Rice is the dominant food staple in southern and eastern Asia. Corn, used primarily as animal food in developed nations, is a staple food in Latin America and Southeast Africa. Potatoes are a basic food in the highlands of South America and in Central and Eastern Europe. Cassava (manioc) is a tropical starch-producing root crop of special dietary importance in portions of lowland South America, the west coast countries of Africa, and sections of South Asia. Barley is an important component of diets in North African, Middle Eastern, and Eastern European countries. Soybeans are an integral part of the diets of those who live in eastern, southeastern, and southern Asia. Sorghums and millets are staple subsistence foods in the savanna regions of Africa and south Asia, while peanuts are a facet of dietary mixes in tropical Africa, Southeast Asia, and South America.

FOOD FROM ANIMALS. Animals have been used as food by humans from the time the earliest people learned to hunt, trap, and fish. However, humans have domesticated only a few varieties of animals. Ranked by tonnage of meat produced, the most commonly eaten animals are cattle, pigs, chickens and turkeys, sheep, goats, water buffalo, camels, rabbits and guinea pigs, yaks, and llamas and alpacas.

Cattle, which produce milk and meat, are important food sources in North Amer-

*Llamas
Page 172*

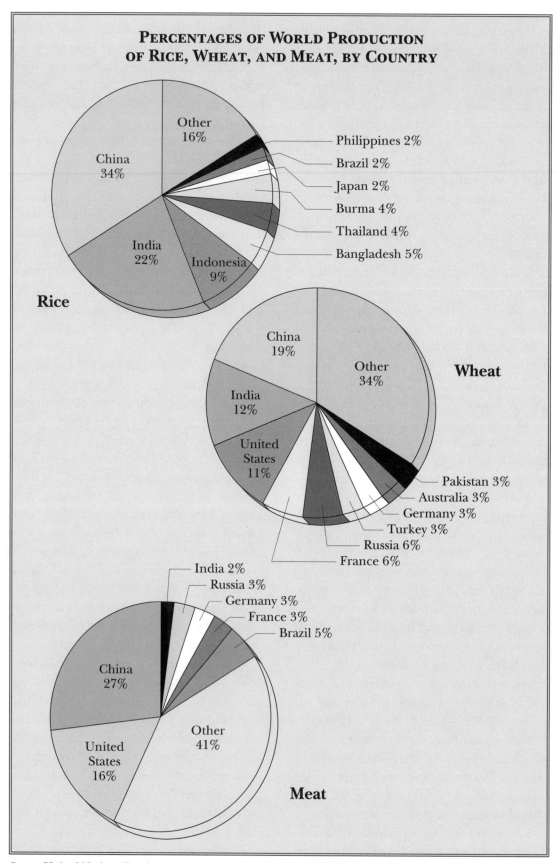

PERCENTAGES OF WORLD PRODUCTION OF RICE, WHEAT, AND MEAT, BY COUNTRY

Rice

Other 16%
China 34%
India 22%
Indonesia 9%
Philippines 2%
Brazil 2%
Japan 2%
Burma 4%
Thailand 4%
Bangladesh 5%

Wheat

China 19%
Other 34%
India 12%
United States 11%
Pakistan 3%
Australia 3%
Germany 3%
Turkey 3%
Russia 6%
France 6%

Meat

India 2%
Russia 3%
Germany 3%
France 3%
Brazil 5%
China 27%
United States 16%
Other 41%

Source: United Nations Food and Agriculture Organization (FAOSTAT Database, 2000).

266

ica, Western Europe, Eastern Europe, Australia and New Zealand, Argentina, and Uruguay. Pigs are bred and reared for food on a massive scale in southern and eastern Asia, North America, Western Europe, and Eastern Europe. Chickens are the most important domesticated fowl used as a human food source and are a part of the diets of most of the world's people. Sheep and goats, as a source of meat and milk, are especially important to the diets of those who live in the Middle East and North Africa, Eastern Europe, Western Europe, and Australia and New Zealand.

Water buffalo, camels, rabbits, guinea pigs, yaks, llamas, and alpacas are food sources in regions of the world where there is low consumption of meat for religious, cultural, or socioeconomic reasons. Fish is an inexpensive and wholesome source of food. Seafood is an important component to the diets of those who live

in southern and eastern Asia, Western Europe, and North America.

THE WORLD'S GROWING POPULATION. The problem of feeding the world is compounded by the fact that population was increasing at a rate of nearly 80 million persons per year at the end of the twentieth century. That rate of increase is roughly equivalent to adding a country the size of Germany to the world every single year.

Also compounding the problem of feeding the world are population redistribution patterns and changing food consumption standards. In the year 2000 the world population was projected to reach approximately ten billion people in 2050—four billion people more than were on the earth in 2000. Most of the increase in world population was expected to occur within the developing nations.

URBANIZATION. Along with an increase in population in developing nations is

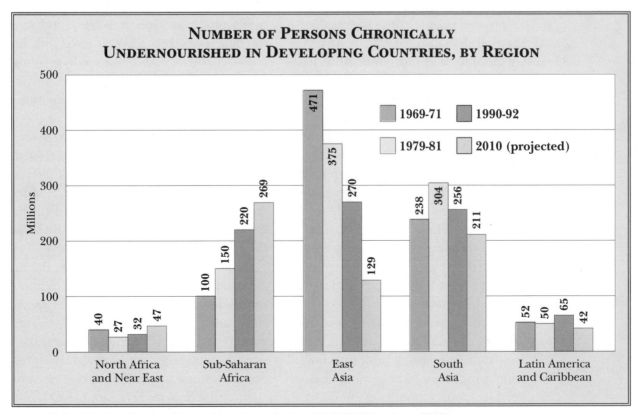

Source: United Nations Food and Agriculture Organization (FAOSTAT Database, 2000).

massive urbanization. City dwellers are food consumers, not food producers. The exodus of young men and women from rural areas has given rise to a new series of megacities, most of which are in developing countries. By the year 2015, twenty-six cities in the world are expected to have populations of ten million people or more.

When rural dwellers move to cities, they tend to change their dietary composition and food-consumption patterns. Qualitative changes in dietary consumption standards are positive, for the most part, and are a result of copying the diets of what is considered a more prestigious group or positive educational activities of modern nutritional scientists working in developing countries. During the last four decades of the twentieth century, a tremendous shift took place in overall dietary habits. Dietary changes and consumption trends have contributed to a decrease in child mortality, an increase in longevity, and a greater resistance to disease. This globalization of people's diets has resulted in increased demands for higher quality, greater quantity, and more nutritious basic foods.

Terraced fields Page 163

STRATEGIES FOR INCREASING FOOD PRODUCTION. To meet the food demands and the food distribution needs of the world's people in the future, a grand strategy has been proposed. Its first step calls for the intensification of agriculture—improving biological, mechanical, and chemical technology and applying proven agricultural innovations to regions of the world where the physical and cultural environments are most suitable for rapid food production increases.

The second step in the strategy is to expand the areas where food is produced so that areas that are empty or underused will be made productive. Reclaiming areas damaged by human mismanagement, expanding irrigation in carefully selected ar-

eas, and introducing extensive agrotechniques to areas not under cultivation could increase the production of inexpensive grains and meats.

Finally, interregional, international, and global commerce should be expanded, in most instances, increasing regional specializations and production of high-quality, high-demand agricultural products for export and importing low-cost basic foods. A disequilibrium of supply and demand for certain commodities will persist, but food producers, regional and national agricultural planners, and those who strive for regional economic integration must take advantage of local conditions and location or create the new products needed by the food-consuming public in a one-world economy.

PERSPECTIVES. Humanity is entering a time of volatility in food production and distribution. The world will produce enough food to meet the demands of those who can afford to buy food. In many developing countries, however, food production is unlikely to keep pace with increases in the demand for food by growing populations. The food gap—the difference between production and demand—could more than double in the first three decades of the twenty-first century. Such a development would increase the dependence of developing countries on food imports. About 90 percent of the rate of increase in aggregate food demand in the early twenty-first century is expected to be the result of population increases.

Factors that could lead to larger fluctuations in food availability include weather variations such as those induced by El Niño and climatic change, the growing scarcity of water, civil strife and political instability, and declining food aid. In developing countries, decision makers need to ensure that policies promote broad-based economic growth—and in particular agri-

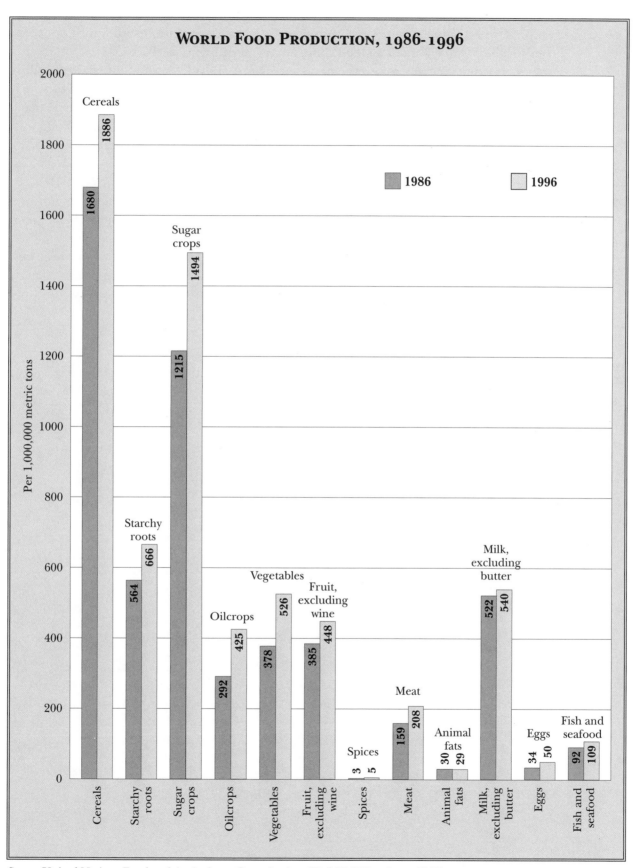

WORLD FOOD PRODUCTION, 1986-1996

Per 1,000,000 metric tons

	1986	1996
Cereals	1680	1886
Starchy roots	564	666
Sugar crops	1215	1494
Oilcrops	292	425
Vegetables	378	526
Fruit, excluding wine	385	448
Spices	3	5
Meat	159	208
Animal fats	30	29
Milk, excluding butter	522	540
Eggs	34	50
Fish and seafood	92	109

Source: United Nations Food and Agriculture Organization (FAOSTAT Database, 2000).

cultural growth—so that their countries can produce enough food to feed themselves or enough income to buy the necessary food on the world market.

William A. Dando

FOR FURTHER STUDY

Castro, José de. *The Geopolitics of Hunger.* New York: Monthly Review Press, 1977.

Dando, William A. *The Geography of Famine.* New York: John Wiley & Sons, 1980.

———, and Caroline Z. Dando. *Food and Famine: A Reference.* Hillside, N.J.: Enslow, 1991.

Johnson, David Gale. *World Food Problems and Prospects.* Washington, D.C.: American Enterprise Institute for Public Policy Research, 1975.

Norse, David. "A New Strategy for Feeding a Crowded Planet." In *Global Issues 93/94,* edited by Robert M. Jackson. 9th ed. Guilford, Conn.: Dushkin, 1993.

Rau, Bill. *From Feast to Famine: Official Cures and Grassroots Remedies to Africa's Food Crisis.* London: Zed Books, 1991.

"Seeds of Change: Genetically Altered Foods." *Consumer Reports* (September, 1999): 41-46.

Stevens, Charles J. *Confronting the World Food Crisis.* Muscatine, Iowa: Stanley Foundation, 1981.

ENERGY AND ENGINEERING

ENERGY SOURCES

Energy is essential for powering the processes of modern industrial society: refining ores, manufacturing products, moving vehicles, heating buildings, and powering appliances. In 1999 energy costs were half a trillion dollars in the United States alone. All technological progress has been based on harnessing more energy and using it more effectively. Energy use has been shaped by geography and also has shaped economic and political geography.

ANCIENT TO MODERN ENERGY. Energy use in traditional tribal societies illustrates all aspects of energy use that apply in modern human societies. Early Stone Age peoples had only their own muscle power, fueled by meat and raw vegetable matter. Warmth for living came from tropical or subtropical climates. Then a new energy source, fire, came into use. It made cold climates livable. It enabled the cooking of roots, grains, and heavy animal bones, vastly increasing the edible food supply. Its heat also hardened wood tools, cured pottery, and eventually allowed metal-working.

Nearly as important as fire was the domestication of animals, which multiplied available muscle energy. Domestic animals carried and pulled heavy loads. Domesticated horses could move as fast as the game to be hunted or large animals to be herded.

Increased energy efficiency was as important as new energy sources in making tribal societies more successful. Cured animal hides and woven cloth were additional factors enabling people to move to cooler climates. Cooking fires also allowed drying meat into jerky to preserve it against times of limited supply. Fire-cured pottery helped protect food against pests and kept water close by. However, energy benefits had costs. Fire drives for hunting may have caused major animal extinctions. Periodic burning of areas for primitive agriculture caused erosion. Trees became scarce near the best campsites because they had been used for camp fires—the first fuel shortage.

ENERGY FUNDAMENTALS. Human use of energy revolves about four interrelated factors: energy sources, methods of harnessing the sources, means of transporting or storing energy, and methods of using energy. The potential energies and energy flows that might be harnessed are many times greater than present use.

The Sun is the primary source of most energy on Earth. Sunlight warms the planet. Plants use photosynthesis to transform water and carbon dioxide into the sugars that power their growth and indirectly power plant-eating and meat-eating animals. Many other energies come indirectly from the Sun. Remains of plants and animals become fossil fuels. Solar heat evaporates water, which then falls as rain, causing water flow in rivers. Regional differences in the amount of sunlight received and reflected cause temperature differences that generate winds, ocean currents, and temperature differences between different ocean layers. Food for muscle power of humans and animals is the most basic energy system.

ENERGY SOURCES. Biomass— wood or other vegetable matter that can be burned—is still the most important energy source in much of the world. Its basic use is to provide heat for cooking and warmth. Biomass fuels are often agricultural or forestry wastes. The advantage of biomass is that it is grown, so it can be re-

placed. However, it has several limitations. Its low energy content per unit volume and unit mass makes it unprofitable to ship, so its use is limited to the amount nearby. Collecting and processing biomass fuels costs energy, so the net energy is less. Biomass energy production may compete with food production, since both come from the soil. Finally, other fuels can be cheaper.

Greater concentration of biomass energy or more efficient use would enable it to better compete against other energy sources. For example, fermenting sugars into fuel alcohol is one means of concentrating energy, but energy losses in processing make it expensive.

Fossil fuels have more concentrated chemical energy than biomass. Underground heat and pressure compacts trees and swampy brush into the progressively more energy-concentrated peat, lignite

After underground oil resources are found by drilling, their contents are pumped to the surface. (PhotoDisc)

coal, bituminous coal, and anthracite or black coal, which is mostly carbon. Industrializing regions turned to coal when they had exhausted their firewood. Like wood, coal could be stored and shoveled into the fire box as needed. Large deposits of coal are still available, but growth in the use of coal slowed by the mid-twentieth century because of two competing fossil fuels, petroleum and natural gas.

Petroleum includes gasoline, diesel fuel, and fuel oil. It forms from remains of one-celled plants and animals in the ocean that decompose from sugars into simpler hydrogen and carbon compounds (hydrocarbons). Petroleum yields more energy per unit than coal, and it is pumped rather than shoveled. These advantages mean that an oil-fired vehicle can be cheaper and have greater range than a coal-fired vehicle.

There are also hydrocarbon gases associated with petroleum and coal. The most common is the natural gas methane. Methane does not have the energy density of hydrocarbon liquids, but it burns cleanly and is a fuel of choice for end uses such as homes and businesses.

Petroleum and natural gas deposits are widely scattered throughout the world, but the greatest known deposits are in an area extending from Saudi Arabia north through the Caucasus Mountains. Deposits extend out to sea in areas such as the Persian Gulf, the North Sea, and the Gulf of Mexico. More exotic sources, such as oil tar sands and shale oil, could be tapped when conventional supplies run low.

Heat engines transform the potential of chemical energies. James Watt's steam engine (1782) takes heat from burning wood or coal (external combustion), boils water to steam, and expands it through pistons to make mechanical motion. In the twentieth century, propeller-like steam turbines were developed to increase efficiency and decrease complexity. Auto and diesel engines burn fuel inside the engine (internal combustion), and the hot gases expand through pistons to make mechanical motion. Expanding them through a gas turbine is a jet engine. Heat engines can create energy from other sources, such as concentrated sunlight, nuclear fission, or nuclear fusion. The electrical generator transforms mechanical motion into electricity that can move by wire to uses far away. Such transportation (or wheeling) of electricity means that one power plant can serve many customers in different locations.

Flowing water and wind are two of the oldest sources of industrial power. The Industrial Revolution began with water power and wind power, but they could only be used in certain locations, and they were not as dependable as steam engines. In the early twentieth century, electricity made river power practical again. Large dams along river valleys with adequate water and steep enough slopes enabled areas like the Tennessee Valley to be industrial centers. In the 1970's wind power began to be used again, this time for generating electricity.

Solar energy can be tapped directly for heat or to make electricity. Although sunlight is free, it is not concentrated energy, so getting usable energy requires more equipment cost. Consequently, fossil-fueled heat is cheaper than solar heat, and power from the conventional utility grid has been much less expensive than solar-generated electricity. However, prices of solar equipment are dropping as technologies improve, and prices of other energy sources may rise.

FUTURE ENERGY SOURCES. Possible future energy sources are nuclear fission, nuclear fusion, geothermal heat, and tides. Fission reactors contain a critical mass of radioactive heavy elements that

Solar panels Page 174

Nuclear fission monument Page 173

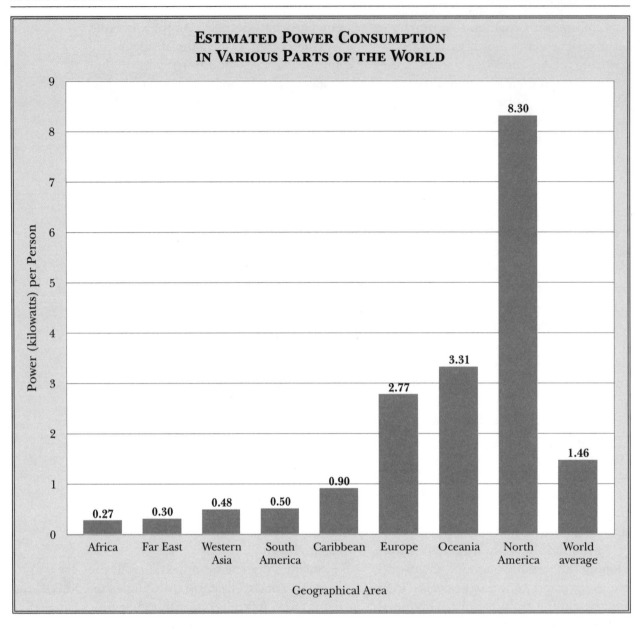

ESTIMATED POWER CONSUMPTION
IN VARIOUS PARTS OF THE WORLD

sustains a chain reaction of atoms splitting (fissioning) into lighter elements—releasing heat to run a steam turbine. Tremendous amounts of fission energy are available, but reactor costs and safety issues have kept nuclear prices higher than that of coal.

Nuclear fusion involves the same reaction that powers the Sun: four hydrogen atoms fusing into one helium atom. However, duplicating the Sun's heat in a small area without damaging the surrounding reactor may be too expensive to allow profitable fusion reactors.

Geothermal power plants, tapping heat energy from within the earth, have operated since 1904, but widespread use depends on cheaper drilling to make them practical in more than highly volcanic areas. Tidal power is limited to the few bays that concentrate tidal energy.

ENERGY AND WARFARE. Much of ancient energy use revolved about herding animals and conducting warfare. Horse riders moved faster and hit harder than warriors on foot. The bow and arrow did not change appreciably for thousands of years. Herders on the plains rode horses and used the bow and arrow as part of tending their flocks, and the small amounts of metal needed for weapons was easily acquired. Consequently, the herders could invade and plunder much more advanced peoples. From Scythians to Parthians to Mongols, these people consistently destroyed the more advanced civilizations.

The geographical effect was that ancient civilizations generally developed only if they had physical barriers separating them from the flat plains of herding peoples. Egypt had deserts and seas. The Greeks and Romans lived on mountainous peninsulas, safe from easy attack. The Chinese built the Great Wall along their northern frontier to block invasions.

Barbarian riders dominated until the advent of an energy system of gunpowder and steel barrels began delivering lead bullets. With them, the Russians broke the power of the Tartars in Eurasia in the late fifteenth century, and various peoples from Europe conquered most of the world. Energy and industrial might became progressively more important in war with automatic weapons, high explosives, aircraft, rockets, and nuclear weapons.

By World War II, oil had become a reason for war and a crucial input for war. The Germans attempted to seize petroleum fields around Baku on the Caspian. Later in the war, major Allied attacks targeted oil fields in Romania and plants in Germany synthesizing liquid fuels. During the Arab-Israeli War of 1973, Arabs countered Western support of Israel with an oil boycott that rocked Western economies. In 1990 Iraq attempted to solve a border dispute with its oil-rich neighbor, Kuwait, by seizing all of Kuwait. An alliance, led by the United States, ejected the Iraqis.

Other wars occur over petroleum deposits that extend out to sea. European nations bordering on the North Sea negotiated a complete demarcation of economic rights throughout that body. There has been no similar negotiation regarding the South China Sea, which may have deposits comparable to those in the Saudi Peninsula. The area is claimed by China, Vietnam, Malaysia, and the Philippines. Turkey and Greece have not resolved ownership division of Aegean waters that might have oil deposits.

ENERGY, DEVELOPMENT, AND ENERGY EFFICIENCY. Ancient civilizations tended to grow and use locally available food and firewood. Soils and wood supplies often were depleted at the same time, which often coincided with declines in those civilizations. The Industrial Revolution caused development to concentrate in new wooded areas where rivers suitable for power, iron ore, and coal were close together, for example, England, Silesia, and the Pittsburgh area. The iron ore of Alsace in France combined with nearby coal from the Ruhr in Germany fueled tremendous growth, not always peacefully.

By the late nineteenth century, the development of Birmingham, Alabama, demonstrated that railroads enabled a wider spread between coal deposits, iron ore deposits, and existing population centers. By the 1920's, the Soviet Union developed entirely new cities to connect with resources. By the 1970's, unit trains and ore-carrying ships transported coal from the thick coal beds in Montana and Wyoming to the United States' East Coast and to countries in Asia.

The mechanized transport of electrical distribution and distribution of natural gas in pipelines also changed settlement patterns. Trains and subway trains allowed cities to spread along rail corridors in the

*Firewood
Page 174*

late nineteenth century and early twentieth century. By the 1940's, cars and trucks enabled cities such as Los Angeles and Phoenix to spread into suburbs. The trend continues with independent solar power that allows houses to be sited anywhere.

Advances in technology have allowed people to get more while using less energy. For example, early peoples stampeded herds of animals over cliffs for food, which was mostly wasted. Horseback hunting was vastly more efficient. Likewise, fireplaces in colonial North America were inefficient, sending most of their heat up the chimney. In the late eighteenth century, inventor and statesman Benjamin Franklin developed a metallic cylinder radiating heat in all directions, which saved firewood.

The ancient Greeks and others pioneered the use of passive solar energy and efficiency after they exhausted available firewood. They sited buildings to absorb as much low winter sun as possible and constructed overhanging roofs to shade buildings from the high summer sun. That siting was augmented by heavy masonry building materials that buffered the buildings from extremes of heat and cold. Later, metal pipes and glass meant that solar energy could be used for water and space heating.

The first seven decades of the twentieth century saw major declines in energy prices, and cars and appliances became less efficient. That changed abruptly with the energy crises and high prices of the 1970's. Since then, countries such as Japan, with few local energy resources, have worked to increase efficiency so they will be less sensitive to energy shocks and be able to thrive with minimal energy inputs. This trend could lead eventually to economies functioning on only solar and biomass inputs.

Solid-state electronics, use of fluorescent lights rather than incandescents, and

fuel cells, which convert fuel directly into electricity more efficiently than combustion engines, all could lead to less energy use. The speed of their adoption depends on the price of competing energies. Predictions that petroleum resources will be exhausted started in 1866; however, this is unlikely to happen before the middle of the twenty-first century. First, drilling will likely go to more exotic locations, and eventually to oil tars, such as those of Venezuela, oil shales in western Canada, and methane hydrates, which are deposits of methane frozen together with water ice on the ocean floors.

ENERGY AND ENVIRONMENT. Energy affects the environment in three major ways. First, firewood gathering in underdeveloped countries contributes to deforestation and resulting erosion. Although more efficient stoves and small solar cookers have been designed, efficiency increases are competing against population increases.

Energy production also frequently causes toxic pollutant by-products. Sulfur dioxide (from sulfur impurities in coal and oil) and nitrogen oxides (from nitrogen being formed during combustion) damage lungs and corrode the surfaces of buildings. Lead additives in gasoline make internal combustion engines run more efficiently, but they cause low-grade lead poisoning. Spent radioactive fuel from nuclear fission reactors is so poisonous that it must be guarded for centuries.

Finally, carbon dioxide from the burning of fossil fuels may be accelerating the greenhouse effect, whereby atmospheric carbon dioxide slows the planetary loss of heat. If the effect is as strong as some research suggests, global temperatures may increase several degrees on average in the twenty-first century, with unknown effects on climate and sea level.

Roger V. Carlson

FOR FURTHER STUDY

Ballanoff, Paul. *Energy: Ending the Never-Ending Crisis.* Washington, D.C.: Cato Institute, 1997.

Berger, John B. *Charging Ahead: The Business of Renewable Energy and What It Means for America.* New York: Henry Holt, 1997.

"Energy for Planet Earth." *Scientific American* 263, no. 3 (September, 1995).

Lee, Thomas H., Ben C. Ball, Jr., and Richard D. Tabors. *Energy Aftermath.* Boston: Harvard Business School Press, 1990.

INFORMATION ON THE WORLD WIDE WEB

The International Energy Agency (IEA) is an energy forum of twenty-three industrialized countries committed to sharing information and taking joint measures to meet oil supply emergencies. The IEA's Web site provides information about the agency's activities and research. (www.iea.org)

ALTERNATIVE ENERGIES

The energy that lights homes and powers industry is indispensable in modern societies. This energy usually comes from mechanical energy that is converted into electrical energy by means of generators—complex machines that harness basic energy captured when such sources as coal, oil, or wood are burned under controlled conditions. This energy, in turn, provides the thermal energy used for heating, cooling, and lighting and for powering automobiles, locomotives, steamships, and airplanes. Because such natural resources as coal, oil, and wood are being used up, it is vital that these nonrenewable sources of energy be replaced by sources that are renewable and abundant. It is also desirable that alternative sources of energy be developed in order to cut down on the pollution that results from the combustion of the hydrocarbons that make the nonrenewable fuels burn.

THE SUN AS AN ENERGY SOURCE. Energy is heat. The Sun provides the heat that makes Earth habitable. As today's commonly used fuel resources are used less, solar energy will be used increasingly to provide the power that societies need in order to function and flourish.

There are two forms of solar energy: passive and active. Humankind has long employed passive solar energy, which requires no special equipment. Ancient cave dwellers soon realized that if they inhabited caves that faced the Sun, those caves would be warmer than those that faced away from the Sun. They also observed that dark surfaces retained heat and that dark rocks heated by the Sun would radiate the heat they contained after the Sun had set. Modern builders often capitalize on this same knowledge by constructing structures that face south in the Northern Hemisphere and north in the Southern

277

Hemisphere. The windows that face the Sun are often large and unobstructed by draperies and curtains. Sunlight beats through the glass and, in passive solar houses, usually heats a dark stone or brick floor that will emit heat during the hours when there is no sunlight. Just as an automobile parked in the sunlight will become hot and retain its heat, so do passive solar buildings become hot and retain their heat.

Solar panels Page 174

Active solar energy is derived by placing specially designed panels so that they face the Sun. These panels, called flat plate collectors, have a flat glass top beneath which is a panel, often made of copper with a black overlay of paint, that retains heat. These panels are constructed so that heat cannot escape from them easily. When water circulated through pipes in the panels becomes hot, it is either pumped into tanks where it can be stored or circulated through a central heating system.

Some active solar devices are quite complex and best suited to industrial use. Among these is the focusing collector, a saucer-shaped mirror that centers the Sun's rays on a small area that becomes extremely hot. A power plant at Odeillo in the French Pyrenees Mountains uses such a system to concentrate the Sun's rays on a concave mirror. The mirror directs its incredible heat to an enormous, confined body of water that the heat turns to steam, which is then used to generate electricity.

Another active solar device is the solar or photovoltaic cell, which gathers heat from the Sun and turns it into energy directly. Such cells help to power spacecraft that cannot carry enough conventional fuel to sustain them through long missions in outer space.

GEOTHERMAL HEATING. The earth's core is incredibly hot. Its heat extends far into the lower surfaces of the planet, at times causing eruptions in the form of gey-sers or volcanoes. Many places on Earth have springs that are warmed by heat from the earth's core.

In some countries, such as Iceland, warm springs are so abundant that people throughout the country bathe in them through the coldest of winters. In Iceland, geothermal energy is used to heat and light homes, making the use of fossil fuels unnecessary.

Hot areas exist beneath every acre of land on Earth. When such areas are near the surface, it is easy to use them to produce the energy that humans require. As dependence on fossil fuels decreases, means will increasingly be found of drawing on Earth's subterranean heat as a major source of energy.

WIND POWER. Anyone who has watched a sailboat move effortlessly through the water has observed how the wind can be used as a source of kinetic energy—the kind of energy that involves motion—whose movement is transferred to objects that it touches. Wind power has been used throughout human history. In its more refined aspects, it has been employed to power windmills that cause turbines to rotate, providing generators with the power they require to produce electricity.

Windmills typically have from two to twenty blades made of wood or of heavy cloth such as canvas. Windmills are most effective when they are located in places where the wind regularly blows with considerable velocity. As their blades turn, they cause the shafts of turbines to rotate, thus powering generators. The electricity created is usually transmitted over metal cables for immediate use or for storage.

Modern vertical-axis wind turbines have two or three strips of curved metal that are attached at both ends to a vertical pole. They can operate efficiently even if they are not turned toward the wind.

Spanish windmills. The simple windmill is one of the oldest and most efficient machines for harnessing alternative energy. (PhotoDisc)

These windmills are a great improvement over the old horizontal axis windmills that have been in use for many years. Although older wind machines did not produce sufficient power for whole communities, the Department of Energy is experimenting with vertical-axis machines that it estimates could some day meet 20 percent of the energy needs of the United States, cheaply and without pollution.

OCEANS AS ENERGY SOURCES. Seventy percent of the earth's surface is covered by oceans. Their tides, which rise and fall with predictable regularity twice a day, would offer a ready source of energy once it becomes economically feasible to harness them and store the electrical energy they can provide. The most promising spots to build facilities to create electrical energy from the tides are places where the tides are regularly quite dramatic, such as Nova Scotia's Bay of Fundy, where the difference between high and low tides averages about 55 feet (17 meters).

Some tidal power stations that currently exist were created by building dams across estuaries. The sluices of these dams are opened when the tide comes in and closed after the resulting reservoir fills. The water captured in the reservoir is held for several hours until the tide is low enough to create a considerable difference between the level of the water in the reservoir and that outside it. Then the sluice gates are opened and, as the water rushes out at a high rate of speed, it turns turbines that generate electricity.

FUTURE OF RENEWABLE ENERGY. As pollution becomes a huge problem

---

OCEAN ENERGY

The oceans have tremendous untapped energy flows in currents and tremendous potential energy in the temperature differences between warmer tropical surface waters and colder deep waters, known as ocean thermal energy conversion. In both cases, the insurmountable cost has been in transporting energy to users on shore.

throughout the world, the race to find nonpolluting sources of energy is accelerating rapidly. New technologies are making renewable energy sources economically practical. As supplies of fossil fuels have diminished, pressure to become less dependent on them has grown worldwide. Alternative energy sources are the wave of the future.

R. Baird Shuman

FOR FURTHER STUDY

Berger, John B. *Charging Ahead: The Business of Renewable Energy and What It Means for America*. New York: Henry Holt, 1997.

Borowitz, Sidney. *Farewell Fossil Fuels: Reviewing America's Energy Policy*. New York: Plenum Trade, 1999.

Chandler, Gary, and Kevin Graham. *Alternative Energy Sources*. New York: Twenty-First Century Books, 1996.

Cole, Nancy, and P. J. Skerrett. *Renewables Are Ready*. White River Junction, Vt.: Chelsea Green, 1995.

Dunn, P. D. *Renewable Energies: Sources, Conversion, and Application*. London: Peter Peregrinus, 1986.

Gipe, Paul. *Wind Power for Home and Business: Renewable Energy for the 1990's and Beyond*. White River Junction, Vt.: Chelsea Green, 1993.

Wrixon, G. T., A.-M. E. Rooney, and W. Palz. *Renewable Energy—2000*. Berlin: Springer-Verlag, 1993.

ENGINEERING PROJECTS

Human beings attempt to overcome the physical landscape by building forms and structures on the earth. Most structures are small-scale, like houses, telephone poles, and schools. Other structures are great engineering works, such as hydroelectric projects, dams, canals, tunnels, bridges, and buildings.

HYDROELECTRIC PROJECTS. The potential for hydroelectricity generation is greatest in rapidly flowing rivers in mountainous or hilly terrain. The moving water turns turbines that, in turn, generate electricity. Hydroelectric power projects also can be built on escarpments and fall lines, where there is tremendous untapped energy in the falling water.

Most of the potential for hydroelectricity remains untapped. Only about one-sixth of the suitable rivers and falls are used for hydroelectric power. Certain areas of the world have used more of their potential than others. The United States, most of Western Europe, Japan, South Korea, and Australia have all tapped about three-fourths of their potential for water power. Brazil, Paraguay, Mexico, and Canada also use significant portions of their hydroelectric potential. Russia, the former Soviet Republics, China, Pakistan, and India have tremendous potential that is not yet fully tapped, but they still produce a significant proportion of the world's hydroelectric power.

Most of the remaining areas of the world have not yet taken advantage of hydroelectric power. In South America, there is great potential for exploiting water power in most areas, especially Colombia, Ecuador, Peru, and Argentina. In Africa, only Zambia, Zimbabwe, and Ghana produce significant hydroelectricity. In the late 1990's, the Democratic Republic of the Congo (formerly Zaire) showed the greatest promise for the future, but it had not yet tapped this resource.

In Southeast Asia, only Thailand and Vietnam have used much of their potential, and even that is not a great amount. The greatest potential in that region lies in Indonesia, New Guinea, and Myanmar (Burma).

DAMS. Dams serve several purposes. One purpose is the generation of hydroelectric power, as discussed above. Dams also provide flood control and irrigation. Rivers in their natural state tend to rise and fall with the seasons. This can cause serious problems for people living in downstream valleys. Flood-control dams also can be used to regulate the flow of water used for irrigation and other projects. A fi-

nal reason to build dams is to reduce swampland, in order to control insects and the diseases they carry.

Famous dams are found in all regions of the world. In North America, two of the most notable dams are Hoover Dam, completed in 1936, on the Colorado River between Arizona and Nevada; and the Grand Coulee Dam, completed in 1942, on the Columbia River in Washington State.

In South America, the most famous dam is the Itaipu Dam, completed in 1983, on the Paraná River between Brazil and Paraguay. In Africa, the Aswan High Dam was completed in 1970, on the Nile River in Egypt, and the Kariba Dam was completed in 1958, on the Zambezi River between Zambia and Zimbabwe. In Asia, the Three Gorges Dam was under construction on the Chiang Jiang (Yangtze River) in China during the late 1990's, with completion scheduled for 2009.

BRIDGES. Bridges are built to span low-lying land between two high places. Most

Tenpozan Watasi bridge in Osaka, Japan. (PhotoDisc)

commonly, there is a river or other body of water in the way, but other features that might be spanned include ravines, deep valleys and trenches, and swamps. A related engineering project is the causeway, in which land in a low-lying area is built up and a road is then constructed on it.

The longest bridge in the world is the Akashi Kaikyo in Japan near Osaka. It was built in 1998 and spans 6,529 feet (1,990 meters), connecting the island of Hōnshū to the small island of Awaji. The Storebælt Bridge in Denmark, also completed in 1998, spans 5,328 feet (1,624 meters), connecting the island of Sjaelland, on which Copenhagen is situated, with the rest of Denmark. Another bridge spanning more than 5,300 feet is the Izmit Bay Bridge in Turkey, which was being built near Istanbul in the late 1990's.

Other long bridges can be found across the Humber River in Hull, England; across the Chiang Jiang (Yangtze River) in China; in Hong Kong, Norway, Sweden, and Turkey and elsewhere in Japan.

The longest bridge in the United States, which was once the longest in the world, is the Verrazano-Narrows Bridge in New York City between Staten Island and Brooklyn. Completed in 1964, it spans 4,260 feet (1,298 meters). Only slightly shorter—at 4,200 feet—is the San Francisco Bay Bridge, which was completed in 1937.

CANALS. Moving goods and people by water is generally cheaper and easier, if a bit slower, than moving them by land. Before the twentieth century, that cost savings overwhelmed the advantages of land travel—speed and versatility. Therefore, human beings have wanted to move things by water whenever possible. To do so, they had two choices: locate factories and people near water, such as rivers, lakes, and oceans, or bring water to where the factories and people are, by digging canals.

Panama Canal Page 236

Dutch canal Page 223

One of the most famous canals in the world is the Erie Canal, which runs from Albany to Buffalo in New York State. Built in 1825 and running a length of 363 miles (584 km.), the Erie Canal opened up the Great Lakes region of North America to development and led to the rise of New York City as one of the world's dominant cities.

Two other important canals in world history are the Panama Canal and the Suez Canal. The Panama Canal connects the Atlantic and Pacific Oceans over a length of 50.7 miles (81.6 km.) on the isthmus of Panama in Central America. Completed in 1914, the Panama Canal eliminated the long and dangerous sea journey around the tip of South America. The Suez Canal in Egypt, which runs for 100 miles (162 km.) and was completed in 1856, eliminates a similar journey around the Cape of Good Hope in South Africa.

The longest canal in the world is the Grand Canal in China, which was built in the seventh century and stretches a length of 1,085 miles (2,904 km.). It connects Tianjin, near Beijing in the north of China, with Nanjing on the Chang Jiang (Yangtze River) in Central China. This canal may eventually be surpassed in length by the Karakum Canal, which runs across the Central Asian desert in Turkmenistan from the Amu Darya River westward to Ashkhabad. That canal was begun in the 1950's and was intended to irrigate the dry lands of Turkmenistan and eventually to reach the Caspian Sea. The project has stalled at a length of 700 miles (1,100 km.) and it is not known if it will ever be completed.

Many canals are found in Europe, particularly in England, France, Belgium, the Netherlands, and Germany, and in the United States and Canada, especially connecting the Great Lakes to each other and to the Ohio and Mississippi Rivers.

ENGINEERING WORKS AND ENVIRONMENTAL PROBLEMS

Although engineering allows humans to overcome natural obstacles, works of engineering often have unintended consequences. Many engineering projects have caused unanticipated environmental problems.

Dams, for instance, create large lakes behind them by trapping water that is released slowly. This water typically contains silt and other material that eventually would have formed soil downstream had the water been allowed to flow naturally. Instead, the silt builds up behind the dam, eventually diminishing the lake's usefulness. As an additional consequence, there is less silt available for soil-building downstream.

Canals also can cause environmental harm by diverting water from its natural course. The river from which water is diverted may dry up, negatively affecting fish, animals, and the people who live downstream.

The benefits of engineering works must be weighed against the damage they do to the environment. They may be worthwhile, but they are neither all good nor all bad: There are benefits and drawbacks in building any engineering project.

TUNNELS. Tunnels connect two places separated by physical features that would make it extremely difficult, if not impossible, for them to be connected without cutting directly through them. Tunnels can be used in place of bridges over water bodies so that water traffic is not impeded by a bridge span. Tunnels of this type are often found in port cities, and cities with them include Montreal, Quebec; New York City; Hampton Roads, Virginia; Liverpool, England; or Rio de Janeiro, Brazil.

Tunnels are often used to go through mountains that might be too tall to climb over. Trains especially are sensitive to changes in slope, and train tunnels are found all over the world. Less common are automobile and truck tunnels, although these are also found in many places. Train and automotive tunnels through mountains are common in the Appalachian Mountains in Pennsylvania, the Rockies in the United States and Canada, Japan, and the Alps in Italy, France, Switzerland, and Austria.

THE CHUNNEL. Arguably the most famous—and one of the most ambitious—tunnels in the world goes by the name Chunnel. Completed in 1994, it connects Dover, England, to Calais, France, and runs 31 miles (50 km.). "Chunnel" is short for the Channel Tunnel, named for the English Channel, the body of water that it goes under. It was built as a train tunnel, but cars and trucks can be carried through it on trains. In the year 2000 plans were underway to cut a second tunnel, to carry automobiles and trucks, that would run parallel to the first Chunnel.

Among undersea tunnels, the Chunnel is exceeded in length only by the Seikan Tunnel in Japan, which connects the large island of Hōnshū with the northern island of Hokkaidō. The Seikan Tunnel is nearly 2.4 miles (4 km.) longer than Europe's Chunnel.

BUILDINGS. Historically, North America has been home to the tallest buildings in the world. Chicago has been called the birthplace of the skyscraper and was at one time home to the world's tallest building. In 1998, however, the two Petronas Towers (each 1,483 feet/452 meters tall) were completed in Kuala Lumpur, Malaysia, surpassing the height of the world's tallest building, Chicago's Sears Tower

The Empire State Building. (Corbis)

East and Southeast Asia. The Jin Mao Building in Shanghai, China, was completed in 1999, at a height of 1,380 feet (421 meters). New York City is home to several of the world's tallest buildings, including the World Trade Center twin towers. They were finished in 1972 and 1973, respectively, at heights of 1,368 feet (417 meters) and 1,362 feet (415 meters). Also in New York are the Empire State Building at a height of 1,250 feet (381 meters) and the Chrysler Building at 1,046 feet (319 meters). These both were finished in the 1930's and were the two tallest buildings in the world until the John Hancock Center was finished in Chicago in 1969, at 1,127 feet (344 meters).

Hong Kong and the Pearl River region of China boast several tall buildings. Citic Plaza in Guangzhou, China, was completed in 1997 at a height of 1,283 feet (391 meters). Shun Hing Square in Shenzhen, China, was completed a year earlier and stands 1,260 feet (384 meters).

Of the twenty tallest buildings standing in the year 2000, New York City is home to four, Chicago and Hong Kong to three each, and Kuala Lumpur to two. Eight other cities boast one each: Shanghai, Guangzhou, and Shenzhen in China; Kaoshiung in Taiwan; Dubai in the United Arab Emirates; Bangkok in Thailand; and Atlanta and Los Angeles in the United States. The tallest building in Europe is Commerzbank Tower in Frankfurt, Germany, completed

(1,450 feet/442 meters), which had been completed in 1974. Not to be outdone, plans were made in Chicago in the late 1990's to build a new skyscraper, called the 7 South Dearborn Building, which would be 1,550 feet (472 meters) in height when it was completed in 2003.

Other famous tall buildings are found primarily in cities of North America and in 1997 at 981 feet (299 meters). In Australia, Rialto Tower in Melbourne was built in 1985 and stands 813 feet (248 meters). There are no buildings over 750 feet (228 meters) in South America or Africa.

Timothy C. Pitts

FOR FURTHER STUDY

Fales, James F. *Construction Technology: Today and Tomorrow.* New York: Glencoe/McGraw-Hill, 1991.

Franck, Irene M., and David M. Brownstone. *Builders.* New York: Facts on File, 1985.

Kingston, Jeremy. *How Bridges Are Made.* New York: Facts on File, 1985.

Toberman, Scott. "The Sky's the Limit." *Popular Mechanics* (March, 2000): 56-59.

Zich, Arthur. "China's Three Gorges." *National Geographic* (September, 1997): 2-33.

New York City's World Trade Center towers. (PhotoDisc)

INDUSTRY AND TRADE

MINERALS

Mineral resources make up all the nonliving matter found in the earth, its atmosphere, and its waters that are useful to humankind. The great ages of history are classified by the resources that were exploited. First came the Stone Age, when flint was used to make tools and weapons. The Bronze Age followed; it was a time when metals such as copper and tin began to be extracted and used. Finally came the Iron Age, the time of steel and other ferrous alloys that required higher temperatures and more sophisticated metallurgy.

*Strip mine
Page 223*

*Mining
operations
Pages 162,
224*

Metals, however, are not the whole story—economic progress also requires fossil fuels such as coal, oil, natural gas, tar sands, or oil shale as energy sources. Beyond metals and fuels, there are a host of mineral resources that make modern life possible: building stone, salt, atmospheric gases (oxygen, nitrogen), fertilizer minerals (phosphates, nitrates, and potash), sulfur, quartz, clay, asbestos, and diamonds are some examples.

MINING AND PROSPECTING. Exploitation of mineral resources begins with the discovery and recognition of the value of the deposits. To be economically viable, the mineral must be salable at a price greater than the cost of its extraction, and great care is taken to determine the probable size of a deposit and the labor involved in isolating it before operations begin. Iron, aluminum, copper, lead, and zinc oc-

cur as mineral ores that are mined, then subjected to chemical processes to separate the metal from the other elements (usually oxygen or sulfur) that are bonded to the metal in the ore.

Some deposits of gold or platinum are found in elemental (native) form as nuggets or powder and may be isolated by alluvial mining—using running water to wash away low-density impurities, leaving the dense metal behind. Most metal ores, however, are obtained only after extensive digging and blasting and the use of large-scale earthmoving equipment. Surface mining or strip mining is far simpler and safer than underground mining.

SAFETY AND ENVIRONMENTAL CONSIDERATIONS. Underground mines can extend as far as a mile into the earth and are subject to cave-ins, water leakage, and dangerous gases that can explode or suffocate miners. Safety is an overriding issue in deep mines, and there is legislation in many countries designed to regulate mine safety and to enforce practices that reduce hazards to the miners from breathing dust or gases.

In the past, mining often was conducted without regard to the effects on the environment. In economically advanced countries such as the United States, this is now seen as unacceptable. Mines are expected to be filled in, not just abandoned after they are worked out, and care must be taken that rivers and streams are not contaminated with mine wastes.

IRON, STEEL, AND COAL. Iron ore and coal are essential for the manufacture of steel, the most important structural metal. Both raw materials occur in many geographic regions. Before the mid-nineteenth century, iron was smelted in the eastern United States—New Jersey, New York, and Massachusetts—but then huge hematite deposits were discovered near Duluth, Minnesota, on Lake Superior. The ore traveled by ship to steel mills in northwest Indiana and northeast

Contaminated water caused by mining in Idaho Springs, Colorado, in the early 1980's. (U.S. Geological Survey)

Illinois, and coal came from Illinois or Ohio. Steel also was made in Pittsburgh and Bethlehem in Pennsylvania, and in Birmingham, Alabama.

After World War II, the U.S. steel industry was slow to modernize its facilities, and after 1970 it had great difficulty producing steel at a price that could compete with imports from countries such as Japan, Korea, and Brazil. In Europe, the German steel industry centered in the Ruhr River valley in cities such as Essen and Düsseldorf. In Russia, iron ore is mined in the Urals, in the Crimea, and at Krivoi Rog in Ukraine. Elsewhere in Europe, the French "minette" ores of Alsace-Lorraine, the Swedish magnetite deposits near Kiruna, and the British hematite deposits in Lancashire are all significant. Hematite is also found in Labrador, Canada, near the Quebec border.

Coal is widely distributed on earth. In the United States, Kentucky, West Virginia, and Pennsylvania are known for their coal mines, but coal is also found in Illinois, Indiana, Ohio, Montana, and other states. Much of the anthracite (hard coal) is taken from underground mines,

where networks of tunnels are dug through the coal seam, and the coal is loosened by blasting, use of digging machines, or human labor. A huge deposit of brown coal is mined at the Yallourn open pit mine west of Melbourne, Australia. In Germany, the mines are near Garsdorf in Nord-Rhein/Westfalen, and in the United Kingdom, coal is mined in Wales. South Africa has coal and is a leader in manufacture of liquid fuels from coal. There is coal in Antarctica, but it cannot yet be mined profitably. China and Japan both have coal mines, as does Russia.

ALUMINUM. Aluminum is the most important structural metal after iron. It is extremely abundant in the earth's crust, but the only readily extractable ore is bauxite, a hydrated oxide usually contaminated with iron and silica. Bauxite was originally found in France but also exists in many other places in Europe, as well as in Australia, India, China, the former Soviet Union, Indonesia, Malaysia, Suriname, and Jamaica.

Much of the bauxite in the United States comes from Arkansas. After purification, the bauxite is combined with the mineral cryolite at high temperature and subjected to electrolysis between carbon electrodes (the Hall-Héroult process), yielding pure aluminum. Because of the enormous electrical energy requirements of the Hall-Héroult method, aluminum can be made economically only where cheap power (preferably hydroelectric) is available. This means that the bauxite often must be shipped long distances—Jamaican bauxite comes to the United States for electrolysis, for example.

COPPER, SILVER, AND GOLD. These coinage metals have been known and used since antiquity. Copper came from Cyprus and takes its name from the name of the island. Copper ores include oxides or sulfides (cuprite, bornite, covellite, and others). Not enough native copper occurs to be commercially significant. Mines in Bingham, Utah, and Ely, Nevada, are major sources in the United States. The El Teniente mine in Chile is the world's largest copper mine, and major amounts of copper also come from Canada, the former Soviet Union, and the Katanga region mines in Congo-Kinshasa and Zambia.

Silver often occurs native, as well as in combination with other metals, including lead, copper, and gold. Famous silver mines in the United States include those near Virginia City (the Comstock lode) and Tonopah, Nevada, and Coeur d'Alene, Idaho. Silver has been mined in the past in Bolivia (Potosi mines), Peru (Cerro de Pasco mines), Mexico, and Ontario and British Columbia in Canada.

Gold occurs native as gold dust or nuggets, sometimes with silver as a natural alloy called electrum. Other gold minerals include selenides and tellurides. Small amounts of gold are present in sea water, but attempts to isolate gold economically from this source have so far failed. Famous gold rushes occurred in California and Colorado in the United States, Canada's Yukon, and Alaska's Klondike region. Major gold-producing countries include South Africa, Siberia, Ghana (once called the Gold Coast), the Philippines, Australia, and Canada.

PETROLEUM AND NATURAL GAS. Petroleum has been found on every continent except Antarctica, with 600,000 producing wells in one hundred different countries. In the United States, petroleum was originally discovered in Pennsylvania, with more important discoveries being made later in west Texas, Oklahoma, California, and Alaska. New wells are often drilled offshore, for example in the Gulf of Mexico or the North Sea. The United States depends heavily on oil imported from Mexico, South America, Saudi Arabia and the Persian Gulf states, and Canada.

Over the years, the price of oil has varied dramatically, particularly due to the attempts of the Organization of Petroleum Exporting Countries (OPEC) to limit production and drive up prices. In Europe, oil is produced in Azerbaijan near the Caspian

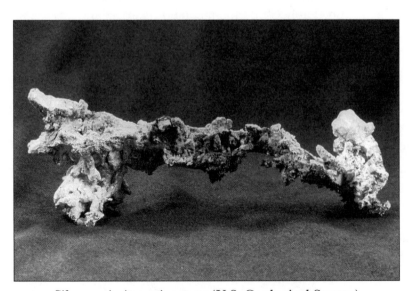

Offshore drill Page 172

Silver ore in its native state. (U.S. Geological Survey)

Sea, where a pipeline is planned to carry the crude to the Mediterranean port of Ceyhan, in Turkey. In Africa, there are oil wells in Gabon, Libya, and Nigeria; in the Persian Gulf region, oil is found in Kuwait, Qatar, Iran, and Iraq. Much crude oil travels in huge tankers to Europe, Japan, and the United States, but some supplies refineries in Saudi Arabia at Abadan. Tankers must exit the Persian Gulf through the narrow Gulf of Hormuz, which thus assumes great strategic importance.

After oil was discovered on the shores of the Beaufort Sea in northern Alaska (the so-called North Slope) in the 1960's, a pipeline was built across Alaska, ending at the port of Valdez. The pipeline is heated to keep the oil liquid in cold weather and elevated to prevent its melting through the permanently frozen ground (permafrost) that supports it. From Valdez, tankers reach Japan or California.

A section of the Alaska Pipeline, which carries crude oil from the state's northern slopes to Valdez on Alaska's southern coast. (PhotoDisc)

THE EXXON VALDEZ OIL SPILL

On March 24, 1989, the tanker *Exxon Valdez*, with a cargo of fifty-three million gallons of crude oil, ran aground on Bligh Reef in Prince William Sound, Alaska. Approximately eleven million gallons of oil were released into the water, in the worst environmental disaster of this type recorded to date. Despite immediate and lengthy efforts to contain and clean up the spill, there was extensive damage to wildlife, including aquatic birds, seals, and fish. Lawsuits and calls for new regulatory legislation on tankers continued a decade later. Such regrettable incidents as these are the almost inevitable result of attempting to transport the huge oil supplies demanded in the industrialized world.

Drilling activities occasionally result in discovery of natural gas, which is valued as a low-pollution fuel. Vast fields of gas exist in Siberia, and gas is piped to Western Europe through a pipeline. Algerian gas is shipped in the liquid state in ships equipped with refrigeration equipment to maintain the low temperatures needed. Late 1990's gas finds in Alberta, Canada, were expected to help supply the energy needs of the central United States when a pipeline is built. Britain and Northern Europe also benefit from gas produced in the

North Sea, between Norway and Scotland.

Shale oil, a plentiful but difficult-to-exploit fossil fuel, exists in enormous amounts near Rifle, Colorado. A form of oil-bearing rock, the shale must be crushed and heated to recover the oil, a more expensive proposition than drilling conventional oil wells. In spite of ingenious schemes such as burning the shale oil in place, this resource is likely to remain largely unused until conventional petroleum is used up. A similar resource exists in Alberta, Canada, where the Athabasca tar sands are exploited for heavy oils.

John R. Phillips

FOR FURTHER STUDY

Alexander, William O., and Street, Arthur C. *Metals in the Service of Man*. Baltimore: Penguin, 1964.

Jones, W. R. *Minerals in Industry*. Baltimore: Penguin, 1963.

Pearl, Richard M. *Gems, Minerals, Crystals and Ores*. New York: Odyssey, 1964.

Robinson, G. W., and T. A. Scovil. *Minerals*. New York: Simon & Schuster, 1994.

Scalisi, P., and Cook, D. *Classic Mineral Localities of the World: Asia and Australia*. New York: Van Nostrand Reinhold, 1983.

Strahler, Alan H. *Introducing Physical Geography*. New York: Wiley, 2000.

MANUFACTURING

Manufacturing is the process by which value is added to materials by changing their physical form—shape, function, or composition. For example, an automobile is manufactured by piecing together thousands of different component parts, such as seats, bumpers, and tires. The component parts in unassembled form have little or no utility, but pieced together to produce a fully functional automobile, the resulting product has significant utility. The more utility something has, the greater its value. In other words, the value of the component parts increases when they are combined with the other parts to produce a useful product.

EMPLOYMENT IN MANUFACTURING. On a global scale, only 20 percent of the world's working population had jobs in the manufacturing sector at the end of the twentieth century. The rest worked in agriculture and mining (49 percent) and services (31 percent). The importance of each of these sectors varies from country to country and from time period to time period. High-income countries have a higher percentage of their labor force employed in manufacturing than low-income countries do. For example, in the United States 18 percent of the labor force worked in manufacturing in the late 1990's, whereas the African country of Tanzania had only 5 percent of its labor force employed in the manufacturing sector at that time.

At the end of the twentieth century, the vast majority of the U.S. labor force (81 percent) worked in services, a sector that includes jobs such as computer programmers, lawyers, and teachers. Only 1 percent worked in agriculture and mining.

This employment structure is typical for a high-income country. In low-income countries, in contrast, the majority of the labor force have agricultural jobs. In Tanzania, for example, 84 percent of the labor force worked in agriculture, while services accounted for 11 percent of the jobs.

The importance of manufacturing as an employer changes over time. In 1950 manufacturing accounted for 38 percent of all jobs in the United States. The percentage of jobs accounted for by the manufacturing sector in high-income countries has decreased in the post-World War II period. The decreasing share of manufacturing jobs in high-income countries is partly attributable to the fact that many manufacturing companies have replaced people with machines on assembly lines. Because one machine can do the work of many people, manufacturing has become less labor-intensive (uses fewer people to perform a particular task) and more capital-intensive (uses machines to perform tasks formerly done by people). In the future, manufacturing in high-income countries is expected to become increasingly capital-intensive. It is not inconceivable that manufacturing's share of the U.S. labor force could fall below 10 percent in the twenty-first century.

Although the importance of manufacturing as an employer is decreasing, it should be noted that manufacturing jobs tend to pay higher wages than jobs in many other sectors. For example, the average manufacturing job in the United States paid more than $35,000 per year, while the average construction job paid just over $31,000 and the average retail job just over $20,000.

GEOGRAPHY OF MANUFACTURING. Every country produces manufactured goods, but the vast bulk of manufacturing activity is concentrated geographically in three major manufacturing regions—eastern North America, Europe, and eastern Asia. Together, these three regions produce more than 85 percent of the world's manufacturing output. In fact, three countries—the United States, Japan, and Germany—produce almost 60 percent of the world's manufactured goods. The concentration of manufacturing activity in a small number of regions means that there are other regions where very little manufacturing occurs. Africa is a prime example of a region with little manufacturing.

Different countries tend to specialize in the production of different products. For example, 50 percent of the automobiles that were produced in that late 1990's were produced in three countries–Germany, Japan, and the United States. In the production of television sets, the top three countries were China, Japan, and South Korea, which together produced 48 percent of the world's television sets. It is important to note that these patterns change over time. For example, in 1960 the top three automobile-producing countries were Germany, the United Kingdom, and the United States, which together produced 76 percent of the world's automobiles.

MULTINATIONAL CORPORATIONS. A multinational corporation is a corporation that is headquartered in one country but owns business facilities, for example, manufacturing plants, in other countries. Some examples of multinational corporations from the manufacturing sector include the automobile maker Ford, whose headquarters are the in the United States, the pharmaceutical company Bayer, whose headquarters are in Germany, and the candy manufacturer Nestle, whose headquarters are in Switzerland. Since the end of World War II, multinational corporations have become increasingly important in the world economy. Most multinational corporations are headquartered in

DAIMLERCHRYSLER: A GLOBAL COMPANY

DaimlerChrysler, the world's fifth-largest auto maker, is a representative example of a large multinational corporation. DaimlerChrysler's automobiles, including brands such as Mercedes Benz and Jeep, are sold to consumers in more than two hundred countries. From its dual headquarters in Auburn Hills, Michigan, and Stuttgart, Germany, DaimlerChrysler owned and operated 141 automobile assembly plants in 34 countries in the late 1990's. Worldwide, the corporation employed over 400,000 people. The following table shows the late-1990's geographic distribution of DaimlerChrysler automobile assembly plants and employees.

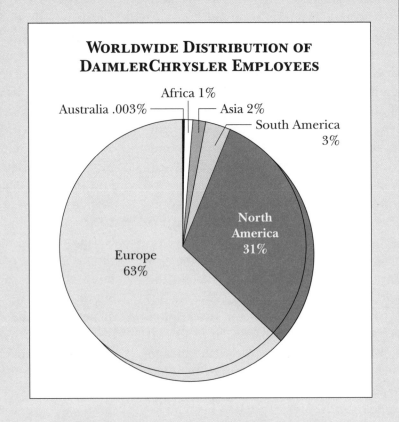

WORLDWIDE DISTRIBUTION OF DAIMLERCHRYSLER EMPLOYEES

Region	Plants	Employees
Europe	95	259,937
North America	19	125,551
Asia	15	7,285
South America	5	13,361
Africa	4	4,063
Australia	3	1,226
Worldwide	141	411,423

high-income countries, such as Japan, the United Kingdom, and the United States.

Companies open manufacturing plants in other countries for a variety of reasons. One of the most common reasons is that it allows them to circumvent barriers to trade that are imposed by foreign governments, especially tariffs and quotas. A tariff is an import tax that is imposed upon foreign-manufactured goods as they enter a country. A quota is a limitation imposed on the volume of a particular good that a particular country can export to another country. The net effect of tariffs and quotas is to increase the cost of imported goods for consumers.

Governments impose tariffs and quotas partly to raise revenue and partly to encourage consumers to purchase goods manufactured in their own country. Foreign manufacturers faced with tariffs and quotas often begin manufacturing their product in the country imposing the tariffs and quotas. As tariffs and quotas apply to imported goods only, producing in the country imposing the quotas or tariffs effectively makes these trade barriers obsolete.

Companies also open manufacturing plants in other countries because of differences in labor costs among countries. While most manufacturing takes place in high-income countries, some low-income countries have become increasingly attractive as production locations because their workers can be hired much more cheaply than in high-income countries. For example, in the late 1990's, the average manufacturing job in the United States paid more than $17 per hour. By comparison, manufacturing employees in the Asian country of Sri Lanka earned less than $1 per hour.

This dramatic differences in labor costs have prompted some companies to close down their manufacturing plants in high-income countries and open up new plants in low-income countries. This has resulted in high-income countries purchasing more manufactured goods from low-income countries. In 1988, for example, 28 percent of the clothing purchased by U.S. consumers was imported from other countries, while in 1998, 48 percent of the clothing purchased in the United States was imported.

More than half the clothing imported into the United States came from Asian countries, for example, China, Taiwan, and South Korea, where labor costs were much lower than in the United States. Much of this clothing was made in factories where workers were paid by companies headquartered in the United States. For example, most of the Nike sports shoes that were sold in the United States were made in China, Indonesia, Vietnam, and Pakistan.

TRANSPORTATION AND COMMUNICATIONS TECHNOLOGY. The ability of companies to have manufacturing plants in other countries stems from the fact that the world has a sophisticated and efficient transportation and communications system. An advanced transportation and communications system makes it relatively easy and relatively cheap to transfer information and goods between geographically distant locations. Thus, Nike can manufacture soccer balls in Pakistan and transport them quickly and cheaply to customers in the United States.

The extent to which transportation and communications systems have improved during the last two centuries can be illustrated by a few simple examples. In 1800, when the stagecoach was the primary method of overland transportation, it took twenty hours to travel the ninety miles from Lansing, Michigan, to Detroit, Michigan. Today, with the automobile, the same journey takes approximately ninety minutes. In 1800 sailing ships traveling at an average speed of ten miles per hour were used to transport people and goods between geographically distant countries. In the year 2000 jet-engine aircraft could traverse the globe at speeds in excess of six hundred miles per hour. Communications technology has also improved over time.

In 1930 a three-minute telephone call between New York and London, England, cost more than $250 in 1998 dollars. In the year 2000 the same telephone call could be made for less than a dollar. In addition to modern telephones, there are fax machines, e-mail, video conferencing capabilities, and a host of other technologies that make communication with other parts of the world both inexpensive and swift

FUTURE PROSPECTS. The global economy of the twenty-first century presents a wide variety of opportunities and challenges. Sophisticated communications and transportation networks provide increasing numbers of manufacturing companies with more choices as to where to locate their factories. However, high-income countries like the United States are increasingly in competition with other countries (both high- and low-income) to maintain existing and manufacturing investments and attract new ones. Persuading existing companies to keep their U.S. factories open and not move overseas has been a major challenge. Likewise, making the United States as an attractive place for foreign companies to locate their manufacturing plants is an equally challenging task.

Neil Reid

FOR FURTHER STUDY

Dicken, Peter. *Global Shift: Transforming the World Economy.* 3d ed. New York: Guilford Press, 1998.

Ferdows, Kasra. "Making the Most of Foreign Factories." *Harvard Business Review* (March-April, 1997): 73-88.

Hounshell, David A. *From the American System to Mass Production, 1800-1932.* Baltimore: Johns Hopkins University Press, 1984.

Marcus, Alan I., and Howard P. Segal. *Technology in America: A Brief History.* New York: Harcourt Brace Jovanovich, 1989.

Noble, David F. *Forces of Production: A Social History of Automation.* New York: Oxford University Press, 1986.

Piore, Michael. *The Second Industrial Divide.* New York: Basic Books, 1984.

Wheeler, James O., Peter O. Mueller, Grant I. Thrall, and Timothy J. Fik. *Economic Geography.* 3d ed. New York: John Wiley & Sons, 1998.

GLOBALIZATION OF MANUFACTURING AND TRADE

Why are most of the patents issued worldwide assigned to U.S. corporations? How did a Taiwanese earthquake prevent millions of Americans from purchasing memory upgrades for their computers? Why have personal incomes in Beijing nearly doubled in less than a decade?

Answers to these questions can be found in the geography of globalization. Globalization is an economic, political, and social process characterized by the integration of the world's many systems of manufacturing and trade into a single and increasingly seamless marketplace. The result: a new world geography.

This new geography is associated with the expansion of manufacturing and trade as capitalist principles replace old ideologies and state-controlled economies. With expanded free markets, the process of manufacturing and trading is constantly changing. Globalization delivers economic growth through improved manufacturing processes, newly developed goods, foreign investment in overseas manufacturing, and expanded employment.

The economies of developing countries are slowly transitioning from agricultural to industrial activities. Nevertheless, more than 70 percent of workers in these countries continue to work in agriculture. Meanwhile, developed countries, such as Australia and Germany, are experiencing high-technology service sector growth and reduced manufacturing employment. In the United States, nearly 50 percent of all workers were employed in manufacturing during the 1950's, but by the late 1990's, less than 20 percent were.

In between these extremes, former state-controlled economies, like Romania, are adopting more efficient economic development strategies. Other nations and economic models, such as Indonesia and China, are pulled into the global marketplace by the growth and expansion of market economies. Despite the different economic paths of developing, transitioning, and developed nations, manufacturing and trade link all nations together and represent an economic convergence with important implications for political, business, and labor leaders—as well as all the world's citizens.

The geographies of manufacturing and trade can be examined as the distribution and location of economic activities in response to technological change and political and economic change.

DISTRIBUTION AND LOCATION. Questions about where people live, work, and spend their money can be answered by reading product labels in any shopping mall, supermarket, or automobile dealership. They reveal the fact that manufacturing is a multistage process of component fabrication and final product assembly that can occur continents apart. For example, a shirt may be designed in New Jersey, assembled in Costa Rica from North Carolina fabric, and sold in British Columbia. To understand how goods produced in faraway locations are sold at neighborhood stores, geographers investigate the spatial, or geographic, distribution of natural resources, manufacturing plants, trading patterns, and consumption.

Historically, the geography of manufacturing and trade has been closely linked to the distribution of raw materials, workers, and buyers. In earlier times, this meant that manufacturing and trade were highly localized functions. In the eighteenth cen-

tury, every North American town had cobblers or blacksmiths who produced goods from local resources for sale in local markets. By the start of the Industrial Revolution, improved transportation and manufacturing techniques had significantly enlarged the geography of manufacturing and trade. As distances increased, new manufacturing and trading centers developed. The location of these centers was contingent upon site and situation. Site and situation refer to a physical location, or site, relative to needed materials, transportation networks, and markets. For example, Pittsburgh, Pennsylvania, became the site of a major steel industry because it was near coal and iron resources. Pittsburgh also benefited from its historical role as a port town on a major river system that provided access to both western and eastern markets.

While relative location and transportation costs continue to be important factors, the geographic distribution of production and movement of goods across space is more complex than the simple calculus of site and situation. New global and local geographies of manufacturing and trade have been fueled by two major factors: technology and political change.

TECHNOLOGICAL CHANGE. The old saying that time is money partially explains where goods are manufactured and traded. By compressing time and space, technology has enabled people, goods, and information to go farther more quickly. In the process, technology has reduced interaction costs, such as telecommunications. Just as steel enabled railroads to push farther westward, new technologies reduce the distance between places and people.

By increasing physical and virtual access to people, places, and things, technology has eliminated many barriers to global trade. However, improved telecommunications and transportation are only part of technology's contribution to globalization. If time is money, new efficient manufacturing processes also have reduced costs and facilitated globalization.

Armed with more efficient production processes, reliable telecommunications infrastructures, and transportation improvements, businesses can increase profits and remain competitive by seeking out lower-cost labor markets thousands of miles from consumers. As trade and manufacturing are increasingly spatially separate activities, the geographic distribution of manufacturing promotes an uneven distribution of income. The global distribution of manufacturing plants is closely related to industry-specific skill and wage requirements. For example, low-wage and low-skill jobs tend to concentrate in the developing regions of Asia, South America, and Africa. Alternately, high-technology and high-wage manufacturing activities concentrate in more developed regions.

In some cases, high wages and global competition force corporations to move their manufacturing plants to save costs and remain competitive. During the early 1990's, this byproduct of globalization was a major issue during the U.S. and Canadian debates to ratify the North American Free Trade Agreement (NAFTA). Focusing on primarily U.S. and Canadian companies that moved jobs to Mexico, the debate contributed to growing anxiety over job security as plants relocate to low-cost labor markets in South America and around the world.

As global competition increases, the geography of manufacturing and trade is increasingly global and rapidly changing. One company that has adapted to the shifting nature of global trade and manufacturing is Nike. Based in Beaverton, Oregon, Nike designs and develops new products at its Oregon world headquarters.

However, Nike has internationalized much of its manufacturing capacity to compete in an aggressive athletic apparel industry. Over the last twenty-five years, Nike's strategy has meant shifts in production from high-wage U.S. locations to numerous low-wage labor markets around Pacific Rim.

POLITICAL AND ECONOMIC CHANGE: A NEW WORLD ORDER. In order for companies such as Nike to successfully adapt to changing global dynamics, a stable international, or multilateral, trading system must be in place. In 1948 the General Agreement on Tariffs and Trade (GATT) was the first major step toward developing this stable global trading infrastructure. During that same period, the World Bank and International Monetary Fund were created to stabilize and standardize financial markets and practices. However, Cold War politics postponed complete economic integration for nearly half a century. Since the collapse of communism, globalization has accelerated as economies coalesce around the principles of free markets and capitalism. These important changes have become institutionalized through multilateral trade agreements and international trading organizations.

International trading organizations try to minimize or eliminate barriers to free and fair trade between nations. Trade barriers include tariffs (taxes levied on imported goods), product quotas, government subsidies to domestic industry, domestic content rules, and other regulations. Barriers prevent competitive access to domestic markets by artificially raising the prices of imported goods too high or preventing foreign firms from achieving economies of scale. In some cases, tariffs can also be used to promote fair trade by effectively leveling the playing field.

Because tariffs can be used both to promote fair trade and to unfairly protect

THE WORLD TRADE ORGANIZATION AND GLOBAL TRADING

In 1998 domestic political pressures and an expected domestic surplus of rice prompted the Japanese government to unilaterally implement a 355 percent tariff on foreign rice, violating the United Nations' General Agreement on Tariffs and Trade (GATT). On April 1, 1999, Japan agreed to return to GATT import levels and imposed new over-quota tariffs. While domestic Japanese politics could have prompted a trade war with rice-exporting countries, the crisis demonstrates how multilateral trading initiatives promote stability. Without an agreement, rice exporters might not have gained access to Japanese markets. By returning to GATT minimum quotas and implementing over-quota taxes, the compromise addressed the interests of both domestic and foreign rice growers.

markets, trading organizations are responsible for distinguishing between the two. For example, the Asian Pacific Economic Cooperation (APEC) forum has established guidelines to promote fair trade and attract foreign investment. APEC initiatives include a public Web-based database of member state tariff schedules and related links. Through programs such as the APEC information-sharing project, trading organizations are streamlining the international business process and promoting the overall stability of international markets.

THE FUTURE. As the globalization of manufacturing and trade continues, a new world geography is emerging. Unlike the Cold War's east-west geography and politics of ideology, an economic politics divides the developed and developing world along a north-south axis. While the types of conflicts associated with these new politics and the rules of engagement are unclear, it is evident that a new hierarchy of nations is emerging.

Globalization will raise the economic standard of living in most nations, but it has also widened the gap between richer and poorer countries. A small group of nations generates and controls most of the world's wealth. Conversely, the poorest countries account for roughly two-thirds of the world's population and less than 10 percent of its wealth.

This fundamental question of economic justice was a motive behind globalization's first major political clash. During the 1999 World Trade Organization (WTO) meetings in Seattle, Washington, approximately fifty thousand environmentalists, labor unionists, and human and animal rights activists protested against numerous issues, including cultural intolerance, economic injustice, environmental degradation, political repression, and unfair labor practices they attribute to free trade. While the protesters managed to cancel the opening ceremonies, the United Nations secretary-general, Kofi Annan, expressed the general sentiment of most WTO member states. Agreeing that the protesters' concerns were important, Annan also asserted that the globalization of manufacturing and trade should not be used as a scapegoat for domestic failures to protect individual rights. More important, the secretary-general feared that those issues could be little more than a pretext for a return to unilateral trade policies, or protectionism.

Like the Seattle protesters, supporters of multilateral trade advocate political and economic reforms. Proponents emphasize that open markets promote open societies. Free traders earnestly believe economic engagement encourages rogue nations to improve poor human rights, environmental, and labor records. It is argued that economic engagement raises the expectations of citizens, thereby promoting change. This phenomenon has been partially credited with the fall of the Berlin Wall. It remains to be seen if free trade equals freedom in all places and under all circumstances, and globalization continues to be controversial.

CONCLUSION. Technological and political change have made global labor and consumer markets more accessible and established an economic world hierarchy. At the top, one-fifth of the world's population consumes the vast majority of produced goods and controls more than 80 percent of the wealth. At the bottom of this hierarchy, poor nations are industrializing but possess less than 10 percent of the world's wealth. In political, social, and cultural terms, this global economic reality defines the contours and cleavages of a changing world geography. Whether geographers calculate the economic and political costs of a widening gap between rich and poor or chart the flow of funds from Tokyo to Toronto, the globalization of manufacturing and trade will remain central to the study of geography well into the twenty-first century.

Jay D. Gatrell

FOR FURTHER STUDY

Carrel, T. "Beijing." *National Geographic* (March, 2000): 117-137.

Freidman, T. *The Lexus and the Olive Tree: Understanding Globalization.* New York: Farrar, Straus & Giroux, 1999.

"Rosey Prospects, Forgotten Dangers: A Testing Time for the World Economy." *The Economist* (April 15, 2000): 15-16.

Samuelson, R. "Judgment Calls: Economic Statecraft." *Newsweek* (November 29, 1999): 58.

"Special Report. Trade Wars: The Meeting." *Time* 154, no. 22 (November, 29, 1999): 40-44.

Vulliamy, E. "A New Day for Romania." *National Geographic* 3 (September, 1998): 34-59.

INFORMATION ON THE WORLD WIDE WEB

The Web site of the World Trade Organization (WTO) details the history of the organization and free trade as well as the implications of global trading. (www.wto.org)

MODERN WORLD TRADE PATTERNS

Trade, its routes, and its patterns are an integral part of modern society. Trade is primarily based on need. People trade the goods that they have, including money, to obtain the goods that they don't have. Some nations are very rich in agriculture or natural resources, while others are centers of industrial or technical activity. Because nations' needs change only slowly, trade routes and trading patterns develop that last for long periods of time.

TYPES OF TRADE. The movement of goods can occur among neighboring countries, such as the United States and Mexico, or across the globe, as between Japan and Italy. Some trade routes are well established with regularly scheduled service connecting points. Such service is called liner service. Liners may also serve intermediate points along a trade route to increase their revenue.

Some trade occurs only seasonally, such as the movement of fresh fruits from Chile to California. Some trade occurs only when certain goods are demanded, such as special orders of industrial goods. This type of service is provided by operators called tramps. They go where the business of trade takes them, rather than along fixed liner schedules and routes.

Many people think of international trade as being carried on great ships plying the oceans of the world. Such trade is important; however, a considerable amount of trade is carried by other modes of transportation. Ships and airplanes carry large volumes of freight over large distances, while trucks, trains, barges, and even animal transport are used to move goods over trade routes among neighboring or landlocked countries.

TRADE ROUTES. Through much of human history, trade routes were limited. Shipping trade carried on sailing vessels, for example, was limited by the prevailing winds that powered the ships. Land routes were limited by the location of water, mountain ranges, and the slow development of roads through thick forests and difficult terrain. The mechanization of transportation eventually freed ships and other forms of transport to follow more direct trade routes. Also, the development of canals and transcontinental highway sys-

tems allowed trade routes to develop based solely upon economic requirements.

Other changes in trade routes have occurred with industrialization of transport systems. The world began to have a great need for coal. Trade routes ran to the countries in which coal was mined. Ships and trains delivered coal to the power industry worldwide. Later, trade shifted to locations where oil (petroleum) was drilled. Now, oil is delivered to those same powerplants and industrial sites around the world.

NONECONOMIC FACTORS. Some trade is not purely economic in nature. Political relationships among countries can play an important part in their trade relations. For example, many national governments try to protect their countries' automobile and electronics industries from outside competition by not allowing foreign goods to be imported easily. Governments control imports by assessing duties, or tariffs, on selected imports.

Some national governments use the concept of cabotage to protect their home transportation industries by requiring that certain percentages of imported and exported trade goods be carried by their own carriers. For example, the U.S. government might require that 50 percent of its trade use American ships, planes or trucks. The government might also require that all American carriers employ only American citizens.

Nations also can exert pressure on their trading partners by limiting access to port or airport facilities. Stronger nations may force weaker nations into accepting unequal trade agreements. For example, the United States once had an agreement with Germany concerning air passenger service between the two countries. The agreement allowed United States carriers to carry 80 percent of the

Dutch canal Page 223

passengers, while German carriers were permitted to carry only 20 percent of the passengers.

MULTILATERAL TRADE. In situations in which pairs of trading nations do not have direct diplomatic contact with each other, they make their trade arrangements through other nations. Such trade is referred to as multilateral. Certain carriers cater to this type of trade. They operate their ships or planes in around-the-world service. They literally travel around the globe picking up and depositing cargo along the way for a variety of nations.

TRADE PATTERNS. For many years, world populations were coast centered. This means that most of the people in the country lived close to the coast. This was due primarily to the availability of water transportation systems to move both goods and people. At this time, major railroad, highway and airline systems did not exist. As railway and highway systems pushed into the interiors of nations, the population followed, and goods were needed as well as produced in these areas. Thus, over the years many inland population centers have developed that require transportation systems to move goods into and away from this area.

In these cases, international trade to these inland centers required the use of a number of different modes of transportation. Each of the different modes required additional paperwork and time for repackaging and securing of the cargo. For example, cargo coming off ships from overseas was unloaded and placed in warehouse storage. At some later time, it was loaded onto trucks that carried it to railyards. There it would be unloaded, stored, and then loaded onto railcars. At the destination, the cargo would once again be shifted to trucks for the final delivery. During the course of the trip, the cargo would have been handled a number

of times, with the possibility of damage or loss occurring each time.

CONTAINERIZATION. As more goods began to move in international trade, the systems for packaging and securing of cargo became more standardized. In the 1960's, shipments began to move in containers. These are highway truck trailers which have been removed from the chassis leaving only the box. Container packaging has become the standard for most cargos moving today in both domestic and international trade. With the advent of containerization of cargo in international trade, cargo movements could quickly move intermodally. Intermodal shipping involves the movement of cargo by using more than a single mode of transportation.

Land, water, and air carriers have attempted to make the intermodal movement of cargo in international trade as seamless as possible. They have not only standardized the box for carrying cargo, but they have also standardized the handling equipment, so that containers move quickly from one mode to another. Advances in communications and electronic banking allow the paperwork and payments also to be completed and transferred rapidly.

As the demands for products have grown and as the size of industrial plants has grown, the size of movements of raw materials and containerized cargo has also grown. Thus, the sizes of the ships and trains required to move these large volumes of cargo have also increased.

The development of VLCC's (very large crude carriers) has allowed shippers to move large volumes of oil products. The development of large bulk carriers has allowed for the carriage of large volumes of dry raw materials such as grains or iron ore. These large vessels take advantage of what is known as economies of scale.

Crane lifting cargo containers at an Asian shipyard. (PhotoDisc)

Goods can moved more cheaply when large volumes of them are moved at the same time. This is because the doubling of the volume of cargo moved does not double the cost to build or operate the vessels in which it is carried. This savings reduces the cost to move large volumes of cargo.

INTERMODAL TRANSPORTATION. Intermodal transportation has allowed cargo to move seamlessly across both international boundaries and through different modes of transportation. This seamless movement has changed ocean trade routes over recent years.

The development of the Pacific Rim nations created a demand for trade between East Asia and both the United States and Europe. This trade has usually taken the all-water routes between Asia and Europe.

Ships moving from East Asia across the Pacific Ocean pass through the Panama Canal and cross the Atlantic Ocean to reach Western Europe. This journey is in excess of 10,000 miles (16,000 km.) and usually takes about thirty days for most ships to complete. The all-water route from Asia to New York is similar. The distance is almost as great as that to Europe and requires about twenty-one to twenty-four days to complete.

Intermodal transportation has given shippers alternatives to all-water routes. A great volume of Asian goods is now shipped to such western U.S. ports as Seattle, Oakland, and Los Angeles, from which these goods are carried by trains across the United States to New York. The overall lengths of these routes to New York are only about 7,400 miles (12,000 km.) and

Panama Canal Page 236

301

take between only fifteen and nineteen days to complete. Cargos continuing to Europe are put back on ships in New York and complete their journeys in an additional seven to ten days. Such intermodal shipping can save as much as a week in delivery time.

AIR FREIGHT. Another changing trend in trade patterns is the development of airfreight as an international competitor. Modern aircraft have improved dramatically both in their ability to lift large weights of cargo as well as their ability to carry cargos over long distances. Because of the speed at which aircraft travel in comparison to other modes of transportation, goods can be moved quickly over large distances. Thus, high-value cargos or very fragile cargos can move very quickly by aircraft.

The drawback to airfreight movement of cargo is that it is more expensive than other modes of travel. However, for businesses that need to move perishable commodities, such as flowers of the Netherlands, or expensive commodities, such as Paris fashions or Singapore-made computer chips, airfreight has become both economic and essential.

Robert J. Stewart

FOR FURTHER STUDY

Grimwade, N. *New Patterns of International Trade.* Beckenham, England: Croom Helm, 1988.

Hardin, Garrett. *Living Within Limits: Ecology, Economics and Population Taboos.* New York: Oxford University Press. 1993.

Kennedy, Malcolm J., and Michael J. O'Connor. *Safely by Sea.* Landham, Md.: University Press of America, 1990.

Rosenthal, Paul. *Where on Earth? A Geografunny Guide to the Globe.* New York: Alfred A. Knopf. 1992.

Zimolzak, Chester E., and Charles Stansfield. *The Human Landscape: Geography and Culture.* 2d ed. Columbus, Ohio: Merrill, 1983

INFORMATION ON THE WORLD WIDE WEB

The International Trade Administration (ITA) of the U.S. Department of Commerce encourages the export of U.S. goods into foreign markets. The ITA's Web site contains analyses and reports, searchable by region and country, as well as links to other sites. (www.ita.doc.gov)

POLITICAL
GEOGRAPHY

FORMS OF GOVERNMENT

Philosophers and political scientists have studied forms of government for many centuries. Ancient Greek philosophers such as Plato and Aristotle wrote about what they believed to be good and bad forms of government. According to Plato's famous work, *The Republic*, the best form of government was one ruled by philosopher-kings. Aristotle wrote that good governments, whether headed by one person (a kingship), a few people (an aristocracy), or many people (a polity), were those that ruled for the benefit of all. Those that were based on narrow, selfish interests were considered bad forms of government, whether ruled by an individual (a tyranny), a few people (an oligarchy), or many people (a democracy). Thus, democracy was not always considered a good form of government.

CONSTITUTIONS AND POLITICAL INSTITUTIONS. All governments have certain things in common: institutions that carry out legislative, executive, and judicial functions. How these institutions are supposed to function is usually spelled out in a country's constitution, which is a guide to organizing a country's political system. Most, but not all, countries have written constitutions. Great Britain, for example, has an unwritten constitution based on documents such as the Magna Carta, the English Bill of Rights, and the Treaty of Rome and on unwritten codes of behavior expected of politicians and members of the royal family.

The world's oldest written constitution still in use is that of the United States. All countries have written or unwritten constitutions, and most follow them most of the time. Some countries do not follow their constitutions—for example, the Soviet Union did not; other countries, for example France, change their constitutions frequently.

Greek philosopher Aristotle, who lived in the fourth century B.C.E., laid down some of the earliest recorded principles of government. (Library of Congress)

Constitutions usually first specify if the country is to be a monarchy or a republic. Few countries still have monarchies, and those that do usually grant the monarch only ceremonial powers and duties. Countries with monarchies at the beginning of the twenty-first century included Spain, Great Britain, Lesotho, Swaziland, Sweden, Saudi Arabia, and Jordan. Most countries that do not have monarchies are republics.

Constitutions also specify if power is to be concentrated in the hands of a strong national government, which is a unitary system; if it is to be divided between a national and various subnational governments such as states, provinces, or territories, which is a federal system; or if it is to be spread among various subnational governments that might delegate some power to a weak national government, which is a confederate system.

Examples of countries with unitary systems include Great Britain, France, and China; federal systems include the United States, Germany, Russia, Canada, India, and Brazil. There were no confederate systems in the late 1990's, although there are examples from history. The United States under its eighteenth-century Articles of Confederation and the nineteenth-century Confederate States of America, made up of the rebelling Southern states, were confederate systems. Switzerland was a confederation for much of the nineteenth century. The concept of dividing power between the national and subnational governments is called the vertical axis of power.

Whether governments share power with subnational governments or not, there must be institutions to make laws, enforce laws, and interpret laws: the legislative, executive, and judicial branches of government. How these branches interact is what determines whether governments

are parliamentary, presidential, or mixed parliamentary-presidential. In a presidential system, such as in the United States, the three branches—legislative, executive, and judicial—are separate, independent, and designed to check and balance each other according to a constitution. In a parliamentary system, the three branches are not entirely separate, and the legislative branch is much more powerful than the executive and judicial branches.

Great Britain is a good example of a parliamentary system. Some countries, such as France and Russia, have created a mixed parliamentary-presidential system, wherein the three branches are separate but are not designed to check and balance each other. In a mixed parliamentary-presidential system, the executive (led by a president) is the most powerful branch of government.

Looking at political systems in this way—how the legislative, executive, and judicial branches of government interact—is to examine the horizontal axis of power. All governments are unitary, federal, or confederate, and all are parliamentary, presidential, or mixed parliamentary-presidential. One can find examples of different combinations. Great Britain is unitary and parliamentary. Germany is federal and parliamentary. The United States is federal and presidential. France is unitary and mixed parliamentary-presidential. Russia is federal and mixed parliamentary-presidential. Furthermore, virtually all countries are either republics or monarchies.

TYPES OF GOVERNMENT. Constitutions describe how the country's political institutions are supposed to interact and provide a guide to the relationship between the government and its citizens. Thus, while governments may have similar political institutions—for example, Germany and India are both federal, parliamentary

MONARCHIES OF THE WORLD

Country	Monarch	Type of monarchy
Australia	Queen Elizabeth II	Constitutional
Bahrain	Sheikh Hamad ibn 'Isa Al Khalifah	Traditional
Belgium	King Albert II	Constitutional
Bhutan	King Jigme Singye Wangchuk	Constitutional
Brunei	Sultan Haji Hassanal Bolkiah	Constitutional
Cambodia	King Norodom Sihanouk	Constitutional
Canada	Queen Elizabeth II	Constitutional
Denmark	Queen Margrethe II	Constitutional
Japan	Emperor Akihito	Constitutional
Jordan	King Abdullah II	Constitutional
Kuwait	Sheik Jaber al-Ahmad al-Sabah	Constitutional
Lesotho	King Letsie III	Constitutional
Liechtenstein	Prince Hans Adam II	Constitutional
Luxembourg	Grand Duke Jean	Constitutional
Malaysia	Salehuddin Abdul Aziz Shah	Constitutional
Monaco	Prince Rainier III	Constitutional principality
Morocco	King Muhammad VI	Constitutional
Nepal	King Birendra Bir Bikram Shah Deva	Constitutional
Netherlands	Queen Beatrix	Constitutional
New Zealand	Queen Elizabeth II	Constitutional
Norway	King Harald V	Constitutional
Oman	Sultan Qabus ibn Sa'id	Absolute
Qatar	Emir Sheikh Hamad ibn Khalifah Al Thani	Traditional
Saudi Arabia	King Fahd bin 'Abdulaziz	Absolute
Spain	King Juan Carlos I	Parliamentary
Swaziland	King Mswati III	Near-absolute
Sweden	King Carl XVI Gustaf	Constitutional
Thailand	King Bhumibol Adulyadej	Constitutional
Tonga	King Taufa'ahau Tupou IV	Constitutional
United Kingdom	Elizabeth II	Constitutional

republics—how the leaders treat their citizens can vary widely. However, governments may have political systems that function similarly although they have different forms of constitutions and institutions. For example, Great Britain, a unitary, parliamentary monarchy with an unwritten constitution, treats its citizens very similarly to the United States, which is a federal, presidential republic with a written constitution.

The three most common terms used to describe the relationships between those who govern and those who are governed are democratic, authoritarian, and totalitarian. Characteristics of democracies are free, fair, and meaningfully contested elections; majority rule and respect for minority rights and opinions; a willingness to hand power to the opposition after an election; the rule of law; and civil rights and liberties, including freedom of speech and press, freedom of association, and freedom to travel. The United States, Canada, Japan, and most European countries are democratic.

An authoritarian system is one that curtails some or all of the characteristics of a democratic regime. For example, authoritarian regimes might permit token electoral opposition by allowing other political parties to run in elections, but they do not allow the opposition to win those elections. If the opposition did win, the authoritarian regime would not hand over power. Authoritarian regimes do not respect the rule of law, the rights of minorities to dissent, or freedom of the press, speech, or association. Authoritarian governments use the police, courts, prisons, and the military to intimidate and threaten their citizens, thus preventing people from uniting to challenge the existing political rulers. Cuba, Mexico, Peru, Libya, Serbia, Belarus, and China are examples of countries with authoritarian regimes.

Totalitarian regimes are similar to authoritarian regimes but are even more extreme. Under a totalitarian regime, there is no legal opposition, no freedom of speech, and no rule of law whatsoever. Totalitarian regimes attempt to control totally all members of the society to the point where everyone always must actively demonstrate their loyalty to and support for the regime. Nazi Germany under Adolf Hitler's rule (1933-1945) and the Soviet Union under Joseph Stalin's rule (1928-1953) are examples of totalitarian regimes.

FORMS OF GOVERNMENT: PUTTING IT ALL TOGETHER. In *The Republic*, Plato asserts that people have varied dispositions, and, therefore, there are various types of governments. In recent years, regimes have been created that some call mafiacracies (rule by criminal mafias), narcocracies (rule by narcotics gangs), gerontocra-

cies (rule by very old people), theocracies (rule by religious leaders), and so forth. Such variations show the ingenuity of the human mind in devising forms of government.

Whatever labels that are given to a political system, there are several basic questions to be asked about that regime: Is it a monarchy or a republic? Is all power concentrated in the hands of a national government, or is power shared between a national government and the states or provinces? Are its institutions those of a parliamentary, presidential, or mixed parliamentary-presidential system? Is it democratic, authoritarian, or totalitarian? Finally, does it live up to its constitution, both in terms of how power is supposed to be distributed among institutions and in its relationship between the government and the people? To paraphrase Aristotle, how many rulers are there, and in whose interests do they rule?

Nathaniel Richmond

FOR FURTHER STUDY

Aristotle. *The Politics*. Translated with an introduction by T. A. Sinclair. New York: Penguin Classics, 1962.

Baradat, Leon P. *Political Ideologies: Their Origin and Impact*. 7th ed. Englewood Cliffs, N.J.: Prentice-Hall, 1999.

Cohen, Carl, ed. *Communism, Fascism and Democracy: The Theoretical Foundations*. 3d. ed. New York: McGraw-Hill, 1997.

Love, Nancy S., ed. *Dogmas and Dreams: A Reader in Modern Political Ideologies*. 2d. ed. Chatham, N.J.: Chatham House, 1998.

Mahler, Gregory S. *Comparative Politics: An Institutional and Cross-National Approach*. 3d ed. Englewood Cliffs, N.J.: Prentice-Hall, 2000.

Plato. *The Republic*. Translated and edited by Raymond Larson. Arlington Heights, Ill.: AHM, 1979.

POLITICAL GEOGRAPHY

Students of politics have been aware that there is a significant relationship between physical and political geography since the time of ancient Greece. The ancient Greek philosopher Plato argued that a *polis* (politically organized society) must be of limited geographical size and limited population or it would lack cohesion. The ideal *polis* would be only as geographically large as required to feed about five thousand people, its maximum population.

Plato's illustrious pupil, Aristotle, agreed that stable states must be small. "One can build a wall around the Hellespont," the main territory of ancient Greece, he wrote in his treatise *Politics*, "but that will not make it a polis." Today human ideas differ about the maximum area of a successful state or nation-state, but the close influence of physical geography on political geography and their profound mutual effects on politics itself are not in question.

GEOGRAPHICAL INFLUENCES ON POLITICS. The physical shape and contours of states may be called their physical geography; the political shape and contours of states, starting with their basic structure as unified state, federation, or confederation, are primary features of their political geography. The idea of "political geography" also can refer to variations in a population's political attitudes and behavior that are influenced by geographical features. Thus, the combination of plentiful land and sparse population tend toward an independent spirit, especially where the economy is agriculturally based. This has historically been the case in the western United States; in the Pampas region of Argentina, where cattle are raised by inde-

pendent-mined gauchos (cowboys); and on the Brazilian frontier, where government regulation is routinely resisted.

Likewise, where physical geography presents significant difficulties for inhabitants in earning a living or associating, as where there is rough terrain and poor soil or inhospitable climate, the populace is likely to exhibit a hardy, self-reliant character that strongly influences political preferences. Thus, physical geography helps to shape national character, including aspects of a nation's politics.

Furthermore, it is well known that where physical geography isolates one part of a country's population from the rest, political radicalism may take root. This tendency is found in coastal cities and remote regions, where labor union radicalism has often been pronounced. Populations in coastal locations with access to foreign trade often show a more liberal, tolerant, and outgoing spirit, as reflected in their political opinions. In ancient Greece, the coastal access enjoyed by Athens through a nearby port in the fifth century B.C.E. had a strong influence on its liberal and democratic political order. In modern times, China's coastal cities, such as Tientsin, and North American cities such as San Francisco, show similar influences.

THE GEOGRAPHICAL IMPERATIVE. In many instances, political geography is shaped by what may be called the "geographical imperative." Physical geography in these instances demands, or at least strongly suggests, that political geography follow its course. The numerous valleys of mountainous Greece strongly influenced the emergence of the small, often fiercely independent, polis of ancient times. The formation and borders of Asian states such as Bhutan, Nepal, and Tibet have been strongly influenced by the Himalaya Mountains, and the Alps shape Switzerland.

As another example, physical geography demands that the land between the Pacific Ocean and the Andes Mountains along the western edge of South America be organized as a separate country—Chile. Island geography often plays a decisive role in its political geography. The qualified political unity of Great Britain can be directly traced to its insular status. Small islands often find themselves combined into larger units, such as the Hawaiian Islands.

The absence of the geographical imperative, however, leaves political geography an open question. For example, Indonesia comprises some thirteen hundred islands stretching three thousand miles in bodies of water such as the Indian Ocean and the Celebes Sea. With so many islands, Indonesia lacks a geographical imperative to be a unified state. It also lacks the imperative of ethnic and cultural homogeneity and cohesion, a circumstance mirrored in its political life, since it has remained unified only through military force. As control by the military waned after the fall of the authoritarian General Suharto in 1998, conflicts among the nation's diverse peoples have threatened its breakup. No such threat, however, confronts Australia, an immense island continent where a European majority dominates a fragmented and primitive aboriginal minority. In Australia, the geographical imperative suggests a unity supported by the cultural unity of the majority.

Lithuania Page 224

As many examples show, the geographical imperative is not absolute. For example, mountainous Greece is politically united in the twentieth century. Although

Tenzin Gyatso, the Dalai Lama, is the secular and ecclesiastical ruler of Tibet. In 1950 he went into exile to protest China's violent occupation of Tibet. (Nobel Foundation)

long shielded geographically, Tibet lost its political independence after it was successfully invaded by China. The formerly independent Himalayan state Sikkim was taken over by India. Thus, political will trumps physical geography.

The frequency of exceptions to the geographical imperative illustrates that human freedom, while not unlimited, often plays a key role in shaping political geography. As one example, the Baltic Republics—Lithuania, Latvia, and Estonia—historically have been dominated, or largely swallowed up, by neighboring Russia. At the start of the twenty-first century, however, they had regained their independence through the political will to self-rule and the drive for cultural survival.

STRATEGICALLY SIGNIFICANT LOCATIONS. Locations of great economic or military significance become focal points of political attention and, potentially, of military conflict. There are innumerable such places in the world, but several stand out as models of how important physical geography can be for political geography in the context of international politics.

One significant example is the Panama Canal, without which ships must sail around South America. The Suez Canal, which connects European and Asian shipping, is a similar waterway, saving passage around Africa. The canal's significance was reduced after 1956, however, when its blockage after the Arab-Israeli war of that year led to the building of supertankers too large to traverse it. Another example is Gibraltar, whose fortifications command the entrance to the Mediterranean Sea from the Atlantic Ocean. A final example is the Bosporus, the tiny entrance from the Black Sea to waters leading to the Mediterranean Sea. It is the only warm-water route to and from Eastern Russia and therefore is of great military and economic importance for regional and world power politics.

Charles F. Bahmueller

FOR FURTHER STUDY

Donnan, Hastings, and Thomas M. Wilson. *Borders: Frontiers of Identity, Nation and State.* Oxford, England: Berg Publishers, 2000.

A Gazetteer of the World: Physical, Political, Statistical, Historical, and Ethnological Geography. Columbia, Mo.: South Asia Books, 1988.

Glassner, Martin Ira. *Political Geography.* New York: John Wiley & Sons, 1995.

McKnight, Tom. *Physical Geography: A Landscape Appreciation.* Englewood Cliffs, N.J.: Prentice-Hall, 1998.

Taylor, Peter J., and Colin Flint. *Political Geography: World-Economy, Nation-State, and Locality.* Reading, Mass.: Addison-Wesley, 1999.

Panama Canal Page 236

GEOPOLITICS

Geopolitics is a concept pertaining to the role of purely geographical features in the relations among states in international politics. Geopolitics is especially concerned with the geographical locations of the states in relationship to one another. Geopolitical relationships incorporate social, economic, political, and historical features of the states that interact with purely geographical elements to influence the strategic thinking and behavior of nations in the international sphere.

Coined in 1899 by the Swedish theorist Rudolf Kjellen, the term "geopolitics" combines the logic of the search for security and competition for dominance among states with geographical methodology. *Geopolitics* must not, however, be confused with *political geography*, which focuses on individual states' territorial sizes, boundaries, resources, internal political relations, and relations with other states.

311

Geopolitical is a term frequently used by military and political strategists, politicians and diplomats, political scientists, journalists, statesmen, and a variety of other government officials, such as policy planners and intelligence analysts.

POWER STRUGGLES AMONG STATES. The idea of geopolitics arises in the course of what might be considered the universal struggle for power among the world's most powerful nations, which compete for political and military leadership. How one state can threaten another, for example, is often influenced by geographical factors in combination with technological, social, economic and other factors. The extent to which individual states can threaten each other depends in no small measure on purely geographical considerations.

By the close of twentieth century the Cold War that had dominated world security concerns was over. Nevertheless, the United States still worried about the danger of being attacked by nuclear missiles fired, not by the former Soviet Union, but by irresponsible, fanatical, or suicidal states. American political leaders and military planners were concerned with the geographical position of so-called "rogue states." or "states of concern." In the year 2000 the two most prominently mentioned states that were potentially of this kind were Iran and North Korea. However, others could emerge.

Geographical factors play prominent roles in assessments of the different threats that those states presented to American interests. How far those states are located from American territory determines whether their missiles might pose a serious threat. A missile may be able to reach only the periphery of U.S. soil, or it might be able to carry only a small payload. Similar considerations determine the threat such states pose for U.S. forces stationed abroad, as well as for such impor-tant U.S. allies as Japan, Western Europe, or Israel. Such questions are thus said to constitute geopolitical, or geostrategic, considerations.

There are many examples of the influence of geopolitical factors on international relations among nations in the past. For example, the Bosporus, the narrow sea lane linking the Black Sea and the Mediterranean where Istanbul is situated, has for long been considered of great strategic importance. In the nineteenth century, the Bosporus was the only direct route through which the Russian navy could reach southern Europe and the Mediterranean Sea.

Because of Russia's nineteenth century history of expansionism and its integration into the pre-World War I European state system, with its networks of competing military alliances, the Bosporus took on added geopolitical meaning. It was the congested (and therefore vulnerable) space through which Russian naval power had to pass to reach the Mediterranean.

HISTORICAL ORIGINS OF GEOPOLITICS. Although political geography was a well-established field by the late nineteenth century, geopolitics was just beginning to emerge as a field of study and political analysis at the end of the century. In 1896 the German theorist Friedrich Ratzel published his *Political Geography,* which put forward the idea of the state as territory occupied by a people bound together by an idea of the state. Ratzel's theory embraced Social Darwinist notions that justified the current boundaries of nations. Ratzel viewed the state as a biological organism in competition for land with other states. The ethical implication of his theory seemed to be that "might makes right."

That theme set the stage for later German geopolitical thought, especially the notion of the need for *Lebensraum* (living room)—space into which the people of a nation could expand. German dictator Adolf Hitler justified his attack on Russia during World War II partly upon his claim that the German people needed more *Lebensraum* to the east. To some modern geographers, the use of geopolitical theories to serve German fascism and to justify other instances of military aggression tarnished geopolitics itself as a field of study.

HISTORICAL DEVELOPMENT OF GEOPOLITICS. Modern geopolitics has further origins in the work of the Scottish geographer Sir Halford John Mackinder. In 1904 he published a seminal article, "The Geographical Pivot of History," in which he argued that the world is made up of a Eurasian "heartland" and a secondary hinterland (the remainder of the world), which he called the "marginal crescent." According to his theory, international politics is the struggle to gain control of the heartland. Any state that managed that feat would dominate the world.

A major proposition of Mackinder's theory was that geographical factors are not merely causative factors, but coercive. He tried to describe the physical features of the world that he believed directed human actions. In his view, "Man and not nature initiates, but nature in large measure controls." Geopolitical factors were therefore to a great extent determinants of the behavior of states. If this were true, geopolitics as a science could have deep relevance and corresponding influence among governments.

After Mackinder's time, the concept of geopolitics had a double significance. On the one hand it was a purely descriptive theory of geographic causation in history. On the other hand, its purveyors also believed, as Mackinder argued in 1904, that geopolitics has "a practical value as setting into perspective some of the competing forces in current international politics." Mackinder sought to promote this field of study as a companion to British statecraft, a tool to further Britain's national interest. By extension, geopolitical theory could assist any government in forming its political/military strategy.

As applied to the early twentieth-century world of international politics, however, Mackinder's theory had major weaknesses. Among his most glaring oversights were his failure to appreciate the rise of the United States, which attained considerable naval power after the turn of the century. Also, he failed to foresee the crucial strategic role that air power would play in warfare—and with it the immense change that air power could make in geopolitical considerations. Air power moves continents closer together, revolutionizing their geopolitical relationships.

One of Mackinder's chief critics was Nicolas John Spykman. Spykman argued that Mackinder had overvalued the potential economic, and therefore political, power of the Eurasian heartland, which could never reach its full potential because it could not overcome the obstacles to internal transportation. Moreover, the weaknesses of the remainder of the world—in effect, northern, western and southern Europe—could be overcome through forging alliances.

The dark side of geopolitical thought as handmaiden to political and military strategy became apparent in the Germany of the 1920's. At that time German theorists sought the resurrection of a German state broken by failure in World War I, the harsh terms of the Versailles Treaty that ended the war, and the hyperinflation that followed, wiping out the German middle class. In his 1925 article "Why Geopolitik?" Karl Haushofer urged the practical applications of *Geopolitik*. He urged that this form of analysis had not only "come to stay" but could also form important services for German political leaders, who should use all available tools "to carry on the fight for Germany's existence."

Haushofer ominously suggested that the "struggle" for German existence was becoming increasingly difficult because of the growth of the country's population. A people, he wrote, should study the living spaces of other nations so it could be prepared to "seize any possibility to recover lost ground." This discussion clearly implied that, from geopolitical necessity, Germany should seek additional territory to feed itself—a view carried into effect by Hitler in his quest for *Lebensraum* in attacking the Soviet Union, including its wheat-producing breadbasket, the Ukraine.

*Gulf War
Page 225*

After World War II, a chastened Haushofer sought to soft-pedal both the direction and influence of his prewar writings. However, Hitler's morally heinous use of *Geopolitik* left geopolitical theorizing permanently tainted, in some eyes. Nevertheless, there is no necessary connection between geopolitics as a purely analytic description and geopolitics as the basis for a selfish search for power and advantage.

GEOPOLITICS IN THE TWENTY-FIRST CENTURY. Geopolitical considerations were unquestionably of profound relevance to the principal states of the post-World War II Cold War period. After the fall of the Berlin Wall in 1989, however, some theorists thought that the age of geopolitics had passed. In 1990 American strategic theorist Edward N. Luttwak, for example, argued that the importance of military power in international affairs had declined precipitously with the winding down of the Cold War. Military power had been overtaken in significance by economic prowess. Consequently, geopolitics had been eclipsed by what Luttwak called "geoeconomics," the waging of geopolitical struggle by economic means.

The view of Luttwak and various geographers of the declining significance of military power and geopolitical analysis, however, was soon proved to be overdrawn by events. As early as the first months of 1991, before the Soviet Union was officially dismantled, military power asserted itself as a key determinant on the international scene. Led by the United States, a far-flung alliance of nations participated in a war to remove Iraqi dictator Saddam Hussein's forces from neighboring Kuwait, which Iraq had illegally occupied. The decisive and successful use of military power in that war dramatically disproved assertions of its growing irrelevance.

Similarly, at the outset of the twenty-first century, military power retained its preeminence in the dynamics of international politics, even as economic forces were seen to gather momentum. To states throughout Asia and the West (especially Western Europe and the United States), the relative military capability of potential adversaries, and therefore geopolitics, remained a vital feature of the international order. Central to this view of the world scene is the growing military rivalry of the United States and China in East Asia. As China modernizes and expands its nuclear and conventional forces, it may feel itself capable of challenging America's predominant military power and prestige in East Asia. This possibility heightens the use of geopolitical thinking, giving it currency in analyzing this emerging situation.

GEOPOLITICS AS CIVILIZATIONAL CLASH. A recent and sometimes controversial expression of geopolitical analysis has been offered by Samuel Huntington of Harvard University. In his *The Clash of Civilizations and the Remaking of World Order* (1996) Huntington constructs a theory to explain certain tendencies of international behavior. He divides the world into a number of cultural groupings, or "civilizations," and argues that the character of various international conflicts can best be explained as conflicts or clashes of civilizations. In his view, Western civilization differs from the civilization of Orthodox Christianity, with a variety of conflicts erupting between the two. An example is the attack by the North Atlantic Treaty Organization (NATO), the bastion of the West, on Serbia, which is part of the Orthodox East.

Huntington's other civilizations include Islamic, Jewish, Eastern Caribbean, Hindu, Sinic (Chinese), and Japanese. The clash between Israel and its neighbors, the struggle between Pakistan and India over Kashmir, the rivalries between the United States and China and between China and India, for example, can be viewed as civilizational conflicts. Huntington has stated, however, that his theory is not intended to explain all of the historical past, and he does not expect it to remain valid long into the future. Instead, he believes it may remain a relevant tool of analysis only until around 2015, after which it will have become dated.

Charles F. Bahmueller

FOR FURTHER STUDY

Huntington, Samuel. *The Clash of Civilizations and the Remaking of World Order.* New York: Simon & Schuster, 1996.

Luttwak, Edward N. "From Geopolitics to Geo-Economics: Logic of Conflict, Grammar of Commerce." *The National Interest,* 1990.

O'Tuathail, Gearoid, Simon Dalby, and Paul Routledge, eds. *The Geopolitics Reader.* New York: Routledge, 1998.

INTERNATIONAL BOUNDARIES

International boundaries are the marked or imaginary lines traversing natural terrain of land or water that mark off the territory of one politically organized society—a state or nation-state—from other states. In addition, states claim "air boundaries." While satellites circumnavigate the earth without nations' permission, airplanes and other air vessels that fly much lower must gain the permission of states over whose territory they travel.

The existence of international boundaries is a consequence of the "territoriality" that is a feature of modern human societies. All politically organized societies, except for nomadic tribes, claim to rule some exactly defined geographical territory. International boundaries provide the limits that define this territory.

The subject of international boundaries is so complex that an encyclopedia on the subject exists and an academic unit of Durham University in Great Britain is devoted to the subject. Many highly trained individuals devote their professional lives to the subject in universities, government agencies, and other settings.

International boundaries have ancient origins. For example, the oldest sections of the Great Wall of China date back to the Ch'in Dynasty of the second century B.C.E. The Roman Empire also maintained boundaries to its territories, such as Hadrian's Wall in the north of England, built by the Romans in 122 C.E. as a defensive barrier against marauders. In these and other ancient instances, however, there was little thought that borders must be exact.

The existence of precisely drawn boundaries among states is relatively recent. The modern state has existed for no more than a few hundred years. In addition, means to determine many boundaries have come into existence only in the nineteenth and twentieth centuries, with the invention of scientific methods and instruments, along with accompanying vocabulary, for determining exact boundaries. The most basic terms of this vocabulary begin with "latitude" and "longitude" and their subdivisions into the "minutes" and "seconds" used in determining boundaries. In modern times, a new attitude toward states' territory was born, especially with the nineteenth century forms of nationalism, which tend to regard every acre of territory as sacred.

TYPES OF BOUNDARIES. There are several types of international boundaries. Some are geographical features, including rivers, lakes, oceans, and seas. Thus boundaries of the United States include the Great Lakes, which border Canada to the north; the Rio Grande, a river that forms part of the U.S. boundary with Mexico to the south; the Atlantic and Pacific Oceans, to the east and west, respectively; and the Gulf of Mexico, to the south. In Africa, Lake Victoria bounds parts of Tanzania, Uganda, and Kenya; and rivers, such as sections of the Congo and the Zambezi, form natural boundaries among many of the continent's states.

Other geographical features, such as mountains, often form international boundaries. The Pyrenes, for example, separate France and Spain and cradle the tiny state of Andorra. In South America, the Andes frequently serve as a boundary,

such as between Argentina and Chile. The Himalayas in South Central Asia create a number of borders, such as between India, China, and Tibet and between Nepal, Butan, and their neighbors. When there are no clear geographical barriers between states, boundaries must be decided by mutual consent or the threat of force.

CREATION AND CHANGE OF INTERNATIONAL BOUNDARIES. War and conquest often have been used to determine borders. Such wars, however, historically have created hostility among losers. Political pressures to recover lost lands build up among aggrieved losers, and such irredentist claims provide fuel for future wars. A classic example is the fate of the regions of Alsace and Lorraine between France and Germany. Although natural resources in the form of coal played a substantial role in the dispute over this area, national pride was also a potent element.

Whether boundaries are fixed through compelling geographical imperatives or in their absence, states typically sign treaties agreeing to their location. These may be treaties that conclude wars, or boundary commissions set up by those involved may draw up borders to which states give formal agreement. In 1846, for example, negotiators for Great Britain and the United States settled on the forty-ninth parallel as the boundary between the western United States and Canada, although in the United States, "Fifty-four [degrees latitude] Forty [minutes] or Fight" had been a popular motto in the presidential election campaign of 1844.

Sometimes no accepted borders exist because of chronic hostility between states. Thus, maps of the Kashmir region between India and Pakistan, claimed by both countries, show only a "line of control" or cease-fire line to divide the two warring states. Similarly, only a cease-fire line, drawn at the armistice of the Korean

War of 1950-1953, divides North and South Korea; a mutually agreed-upon border remains unfixed.

In rare instances, no true boundary exists to mark where a state's territory begins and ends. Classic cases are found on the Arabian Peninsula, where the land borders of principalities, known as the Gulf Sheikdoms, are vague lines in the sand. Such circumstances usually create no difficulties where nothing is at stake, but when oil is discovered, states must come to agreement or risk coming to blows.

In other instances, negotiations and international arbitration have been effective for determining borders. Perhaps the most important principle for determining the borders of newly created states is found in the Latin phrase, *Uti possidetis iurus*. This principle is used when states become independent after having been colonies or constituent parts of a larger state that has broken up. The principle holds that states shall respect the borders in place when they were colonies. *Uti possidetis* was first extensively used in South America in the nineteenth century, when European colonial powers withdrew, leaving several newly born states to determine their own boundaries. The principle may be used as a basis for border agreements among the fifteen states of the former Soviet Union.

Besides war and negotiation, purchase has sometimes been a means of creating international boundaries. For example, in 1853 the United States purchased territory from Mexico in the southwest; in 1867, it purchased Alaska from Russia.

In rare cases, natural boundaries may change naturally or be changed deliberately by one side, incurring resentment among victims. An example occurred in 1997, when Vietnam complained that China had built an embankment on a border river embankment that caused the river to change its course; China countered that Vietnam had built a dam altering the river's course.

Other border difficulties among states include conflicts over water that flows from one country to another. In the 1990's, for example, Mexico complained of excessive U.S. use of Colorado River waters and demanded adjustment.

BORDER DISPUTES. Border disputes among states in the past two centuries have been numerous and lethal. In the

A PEACEFULLY RESOLVED BORDER DISPUTE

The peaceful resolution of the border dispute between the Southern African states of Botswana and Namibia was hailed by observers of African politics. Instead of resorting to the armed warfare that so often has marked similar disputes on the continent, the two states chose a different course in 1996, when they found negotiations stalemated. They submitted their claims to the International Court of Justice in The Hague and agreed to accept the court's ruling. Late in 1999, by an eleven-to-four vote, the court ruled for Botswana, and Namibia kept its word to embrace the decision. At issue was a tiny island in the Chobe River on Botswana's northern border. An 1890 treaty between colonial rulers Great Britain and Germany had described the border at the disputed point vaguely, as the river's "main channel." The court took the course of the deepest channel to mark the agreed boundary, giving Botswana title to the 1.4-square-mile (3.5-sq.-km.) territory.

twentieth century, numerous such controversies degenerated into violence. In Asia, India and Pakistan fought over Kashmir, beginning in 1947-1949 and recurring in 1965 and 1999. China has been involved in violent border disputes with India, especially in 1962; Vietnam in 1979; and Russia in 1969. In South America, border wars between Ecuador and Peru broke out in 1941, 1981, and 1995. This dispute was settled by negotiation in 1998. In Africa, among numerous recent armed conflicts, the bloody border conflict between Eritrea and Ethiopia in the 1990's was notable.

Other recent disputes have ended peacefully. Eritrea avoided violence with Yemen over several Red Sea islands by accepting arbitration by an international tribunal. In 1995 Saudi Arabia and the United Arab Emirates negotiated a peaceful agreement to their border dispute involving oil rights.

Many unresolved boundary disputes might yet lead to conflicts. Among the most complex is the multinational dispute over the six hundred tiny Spratly Islands in the South China Sea. Uninhabited but potentially valuable because of oil, the Spratlys are claimed by China, Brunei, Malaysia, Indonesia, the Philippines, Taiwan, and Vietnam.

BORDER POLICIES. Problems with international borders are not limited to territorial disputes. Policies regarding how borders should be operated—including the key questions of who and what should be allowed entrance and exit under what conditions—can be expected to continue as long as independent states exist. While the members of the European Union have agreed to allow free passage of people and goods among themselves, this policy does not extent to nonmembers.

The most important purpose of states is to protect the lives and property of their citizens. One of the principal purposes of international boundaries is to further this purpose. Most states insist on controlling their borders, although borders seem increasingly porous. Given the imperatives of control and the increasing difficulties of maintaining it, issues surrounding international borders are expected to continue indefinitely in the twenty-first century.

Charles F. Bahmueller

FOR FURTHER STUDY
Biger, Gideon. *The Encyclopedia of International Boundaries.* New York: Facts On File, 1995.
Donnan, Hastings, and Thomas M. Wilson. *Borders: Frontiers of Identity, Nation, and State.* Oxford, England: Berg Publishers, 2000.
Khan, L. Ali. *The Extinction of Nation-States: A World Without Borders.* New York: Kluwer Law International, 1996.
Lee, Boon Thong, ed. *Vanishing Borders: The New International Order of the 21st Century.* Burlington, Vt.: Ashgate Publishing, 1998.
Sohn, Louis B. *The Movement of Persons Across Borders.* Studies in Transnational Legal Policy 23. Washington, D.C.: American Society of International Law, 1992.

GAZETTEER OF OCEANS AND CONTINENTS

Places whose names are printed in SMALL CAPS *are subjects of their own entries in this gazetteer.*

Aden, Gulf of. Deep-water area between the RED and ARABIAN SEAS, bounded by Somalia, Africa, on the south and Yemen on the north. Water is warmer and saltier in the Gulf of Aden than in the Red and Arabian Seas, because little water enters from rain or land runoff.

Africa. Second largest continent, connected to ASIA by the narrow isthmus of Suez. Bounded on the east by the INDIAN OCEAN and on the west by the ATLANTIC OCEAN. Countries of Africa are Algeria, Angola, Benin, Botswana, Burkina Faso, Burundi, Cameroon, Central African Republic, Chad, Congo, Côte d'Ivoire (Ivory Coast), the Democratic Republic of Congo, Egypt, Ethiopia, Gabon, Gambia, Ghana, Guinea, Kenya, Liberia, Libya, Madagascar, Malawi, Mali, Mauritania, Morocco, Mozambique, Namibia, Niger, Nigeria, Rio Muni (Mbini), Rwanda, Senegal, Sierra Leone, Somalia, South Africa, Sudan, Tanzania, Togo, Tunisia, Uganda, Western Sahara, Zambia, and Zimbabwe. Climate ranges from hot and rainy near the equator, to hot and dry in the huge Sahara Desert in the north and the Kalahari Desert in the south, to warm and fairly mild at the northern and southern extremes. Paleontological evidence indicates that humans originally evolved in Africa.

Agulhas Current. Warm, swift ocean current moving south along East AFRICA's coast. Part moves between AFRICA and MADAGASCAR to form the Mozambique Current. The warm water of the Agulhas Current increases the average temperatures in the eastern part of South Africa.

Agulhas Plateau. Relatively small ocean-bottom plateau that lies south of South AFRICA, at the area where the INDIAN and ATLANTIC OCEANS meet.

Aleutian Islands. Chain of volcanic islands that extends 1,100 miles (1,770 km.) from the tip of the Alaska Peninsula to the Kamchatka Peninsula in Russia and forms the boundary between the North PACIFIC OCEAN and the BERING SEA. The area is hazardous to navigation and has been called the "Home of Storms."

Gulf of Aden Page 225

Aleutian Trench. Located on the northern margin of the PACIFIC OCEAN, stretching 3,666 miles (5,900 km.) from the western edge of the Aleutian Island chain to Prince William Sound, Alaska. Depth is 25,263 feet (7,700 meters).

American Highlands. Elevated region on the ANTARCTIC coast between Enderby Land and Wilkes Land, located far south of India. The Lambert and Fisher glaciers originate in the American Highlands and move down to feed the AMERY ICE SHELF.

Amery Ice Shelf. Year-round shelf of relatively flat ice in a bay of ANTARCTICA, located at approximately longitude 70 degrees east, between MAC. ROBERTSON LAND and the AMERICAN HIGHLANDS. The ice shelf is fed by the Lambert and Fisher glaciers.

Amundsen Sea. Portion of the southernmost PACIFIC OCEAN off the Wahlgreen Coast of ANTARCTICA, approximately longitude 100 to 120 degrees west. Named for the Norwegian explorer Roald Amundsen, who became the first person to reach the SOUTH POLE in 1911.

Antarctic Circle. Latitude of 66.3 degrees south. South of this line, the Sun does not set on the day of the summer solstice, about December 22 in the SOUTHERN HEMISPHERE, and does not rise on the day of the winter solstice, about June 21.

Antarctic Circumpolar Current. Eastward-flowing current that circles ANTARCTICA and extends from the surface to the deep ocean floor. The largest-volume current in the oceans. Extends northward to approximately 40 degrees south latitude and is driven by westerly winds.

Antarctic Convergence. Meeting place where cold Antarctic water sinks below the warmer sub-Antarctic water.

Antarctic Ocean. See SOUTHERN OCEAN.

Antarctica. Fifth largest continent, located at the southernmost part of the world. There are two major regions; western Antarctica, which includes the mountainous Antarctic peninsula, and eastern Antarctica, which is mostly a low continental shield area. An ice cap up to 13,000 feet (4,000 meters) thick covers 95 percent of the continent's surface. Temperatures in the austral summer (December, January, and February) rarely rise above 0 degrees Fahrenheit (–18 degrees Celsius) except on the peninsula. By international treaty, the continent is not owned by any single country, and human access is largely regulated. There has never been a self-supporting human habitation on Antarctica.

Arabian Sea. Portion of the INDIAN OCEAN bounded by India on the east, Pakistan on the north, and Oman and Yemen of the Arabian Peninsula on the west.

Arabian Sea Pages 225, 227

Arctic Circle. Latitude of 66.3 degrees north. North of this line, the Sun does not set on the day of the summer solstice, about June 21 in the NORTHERN HEMISPHERE, and does not rise on the day of the winter solstice, about December 22.

Mt. Everest Page 228

Arctic Ocean. World's smallest ocean. It centers on the geographic NORTH POLE and connects to the PACIFIC OCEAN through the BERING SEA, and to the ATLANTIC OCEAN through the GREENLAND SEA. The Arctic Ocean is covered with ice up to 13 feet (4 meters) thick all year, except at its edges. Norwegian explorers on the ship *Fram* stayed locked in the icepack from 1893 to 1896, in order to study the movement of polar ice. They drifted in the ice a total of 1,028 miles (1,658 km.), from the Bering Sea to the Greenland Sea, proving that there was no land mass under the Arctic ice at the top of the world. Also known as Arctic Sea or Arctic Mediterranean Sea.

Argentine Basin. Basin on the floor of the western ATLANTIC OCEAN, off the coast of Argentina in SOUTH AMERICA. Among ocean basins, this one is unusually circular.

Ascension Island. Isolated volcanic island in the South ATLANTIC OCEAN, about midway between SOUTH AMERICA and AFRICA. One of the islands visited by British biologist Charles Darwin during his five-year voyage on the *Beagle*.

Asia. Largest continent; joins with EUROPE to form the great Eurasian landmass. Asia is bounded by the ARCTIC OCEAN on the north, the western PACIFIC OCEAN on the east, and the INDIAN OCEAN on the south. Its countries include Afghanistan, Bahrain, Bangladesh, Bhutan, Cambodia, China, India, Iran, Iraq, Irian Jaya, Israel, Japan, Jordan, Kalimantan, Kazakhstan, North and South Korea, Kyrgyzstan, Laos, Lebanon, Malaysia, Myanmar, Mongolia, Nepal, Oman, Pakistan, the Philippines, Russia, Sarawak, Saudi Arabia, Sri Lanka, Sumatra, Syria, Tajikistan, Thailand, Asian Turkey, Turkmenistan, United Arab Emirates, Uzbekistan, Vietnam, and Yemen. Climates include virtually all types on earth, from arctic to tropical, desert to rain forest. Asia has the highest (Mount Everest) and lowest (Dead Sea) surface points in the world. Nearly 60 percent of the world's people live in Asia.

Atlantic Ocean. Second largest body of water in the world, covering more than 25 percent of Earth's surface. Bordered by NORTH and SOUTH AMERICA on the west, and EUROPE and East AFRICA on the east. The widest part (5,500 miles/ 8,800 km.) lies between West AFRICA and Mexico, along 20 degrees latitude. Scientists disagree on the north-south boundaries of the Atlantic; if one includes the ARCTIC OCEAN and the SOUTHERN OCEAN, the Atlantic Ocean extends about 13,300 miles (21,400 km.). The deepest spot (28,374 feet/ 8,648 meters) is found in the PUERTO RICO TRENCH. The Atlantic Ocean has been a major route for trade and communications, especially between North America and Europe, for hundreds of years. This is because of its relatively narrow size and favorable currents, such as the GULF STREAM.

Australasia. Loosely defined term for the region, which, at the least, includes AUSTRALIA and New Zealand; at the most, it also includes other South Pacific Islands in the region.

Australia. Smallest continent, sometimes called the "island continent." Located between the INDIAN and PACIFIC OCEANS. It is the only continent occupied by a single nation, the Commonwealth of Australia. Australia is the flattest and driest continent; two-thirds is either desert or semiarid. Geologically, it is the oldest and most isolated continent. Unlike any other place on Earth, large mammals never evolved in Australia. Marsupials (pouched, warm-blooded animals) and unusual birds developed in their place.

Azores. Archipelago (group of islands) in the eastern ATLANTIC OCEAN lying abut 994 miles (1,600 km.) west of Portugal. The islands are of volcanic origin and have been known, fought over, and used by the Europeans since before the fourteenth century. Spanish explorer Christopher Columbus stopped in the Azores to wait for favorable winds before his first trip across the ATLANTIC OCEAN.

Barents Sea. Partially enclosed section of the ARCTIC OCEAN, bounded by Russia and Norway on the south and the Russian island of Navaya Zemlaya on the east. The Barents Sea was important in World War II because Allied convoys had to cross it, through storms and submarine patrols, to deliver war supplies to Murmansk, the only ice-free port in western Russia. It was named for the Dutch explorer Willem Barents.

Bays. See under individual names.

Beaufort Sea. Area of the ARCTIC OCEAN located off the northern coast of Alaska and western Canada. It is usually frozen over and has no islands. Named for British admiral Sir Francis Beaufort, who devised the Beaufort Wind Scale as a means of classifying wind force at sea.

Bengal, Bay of. Northeast arm of the INDIAN OCEAN, bounded by India on the west and Myanmar on the east. The Ganges River empties into the Bay of Bengal. The great ports of Calcutta and Madras in India, and Rangoon in Myanmar lie in the bay, making it a busy and important area for shipping for centuries.

Benguela Current. Northward-flowing current along the western coast of Southern AFRICA. Normally, the Benguela Current carries cold, rich water that wells up from the ocean depths and supports a large fishing industry. A change in winds can reduce the oxygen supply and kill huge numbers of fish, similar to what may happen off the

Atlantic Ocean Page 229

Bermuda Triangle Page 228

coast of Peru during El Niño weather conditions.

Bering Sea. Portion of the northernmost PACIFIC OCEAN that is bounded by the state of Alaska on the east, Russia and the Kamchatka Peninsula on the west, and the BERING STRAIT on the north. It is a valuable fishing ground, rich in shrimp, crabs, and fish. Whales, fur seals, sea otters, and walrus are also found there.

Bering Strait. Narrowest point of connection between the BERING SEA and the ARCTIC OCEAN, located between the easternmost point of Siberia on the west and Alaska on the east. The Bering Strait is 52 miles (84 km.) wide. During the Ice Age, when the sea level was lower, humans and animals were able to walk from the Asian continent across a land bridge—now known as Beringia—to the North American continent across the frozen strait, providing the first human access to the Americas.

California coast Page 232

Bikini Atoll. Small atoll in the Marshall Islands group in the western PACIFIC OCEAN. In the 1940's, the United States began testing nuclear bombs on Bikini and neighboring atolls. The U.S. Army removed the inhabitants of Bikini, and testing occurred from 1946 to 1958. The Bikini inhabitants were allowed to return in 1969, then removed again in 1978 when high levels of radioactivity were found to remain.

Black Sea. Large inland sea situated where southeastern EUROPE meets ASIA; connected to the MEDITERRANEAN SEA through Turkey's Bosporus strait. The sea covers an area of about 178,000 square miles (461,000 sq. km.), with a maximum depth of more than 7,250 feet (2,210 meters).

Brazil Current. Extension of part of the warm, westward-flowing South EQUATORIAL CURRENT, which turns south to the coast of Brazil. The Brazil Current has very salty water because of its long flow across the equator. It joins the WEST WIND DRIFT and moves eastward across the South ATLANTIC OCEAN as part of the SOUTH ATLANTIC GYRE.

California, Gulf of. Branch of the eastern PACIFIC OCEAN that separates Baja California from mainland Mexico. Warm, nutrient-rich water supports a variety of fish, oysters, and sponges. California gray whales migrate to the gulf to give birth and breed, January through March. Fisheries and tourism are important industries in the Gulf of California. Also known as the Sea of Cortés.

California Current. Cool water that flows southeast along the western coast of NORTH AMERICA from Washington State to Baja California. The eastern portion of the NORTH PACIFIC GYRE.

California Current. Surfers at Santa Cruz enjoy some of the best surfing conditions on the California coast. (PhotoDisc)

One of the Caribbean's major attractions to tourists is its many fine sites for scuba diving. (Corbis)

Canada Basin. Part of the ocean floor that lies north of northeastern Canada and Alaska. The BEAUFORT SEA lies above the Canada Basin.

Cape Horn. Southernmost tip of SOUTH AMERICA. It is the site of notoriously severe storms and is hazardous to shipping.

Cape Verde Plateau. ATLANTIC OCEAN plateau lying off the western bulge of the AFRICAN continent. The volcanic Cape Verde Islands lie on the plateau.

Caribbean Sea. Portion of the western ATLANTIC OCEAN bounded by CENTRAL and SOUTH AMERICA to the west and south, and the islands of the Antilles chain on the north and east. Mostly tropical in climate, the Caribbean Sea supports a large variety of plant and animal life. Its islands, including Puerto Rico, the Cayman Islands, and the Virgin Islands, are popular tourist sites.

Caspian Sea. World's largest inland sea. Located east of the Caucasus Mountains at EUROPE's southeasternmost extremity, it dominates the expanses of western Central ASIA. Its basin is 750 miles (1,200 kilometers) long, and its average width is 200 miles (320 kilometers). It covers 149,200 square miles (386,400 sq. km.).

Central America. Region generally understood to constitute the irregularly shaped neck of land linking NORTH and SOUTH AMERICA, containing Belize, Guatemala, Honduras, El Salvador, Nicaragua, Costa Rica, and Panama.

Chukchi Sea. Portion of the ARCTIC OCEAN, bounded by the BERING STRAIT on the south, Siberia on the southwest, and Alaska on the southeast. The Chukchi Sea is the area of exchange between waters and sea life of the PACIFIC and ARCTIC OCEANS, and so is an area of interest to oceanographers and fishermen.

Clarion Fracture Zone. East-west-running fracture zone that begins off the west coast of Mexico and extends approximately 2,500 miles (4,023 km.) to the southwest.

Cocos Basin. Relatively small ocean basin located off the west coast of Sumatra in the northeast INDIAN OCEAN.

Coral Sea. Area of the PACIFIC OCEAN off the northeast coast of AUSTRALIA, between Australia on the southwest, Papua New Guinea and the Solomon Islands on the northeast, and New Caledonia on the east. Site of a naval battle in 1942 that prevented the Japanese invasion of Australia.

Cortés, Sea of. See CALIFORNIA, GULF OF.

Denmark Strait. Channel that separates GREENLAND and ICELAND and connects the North Atlantic Ocean with the ARCTIC OCEAN.

Dover, Strait of. Body of water between England and the European continent, separating the NORTH SEA from the ENGLISH CHANNEL. It is 33 miles (53 km.) wide at its narrowest point. The tunnel between England and France (known as the "Chunnel") was cut into the rock under the Strait of Dover.

Drake Passage. Narrow part of the SOUTHERN OCEAN that connects the ATLANTIC and PACIFIC OCEANS between the southern tip of SOUTH AMERICA and the ANTARCTIC peninsula. Named for sixteenth-century English navigator and explorer Sir Francis Drake, who discovered the passage when his ship was blown into it during a violent storm. Also called Drake Strait.

East China Sea. Area of the western PACIFIC OCEAN bounded by China on the west, the YELLOW SEA on the north, and Japan on the northeast. Large oil deposits were found under the East China Sea floor in the 1980's.

East Pacific Rise. Broad, nearly continuous undersea mountain range that extends from the southern end of Baja California southward, then curves east near ANTARCTICA. It is formed along the southeast side of the Pacific Plate and is part of the RING OF FIRE, a nearly continuous ring of volcanic and tectonic activity around the rim of the Pacific Ocean. Also called East Pacific Ridge.

East Siberian Sea. Portion of the ARCTIC OCEAN bounded by the CHUKCHI SEA on the east, Siberia on the south, and the LAPTEV SEA on the west. Much of the East Siberian Sea is covered with ice year-round.

Eastern Hemisphere. The half of the earth containing EUROPE, ASIA, and AFRICA; generally understood to fall between longitudes 20 degrees west and 160 degrees east.

El Niño. Conditions—also known as El Niño-Southern Oscillation (ENSO) events—that occur every two to ten years and cause weather and ocean temperature changes off the coast of Ecuador and Peru. Most of the time, the PERU CURRENT causes cold, nutrient-rich water to well up off the coast of Ecuador and Peru. During ENSO years, the cold upwelling is replaced by warmer surface water that does not support plankton and fish. Fisheries decline and seabirds starve. Climatic changes of El Niño can bring floods to normally dry areas and drought to wet areas. Effects can extend across NORTH and SOUTH AMERICA, and to the western PACIFIC OCEAN.

During the 1990's, the ENSO event fluctuated but did not go comletely away, which caused tremendous damage to fisheries and agriculture, storms and droughts in North America, and numerous hurricanes.

Emperor Seamount Chain. Largest known example of submerged underwater volcanic ridges, located in the northern PACIFIC OCEAN and extending southward from the Kamchatka Peninsula in Russia for about 2,500 miles (4,023 km.).

Enderby Land. Section of ANTARCTICA that lies between the INDIAN OCEAN and the South Polar Plateau, east of QUEEN MAUD LAND. Enderby Land lies between approximately longitude 45 and 60 degrees east.

English Channel. Strait water separating continental France from Great Britain. Runs for roughly 350 miles (560 km.), from the ATLANTIC OCEAN in the west to the Strait of Dover in the east.

Equatorial Current. Currents just north and south of the equator that flow from east to west. Equatorial currents are found in the PACIFIC and ATLANTIC OCEANS. The equatorial currents and the trade winds, which move in the same direction, greatly aid oceangoing traffic.

Eurasia. Term for the combined landmass of EUROPE and ASIA.

Europe. Sixth largest continent, actually a large peninsula of the Eurasian landmass. Europe is densely populated and includes the countries of Albania, Andorra, Austria, Belarus, Belgium, Bulgaria, Bosnia-Herzegovina, Croatia, the Czech Republic, Denmark, Estonia, Finland, France, Germany, Greece, Hungary, Iceland, Ireland, Italy, Latvia, Lithuania, Macedonia, Malta, Moldova, Monaco, the Netherlands, Norway, Poland, Portugal, Romania, Slovakia, Spain, Switzerland, Turkey, and the United Kingdom (England, Northern Ireland, Scotland, and Wales). Climate ranges from near arctic in the north, to temperate and Mediterranean in the south.

Florida Current. Water moving northward along the east coast of Florida to Cape Hatteras, North Carolina, where it joins the GULF STREAM.

Fundy, Bay of. Large inlet on the North American Atlantic coast, northwest of Maine, separating New Brunswick and Nova Scotia in Canada. Renowned for having the largest tidal change in the world, more than 56 feet (17 meters).

Galapagos Islands. Located directly on the equator, 600 miles (965 km.) west of Ecuador. The islands are volcanic in origin and sit directly in the cold PERU CURRENT, which cools the islands and creates unusual microclimates and fogs. The extreme isolation of the islands allowed unique species to develop. Biologist Charles Darwin visited the Galapagos in the 1830's, and the unusual organisms he observed helped him to conceive the theory of evolution.

Galapagos Rift. Divergent plate boundary extending between the GALAPAGOS ISLANDS and SOUTH AMERICA. The first hydrothermal vent community was discovered in 1977 in the Galapagos Rift. This unusual type of biological habitat is based on energy from bacteria that use heat and chemicals to make food, instead of sunlight.

Grand Banks. Portion of the northwest ATLANTIC OCEAN southeast of Nova Scotia and Newfoundland. The Grand Banks are extremely rich fishing grounds, although in the 1980's and 1990's catches of cod, flounder, and many other fish dropped dramatically due to overfishing and pollution.

Great Barrier Reef Pages 230, 231

Hatteras Abyssal Plain Page 232

Great Barrier Reef. Largest coral reef in the world, lying in the CORAL SEA off the east coast of AUSTRALIA. The reef system and its small islands stretch for more than 1,100 miles (1,750 km.) and is difficult to navigate through. The reefs are home to an incredible variety of tropical marine life, including large numbers of sharks.

Greenland. Largest island in the world that is not rated as a continent; lies between the northernmost part of the ATLANTIC OCEAN and the ARCTIC OCEAN, northeast of the North American continent. About 90 percent of Greenland is permanently covered with an ice sheet and glaciers. Residents engage in limited agriculture, growing potatoes, turnips, and cabbages. Most people live along the southwest coast, where the climate is warmed by the NORTH ATLANTIC CURRENT.

Greenland Sea. Body of water bounded by GREENLAND on the west, ICELAND on the north, and Spitsbergen on the east. It is often ice-covered.

Guinea, Gulf of. Arm of the North ATLANTIC OCEAN below the great bulge of West AFRICA.

Gulf Stream. Current of westward-moving warm water originating along the equator in the ATLANTIC OCEAN. The mass of water moves along the east coast of Florida as the FLORIDA CURRENT, then turns in a northeasterly direction off North Carolina to become the Gulf Stream. The Gulf Stream flows northeast past Newfoundland and the western edge of the British Isles. The warmer water of the Gulf Stream moderates the climate of northwestern EUROPE, causing temperatures in winter to be several degrees warmer than in areas of NORTH AMERICA at the same latitudes. The Gulf Stream decreases the time required for ships to travel from North America to Europe. This was an important factor in trade and communication in American Colonial times and has continued to be significant.

Gulfs. See under individual names.

Hatteras Abyssal Plain. Part of the floor of the northwest ATLANTIC OCEAN Basin, east of North Carolina. It rises to form shallow sandbars around Cape Hatteras, which are a notorious navigational hazard. In the seventeenth and eighteenth centuries, so many ships were lost in the area that Cape Hatteras became known as "The Graveyard of the Atlantic."

horse latitudes. Latitude belts between 30 and 35 degrees north and south latitude, where winds are usually light and variable and the climate mostly hot and dry.

Humboldt Current. See PERU CURRENT

Iceland. Island country bounded by the GREENLAND SEA on the north, the NORWEGIAN SEA on the east, and the ATLANTIC OCEAN on the south and west. Total area of 39,768 square miles (103,000 sq. km.). The nearest land mass is GREENLAND, 200 miles (320 km.) to the northwest. Situated on top of the northern part of the Atlantic Mid-Oceanic Ridge, it is characterized by major volcanic activities, geothermal springs, and glaciers.

Idzu-Bonin Trench. Ocean trench in the western PACIFIC OCEAN, about 6,082 miles (9,810 km.) long and 2,624 feet (800 meters) deep.

Indian Ocean. Third largest of the world's oceans, bounded by the continents of AFRICA to the west, ASIA to the north, AUSTRALIA to the east, and ANTARCTICA to the south. Most of the Indian Ocean lies below the equator. It has an approximate area of 33 million square miles (76 million sq. km.) and an average depth of about 13,120 feet (4,000

Indian Ocean Maldive archipelago Page 233

meters). Its deepest point is 24,442 feet (7,450 meters), in the JAVA TRENCH. The Indian Ocean was the first major ocean to be used as a trade route, particularly by the Egyptians. About 600 B.C.E., the Egyptian ruler Necho sent an expedition into the Indian Ocean, and the ship circumnavigated Africa, probably the first time this feat was accomplished. Warm winds blowing over the northern part of the Indian Ocean from May to September pick up huge amounts of moisture, which falls on India and Sri Lanka as monsoons. Fishing is important and mostly is done by small, family boats. About 40 percent of the world's offshore oil production comes from the Indian Ocean.

Indonesian Trench. See JAVA TRENCH.

Intracoastal Waterway. Series of bays, sounds, and channels, part natural and part humanmade, that extends along the eastern coast of the United States from the Delaware River in New Jersey, south to the tip of Florida, then around the west coast of Florida. It extends around the Gulf Coast to the Rio Grande in Texas. It runs 2,455 miles (3,951 km.) and is an important, protected route for commercial and pleasure boat traffic.

Japan, Sea of. Marginal sea of the western Pacific Ocean that is bounded by Japan on the east and the Russian mainland on the west. Its surface area is approximately 377,600 square miles (978,000 sq. km.). It has an average depth of 5,750 feet (1,750 meters) and a maximum depth of 12,276 feet (3,742 meters).

Japan Trench. Ocean trench approximately 497 miles (800 km.) long, beginning at the eastern edge of the Japanese islands and stretching southward toward the MARIANA TRENCH. Depth is 27,560 feet (8,400 meters).

Java Sea. Portion of the western PACIFIC OCEAN between the islands of Java and Borneo. The sea has a total surface area of 167,000 square miles (433,000 sq. km.) and a comparatively shallow average depth of 151 feet (46 meters).

Java Trench. Ocean trench in the INDIAN OCEAN, 2,790 miles (4,500 km.) long and 24,443 feet (7,450 meters) deep. Also called the Indonesian Trench.

Kermadec Trench. Ocean trench approximately 930 miles (1,500 km.) long, located in the southwest PACIFIC OCEAN, beginning northeast of New Zealand. It has a depth of 32,800 feet (10,000 meters). Its northern end connects with the TONGA TRENCH.

Kurile Trench. Ocean trench approximately 1,367 miles (2,200 km.) long along the northeast rim of the PACIFIC OCEAN, beginning at the north end of the Japanese island chain and extending northeastward. Depth is 34,451 feet (10,500 meters).

Labrador Current. Cold current that begins in Baffin Bay between GREENLAND and northeastern Canada and flows southward. The Labrador Current sometimes carries icebergs into North Atlantic shipping channels; such an iceberg caused the famous sinking of the great passenger ship *Titanic* in 1912.

Laptev Sea. Marginal sea of the ARCTIC OCEAN off the coast of northern Siberia. The Taymyr Peninsula bounds it on the west and the New Siberian Islands on the east. Its area is about 276,000 square miles (714,000 sq. km.). Its average depth is 1,896 feet (578 meters), and the greatest depth is 9,774 feet (2,980 meters).

Lord Howe Rise. Elevation of the floor of the western PACIFIC OCEAN that lies between AUSTRALIA and New Guinea and under the TASMAN SEA.

Mac. Robertson Land. Land near the coast

of ANTARCTICA, located between the INDIAN OCEAN and the south Polar Plateau, east of ENDERBY LAND. Mac.Robertson Land lies between approximately longitude 60 and 65 degrees east.

Macronesia. Loose grouping of islands in the ATLANTIC OCEAN that includes the Azores, Madeira, the Canary Islands and Cape Verde. The term is derived from Greek words meaning "large" and "island" and should not be confused with MICRONESIA, small islands in the central and North PACIFIC OCEAN.

Madagascar. Large island nation, officially called the Malagasy Republic, located in the INDIAN OCEAN about 200 miles from the southeast coast of AFRICA. Although geographically tied to the African continent, it has a culture more closely tied to those of France and Southeast Asia. Area is 226,657 square miles (587,042 sq. km.).

Magellan, Strait of. Waterway connecting the south ATLANTIC OCEAN with the South Pacific. Ships passing through the strait, north of Tierra del Fuego Island, avoid some of the world's roughest seas around CAPE HORN.

magnetic poles. The two points on the earth, one in the NORTHERN HEMISPHERE and one in the SOUTHERN HEMISPHERE, which are defined by the internal magnetism of the earth. Each point attracts one end of a compass needle and repels the opposite end.

Malacca, Strait of. Relatively narrow passage (200 miles/322 kilometers wide) bordered by Malaysia and Sumatra and linking the SOUTH CHINA SEA and the JAVA SEA. It is one of the most heavily traveled waterways in the world, with more than one thounsand ships every week.

Mariana Trench. Lowest point on Earth's surface, with a maximum depth of 36,150 feet (11,022 meters) in the Challenger Deep. The Mariana Trench is located on the western margin of the PACIFIC OCEAN southeast of Japan, and is approximately 1,584 miles (2,550 km.) long.

Marie Byrd Land. Section of ANTARCTICA located at the base of the Antarctic peninsula and shaped like a large peninsula itself. It is bounded at its base by the ROSS ICE SHELF and the Ronne Ice Shelf.

Mediterranean Sea. Large sea that separates the continents of EUROPE, AFRICA, and ASIA. It takes its name from Latin words meaning "in the middle of land"—a reference to its nearly landlocked nature. Covers about 969,100 square miles (2.5 million sq. km.) and extends 2,200 miles (3,540 km.) from west to east and about 1,000 miles (1,600 km.) from north to south at its widest. Its greatest depth is 16,897 feet (5,150 meters).

Melanesia. One of three divisions of the Pacific Islands, along with MICRONESIA and POLYNESIA; located in the western Pacific. The name Melanesia, for "dark islands," was given to the area because of its inhabitants' dark skins

Mexico, Gulf of. Nearly enclosed arm of the western ATLANTIC OCEAN, bounded by the states of Florida, Alabama, Mississippi, Louisiana, and Texas, and Mexico and the Yucatan Peninsula. Cuba is located in the gap between the Yucatan Peninsula and Florida. Most ocean water enters through the Yucatan passage and exits the Gulf of Mexico around the tip of Florida, becoming the FLORIDA CURRENT. Fisheries, tourism, and oil production are important activities.

Micronesia. One of three divisions of the Pacific Islands, along with MELANESIA and POLYNESIA. Micronesia means

*Mediterranean
Pages 233, 234*

*Gulf of
Mexico
Pages 234, 235*

"small islands." Micronesia's islands are mostly atolls and coral islands, but some are of volcanic origin. The more than two thousand islands of Micronesia are located in the Pacific Ocean east of the Philippines, mostly north of the EQUATOR.

Mid-Atlantic Ridge. Steep-sided, underwater mountain range running down the middle of the ATLANTIC OCEAN. Formed by the divergent boundaries, or region where tectonic plates are separating.

Mozambique Current. See AGULHAS CURRENT.

New Britain Trench. Ocean trench in the southwest PACIFIC OCEAN, about 5,158 miles (8,320 km.) long and 2,460 feet (750 meters) deep.

New Hebrides Basin. Part of the CORAL SEA, located east of AUSTRALIA and west of the New Hebrides island chain. The basin contains volcanic islands, both old and recent.

New Hebrides Trench. Ocean trench in the southwest PACIFIC OCEAN, about 5,682 miles (9,165 km.) long and 3,936 feet (1,200 meters) deep.

North America. Third largest continent, usually considered to contain all land and nearby islands in the WESTERN HEMISPHERE north of the Isthmus of Panama, which connects it to SOUTH AMERICA. The major mainland countries are Canada, the United States, Mexico, Guatemala, El Salvador, Honduras, Nicaragua, Costa Rica, and Panama. Island countries include the islands of the CARIBBEAN SEA and GREENLAND. Climate ranges from arctic to tropical.

North Atlantic Current. Continuation of the GULF STREAM, originating near the GRAND BANKS off Newfoundland. It curves eastward and divides into a northern branch, which flows into the NORWEGIAN SEA, a southern branch, which flows eastward, and a branch that forms the Canary Current and flows south along the coast of EUROPE.

North Atlantic Gyre. Large mass of water, located in the ATLANTIC OCEAN in the NORTHERN HEMISPHERE, that rotates clockwise. Warm water moves toward the pole and cold water moves toward the equator.

North Pacific Current. Eastward flow of water in the PACIFIC OCEAN in the NORTHERN HEMISPHERE. It originates as the Kuroshio Current and moves from Japan toward NORTH AMERICA.

North Pacific Gyre. Large mass of water, located in the PACIFIC OCEAN in the NORTHERN HEMISPHERE, that rotates clockwise. Warm water moves toward the pole and cold water moves toward the equator.

North Pole. Northern end of the earth's geographic axis, located at 90 degrees north latitude and longitude zero degrees. The North Pole itself is located on the Polar Abyssal Plain, about 14,000 feet (4,000 meters) deep in the ARCTIC OCEAN. U.S. explorer Robert Edwin is credited with being the first person to reach the North Pole, in 1909, although there is historical dispute over the claim. The North Pole is different from the North MAGNETIC POLE.

North Sea. Arm of the northeastern ATLANTIC OCEAN, bounded by Great Britain on the west and Norway, Denmark, and Germany on the east and south. The North Sea is one of the great fishing areas of the world and an important source of oil.

Northern Hemisphere. The half of the earth above the equator.

Norwegian Sea. Section of the North Atlantic Ocean. Norway borders it on the east and Iceland on the west. A subma-

331

rine ridge linking GREENLAND, ICE-LAND, the Faroe Islands, and northern Scotland separates the Norwegian Sea from the open ATLANTIC OCEAN. Cut by the ARCTIC CIRCLE, the sea is often associated with the ARCTIC OCEAN to the north. Reaches a maximum depth of about 13,020 feet (3,970 meters).

Oceania. Loosely applied term for the large island groups of the central and South Pacific; sometimes used to include AUSTRALIA and New Zealand.

Okhotsk, Sea of. Nearly enclosed area of the northwestern PACIFIC OCEAN bounded by Russia's Kamchatka Peninsula on the east and Siberia on the west. It is open to the Pacific Ocean on the south side only through Japan and the Kuril Islands, a string of islands belonging to Russia.

Pacific Ocean. Largest body of water in the world, covering more than one-third of Earth's surface—an area of about 70 million square miles (181 million sq. km.), more than the entire land area of the world. At its widest point, between Panama in CENTRAL AMERICA and the Philippines, it stretches 10,700 miles (17,200 km.). It runs 9,600 miles (15,450 km.) from the BERING STRAIT in the north to ANTARCTICA in the south. Bordered by NORTH and SOUTH AMERICA in the east, and ASIA and AUSTRALIA in the west. The average depth is about 12,900 feet (3,900 meters). It contains the deepest point on Earth (36,150 feet/11,022 meters), in the Challenger Deep of the MARIANA TRENCH, southwest of Japan. The Pacific Ocean bottom is more geologically varied than the INDIAN or ATLANTIC OCEANS; it has more volcanoes, ridges, trenches, seamounts, and islands. The vast size of the Pacific Ocean was a formidable barrier to travel, communications, and trade well into the nine-

Panama Canal Page 236

Pacific Ocean Pages 235, 236

teenth century. However, evidence shows that people crossed the Pacific Ocean in rafts or canoes as early as 3,000 B.C.E.

Pacific Rim. Modern term for the nations of ASIA and NORTH and SOUTH AMERICA that border, or are in, the PACIFIC OCEAN. Used mostly in discussions of economic growth.

Palau Trench. Ocean trench in the western PACIFIC OCEAN, about 250 miles (400 km.) long and 26,425 feet (8,054 meters) deep.

Palmer Land. Section of ANTARCTICA that occupies the base of the Antarctic peninsula.

Panama, Isthmus of. Narrow neck of land that joins CENTRAL and SOUTH AMERICA. In 1914 the Panama Canal was opened through the isthmus, creating a direct sea link between the PACIFIC OCEAN and the CARIBBEAN SEA. The canal stretches about 50 miles (80 km.) from Panama City on the Pacific to Colon on the Caribbean. More than twelve thousand ships pass through the canal annually.

Persian Gulf. Large extension of the ARABIAN SEA that separates Iran from the Arabian Peninsula in the Middle East. It covers about 88,000 square miles (226,000 sq. km.) and is about 620 miles (1,000 km.) long and 125-185 miles (200-300 km.) wide.

Peru-Chile Trench. Ocean trench that runs along the eastern boundary of the PACIFIC OCEAN, off the western edge of SOUTH AMERICA. It is 3,666 miles (5,900 km.) long and 26,576 feet (8,100 meters) deep.

Peru Current. Cold, broad current that originates in the southernmost part of the SOUTH PACIFIC GYRE and flows up the west coast of SOUTH AMERICA. Off the coast of Peru, prevailing winds usually push the warmer surface water to the west. This causes the nutrient-rich, colder water of the Peru Current to well up to the surface, which provides excellent feeding for fish. At times, the upwelling ceases and biological, economic, and climatic catastrophe can result in EL NIÑO weather conditions. Also known as the Humboldt Current.

Philippine Trench. Ocean trench located on the western rim of the PACIFIC OCEAN, at the eastern margin of the PHILIPPINE ISLANDS. It is about 870 miles (1,400 km.) long and 34,451 feet (10,500 meters) deep.

Polynesia. One of three main divisions of the Pacific Islands, along with MELANESIA and MICRONESIA. The islands are spread through the central and South Pacific. Polynesia means "many islands." Mostly small, the islands are predominantly coral atolls, but some are of volcanic origin.

Puerto Rico Trench. Ocean trench in the western ATLANTIC OCEAN, about 27,500 feet (8,385 meters) deep and 963 miles (1,550 km.) long.

Queen Maud Land. Section of ANTARCTICA that lies between the ATLANTIC OCEAN and the south Polar Plateau, between approximately longitude 15 and 45 degrees east.

Red Sea. Narrow arm of water separating AFRICA from the ARABIAN PENINSULA. One of the saltiest bodies of ocean water on Earth, as a result of high evaporation and little freshwater input. It was used as a trade route for Mediterranean, Indian, and Chinese peoples for centuries before the Europeans discovered it in the fifteenth century. The Suez Canal was opened in 1869 between the MEDITERRANEAN SEA and the Red Sea, cutting the distance from the northern INDIAN OCEAN to northern EUROPE by

Red Sea Pages 225, 238

French Polynesia Page 237

Ship sailing north on the Red Sea as the sun sets over the Northeast African coast. (National Oceanic and Atmospheric Administration)

The Ring of Fire includes California, where San Francisco was devastated by an earthquake in 1906. (National Oceanic and Atmospheric Administration)

California earthquake Page 238

about 5,590 miles (9,000 km.). This greatly increased the economic and military importance of the Red Sea.

Ring of Fire. Nearly continuous ring of volcanic and tectonic activity around the margins of the PACIFIC OCEAN.

Ross Ice Shelf. Thick layer of ice in the ROSS SEA off the coast of ANTARCTICA. The relatively flat ice is attached to and nourished by a continental glacier.

Ross Sea. Bay in the SOUTHERN OCEAN off the coast of ANTARCTICA, located south of New Zealand. Named for English explorer James Clark Ross, the first person to break through the Antarctic ice pack in a ship, in 1841.

St. Peter and St. Paul. Cluster of rocks showing above the surface of the AT-LANTIC OCEAN between Brazil and

West AFRICA. Important landmarks in the days of slave ships.

Sargasso Sea. Warm, salty area of water located in the ATLANTIC OCEAN south and east of Bermuda, formed from water that circulates around the center of the NORTH ATLANTIC GYRE. Named for the seaweed, *Sargassum*, that floats on the surface in large amounts.

Scotia Sea. Area of the southernmost AT-LANTIC OCEAN between the southern tip of SOUTH AMERICA and the ANT-ARCTIC peninsula. The area is known for severe storms.

Seas. See under individual names.

Siam, Gulf of. See THAILAND, GULF OF.

South America. Fourth largest continent, usually considered to contain all land and nearby islands in the Western

Hemisphere south of the Isthmus of Panama, which connects it to NORTH AMERICA. Countries are Argentina, Bolivia, Brazil, Chile, Colombia, Ecuador, French Guiana, Guyana, Paraguay, Peru, Suriname, Uruguay, and Venezuela. Climate ranges from tropical to cold, nearly sub-Antarctic.

South Atlantic Gyre. Large mass of water, located in the ATLANTIC OCEAN in the SOUTHERN HEMISPHERE, that rotates counterclockwise. Warm water moves toward the pole and cold water moves toward the equator.

South China Sea. Portion of the western PACIFIC OCEAN that lies along the east coast of China, Vietnam, and the southeastern part of the Gulf of Thailand. The eastern and southern edges are defined by the Philippine and Indonesian Islands.

South Equatorial Current. Part of the SOUTH ATLANTIC GYRE that is split in two by the eastern prominence of Brazil. One part moves along the northeastern coast of SOUTH AMERICA toward the CARIBBEAN SEA and the North ATLANTIC OCEAN; the other turns southward and forms the BRAZIL CURRENT.

South Pacific Gyre. Large mass of water, located in the PACIFIC OCEAN in the SOUTHERN HEMISPHERE, that rotates counterclockwise. Warm water moves toward the pole and cold water moves toward the equator.

South Pole. Southern end of the earth's geographic axis, located at 90 degrees south latitude and longitude zero degrees. The first person to reach the South Pole was Norwegian explorer Roald Amundsen, in 1911. The South Pole is different from the South MAGNETIC POLE.

Southeastern Pacific Plateau. Portion of the PACIFIC OCEAN floor closest to SOUTH AMERICA.

Southern Hemisphere. The half of the earth below the equator.

Southern Ocean. Not officially recognized as one of the major oceans, but a commonly used term for water surrounding ANTARCTICA and extending northward to 50 degrees south latitude. Also known as the Antarctic Ocean.

Straits. See under individual names.

Sunda Shelf. One of the largest continental shelves in the world, nearly 772,000 square miles (2 million sq. km.). Located in the JAVA SEA, SOUTH CHINA SEA, and Gulf of THAILAND. The area was above water in the Quaternary period, enabling large animals such as elephants and rhinoceros to migrate to Sumatra, Java, and Borneo.

Surtsey Island. Island formed by a volcanic explosion off the coast of ICELAND in 1963. It is valuable to scientists studying how island flora and fauna develop and is a popular tourist site.

Tashima Current. See TSUSHIMA CURRENT.

Tasman Sea. Area of the PACIFIC OCEAN off the southeast coast of AUSTRALIA, between Australia and Tasmania on the west and New Zealand on the east. First crossed by the Morioris people sometime before 1300 C.E. Also called the Tasmanian Sea.

Tasmanian Sea. See TASMAN SEA.

Thailand, Gulf of. Also known as the Gulf of Siam, inlet of the South China Sea, located between the Malay Archipelago and the Southeast Asian mainland. Bounded by Thailand, Cambodia, and Vietnam.

Tonga Trench. Ocean trench in the PACIFIC OCEAN, northeast of New Zealand. It stretches for 870 miles (1,400 km.), beginning at the northern end of the KERMADEC TRENCH. Depth is 32,810 feet (10,000 meters).

Tsushima Current. Warm current in the

335

western PACIFIC OCEAN that flows out of the YELLOW SEA into the Sea of JAPAN in the spring and summer. Also called Tashima Current.

Walvis Ridge (Walfisch Ridge). Long, narrow undersea elevation near the southwestern coast of AFRICA, which extends about 1,900 miles (3,000 km.) in a southwesterly direction under the ATLANTIC OCEAN.

Weddell Sea. Bay in the SOUTHERN OCEAN bounded by the ANTARCTIC peninsula on the west and a northward bulge of ANTARCTICA on the east, stretching from approximately longitude 60 to 10 degrees west. One of the harshest environments on Earth; surface water temperatures stay near 32 degrees Fahrenheit (0 degrees Celsius) all year. The Weddell Sea was the site of much whaling and seal hunting in the nineteenth and twentieth centuries.

West Caroline Trench. See YAP TRENCH.

West Wind Drift. Surface portion of the ANTARCTIC CIRCUMPOLAR CURRENT, driven by westerly winds. Often extremely rough; seas as high as 98 feet (30 meters) have been reported.

Western Hemisphere. The half of the earth containing NORTH and SOUTH AMERICA; generally understood to fall between longitudes 160 degrees east and 20 degrees west.

Wilkes Land. Broad section near the coast of ANTARCTICA, which lies south of AUSTRALIA and east of the AMERICAN HIGHLANDS. Wilkes Land is the nearest landmass to the South MAGNETIC POLE.

Yap Trench. Ocean trench in the western PACIFIC OCEAN, about 435 miles (700 km.) long and 27,900 feet (8,527 meters) deep. Also called the West Caroline Trench.

Yellow Sea. Area of the PACIFIC OCEAN bounded by China on the north and west and Korea on the east. Named for the large amounts of yellow dust carried into it from central China by winds and by the Yangtze, Yalu, and Yellow Rivers. Parts of the sea often show a yellow color from the dust.

Kelly Howard

INDEX TO VOLUME 1

See volume 8 for a comprehensive index to all eight volumes in set.

F L A G S O F T H E W O R L D

Lebanon	Lesotho	Liberia	Libya	Liechtenstein
Lithuania	Luxembourg	Macedonia	Madagascar	Malawi
Malaysia	Maldives	Mali	Malta	Marshall Islands
Mauritania	Mauritius	Mexico	Micronesia	Moldova
Monaco	Mongolia	Morocco	Mozambique	Myanmar
Namibia	Nauru	Nepal	Netherlands	New Zealand
Nicaragua	Niger	Nigeria	Norway	Oman
Pakistan	Palau	Palestine	Panama	Papua New Guinea
Paraguay	Peru	Philippines	Poland	Portugal
Qatar	Romania	Russia	Rwanda	St. Kitts & Nevis

Qatar

Romania

Russia

St. Kitts & Nevis